Theoretical ecology

Theoretical Ecology

PRINCIPLES AND APPLICATIONS

Theoretical Ecology

PRINCIPLES AND APPLICATIONS

EDITED BY

ROBERT M. MAY

Biology Department
Princeton University

W. B. SAUNDERS COMPANY

PHILADELPHIA TORONTO

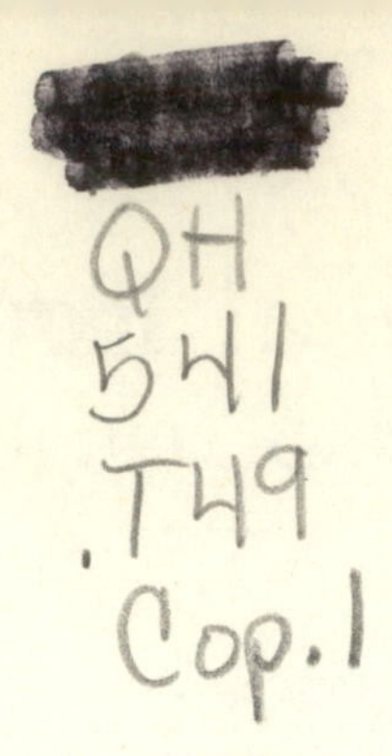

Western Hemisphere
distribution rights assigned to

W.B. SAUNDERS COMPANY

West Washington Square, Philadelphia, Pa. 19105
833 Oxford Street, Toronto M8Z 5T9, Canada

Library of Congress Catalog Card Number: 76-9642

SBN 0-7216-6205-6

Printed in Great Britain

Contents

List of Authors

GRAEME CAUGHLEY Department of Biology, University of Sydney, Sydney, N.S.W., 2006, Australia

JOEL E. COHEN Department of Biology, Rockefeller University, New York, N.Y., 10021, U.S.A.

GORDON CONWAY Environmental Management Unit, Imperial College Field Station, Silwood Park, Ascot, Berks, SL5 7PY, England

JARED M. DIAMOND Physiology Department, U.C.L.A. Medical Center, Los Angeles, California, 90024, U.S.A.

STEPHEN JAY GOULD Department of Paleontology, Harvard University, Cambridge, Massachusetts, 02138, U.S.A.

MICHAEL P. HASSELL Imperial College Field Station, Silwood Park, Ascot, Berks, SL5 7PY, England

HENRY S. HORN Biology Department, Princeton University, Princeton, New Jersey, 08540, U.S.A.

ROBERT M. MAY Biology Department, Princeton University, Princeton, New Jersey, 08540, U.S.A.

ERIC R. PIANKA Department of Zoology, University of Texas, Austin, Texas, 78712, U.S.A.

T. R. E. SOUTHWOOD Department of Zoology and Applied Entomology, Imperial College, London, SW.7, England

EDWARD O. WILSON Department of Biology, Harvard University, Cambridge, Massachusetts, 02138, U.S.A.

Acknowledgements

In a multi-authored volume such as this, the number of people to whom the authors are obliged for stimulating conversation and helpful criticism is unmanagably large. A short list includes Robert Campbell (who suggested this book, and saw it through to publication), J.R.Beddington, J.T.Bonner, W.H.Drury, T.Fenchel, N.G.Hairston, L.R.Lawlor, J.H.Lawton, S.A.Levin, the late R.H.MacArthur, J.Maynard Smith, G.F.Oster, Ruth Patrick, J.Roughgarden, T.W.Schoener, J.M.A.Swan, J.Terborgh, M.Williamson and all the graduate students in the population biology group at Princeton University. My secretary, B.DeLanoy, helped produce order out of chaos.

In choosing the topics to be included in this book, I have exhibited a certain amount of bias and caprice; the guilt is entirely mine.

Princeton University R.M.M.
January 1976

1
Introduction

An increasing amount of study is being devoted to mathematical models which seek to capture some of the essential dynamical features of plant and animal populations. Some of these models describe specific systems in a very detailed way, and others deal with general questions in a relatively abstract fashion: all share the common purpose of helping to construct a broad theoretical framework within which to assemble an otherwise indigestible mass of field and laboratory observations.

The present book aims to review and to draw together some of these theoretical insights, to show how they can shed light on empirical observations, and to examine some of the practical implications. In so doing, the book seeks to occupy a useful niche intermediate between the compendious and general text (of which there are an increasing number of excellent examples) and the often highly technical journal and monograph literature on theoretical ecology (which many people will find impenetrable). The book is directed to an audience of upper level undergraduates, graduate students, or general readers with an educated interest in the discipline of ecology.

Attention is focussed on the biological assumptions which underlie the various models, and on the way the consequent mathematical behaviour of the models explains aspects of the dynamics of populations or of entire communities. That is, the approach is descriptive, with emphasis on the biological inputs in constructing the models, and on the emergent biological understanding. The intervening mathematical details are, by and large, glossed over; this is a book for people who did not get beyond a freshman course or A levels in calculus. Those readers who dislike ex cathedra pronouncements, or who wish to savour the detailed mathematical development, will find signposts to guide them to the more technical literature. Other people may be content to follow the advice given by St Thomas Aquinas (concerning technical details of proofs of the existence of God): 'Truths which can be proved can also be known by faith. The proofs are difficult, and can only be understood by the learned; but faith is necessary also to the young, and to those who, from practical preoccupations, have not the leisure to learn philosophy. For them, revelation suffices' (From Russell, 1946, p. 46).

As will be seen from the chapter headings, the first two-thirds of the book (chapters 2 to 10) deals with plant and animal ecology as such: theoretical and empirical aspects of the dynamics of single populations, of pairs of interacting populations, and of whole communities of different species. The last third of the book (chapters 11 to 14) is devoted to various subjects which are having fruitful reciprocal interaction with theoretical ecology. Although primarily aimed at review and synthesis, the book does contain a significant amount of new material.

Chapter 2 outlines the dynamical behaviour of single species in which population change takes place either in discrete steps, or continuously, subject to density dependent regulatory mechanisms with time-delays; the environment may be constant, or it may vary in time. The consequent population behaviour may be a stable equilibrium point (with disturbances damped in a monotonic or an oscillatory manner), or stable cycles, or even apparently random fluctuations, depending on the relations among the various natural time scales in the system. In chapter 2, these ideas are applied narrowly to the dynamical trajectories of particular laboratory and field populations, and in chapter 3 they are applied much more broadly to discuss the way an organism's bionomic strategy (size, longevity, fecundity, range and migration habit) is fashioned by its general environment.

Chapter 4 gives a brief survey of the basic dynamical features of two species interacting as prey-predator, as competitors, or as mutualists. This serves as a background to the next three chapters, which focus upon two special cases of the general prey-predator relationship, and upon competition. In chapter 5, mathematical models are combined with field and laboratory studies to elucidate the components of predation in arthropod systems. Chapter 6 does a similar thing for plant-herbivore systems. Competition is addressed in chapter 7, which shows how theory and empirical observation can illuminate such questions as the meaning of the ecological niche, or the limits to similarity among coexisting competitors.

Quantitative understanding of the populations' dynamics, of the sort which enlivens some of the earlier chapters, is rarely feasible for complex communities of interacting species. Here the search is more for broad patterns of community organization. Chapter 8 discusses some of these patterns: energy flow (where much progress has accrued from the work of the International Biological Program); relative abundances of the different species in the community; and convergence in the structure of geographically distinct communities in similar environments. Some of these notions can be developed further and applied to the biogeography of islands, and thence to the design and management of floral and faunal conservation areas; this is done in chapter 9. Chapter 10 presents a somewhat revisionist account of succession, arguing from theory and observation that ecosystems require occasional (neither too frequent, nor too infrequent) disturbance back to the zeroth successional state, if they are to maintain their potential diversity.

The final four chapters deal with areas on the edge of mainstream ecology. Chapter 11 summarizes some of the outstanding problems in sociobiology (the recently christened subject dealing with the structure and organization of animal societies). The way recent ecological advances may shed light on such long debated riddles as the great waves of extinction which mark the end of many of the conventional geological epochs is the subject of chapter 12, devoted to palaeobiology. Chapter 13 pursues the population biology of human host-parasite systems, such as schistosomiasis, while chapter 14 draws together ecology and economics in a discussion of optimal strategies for pest control.

I hope that the selection of topics has provided a representative range of examples where theoretical models have been successfully intermeshed with real world observations. More than this, I hope that the collection may convey a sense of excitement, and may, to some small degree, serve to indicate unanswered questions and future directions for research.

2
Models for Single Populations

ROBERT M. MAY

2.1 Introduction

One broad aim in constructing mathematical models for populations of plants and animals is to understand the way different kinds of biological and physical interactions affect the dynamics of the various species. In this enterprise, we are relatively uninterested in the algebraic details of any one particular formula, but are instead interested in questions of the form: which factors determine the numerical magnitude of the population; which parameters determine the time scale on which it will respond to natural or man-made disturbances; will the system track environmental variations, or will it average over them? Accordingly, attention is directed to the biological significance of the various quantities in the equations, rather than to the mathematical details; to do otherwise is to risk losing sight of the real wood in contemplation of the mathematical trees.

In this use of mathematical models to grasp at general principles, it is helpful to begin with models for a single species. Models of this kind seek to elucidate the behaviour of a single population, $N(t)$, as a function of time, t.

Many isolated laboratory populations, carefully maintained in a controlled environment, may realistically be modelled by such a single equation.

On the other hand, there are few, if any, truly single species situations in the natural world. Populations will tend to interact with their food supply (below them on the trophic ladder), with their competitors for these resources (on the same trophic level), and with their predators (above them on the ladder). In addition, populations will be influenced by various factors in their physical environment. Even so, it is often useful to regard these biological and physical interactions as passive parameters in an equation for the single population, summarizing them as some overall 'intrinsic growth rate', 'carrying capacity', or the like.

Section 2.2 discusses models where generations completely overlap and population growth is a continuous process (first order differential equations), and section 2.3 treats models where generations are non-overlapping and growth is a discrete process (first order difference equations). Some of the emergent insights are applied to field and laboratory data in section 2.4, and extended to encompass time-varying environments in section 2.5. Section 2.6 briefly discusses the more complicated case of many distinct but overlapping age classes. Section 2.7 concludes the chapter by reechoing the major themes.

2.2 Continuous growth (differential equations)

In situations where there is complete overlap between generations (as in human populations), the population changes in a continuous manner. Study of the dynamics of such systems thus involves differential equations, which relate the rate at which the population is changing, dN/dt, to the population value at any time, $N(t)$.

2.2.1 *Density independent growth*

The simplest such model has a constant per capita growth rate, r, which is independent of the population density:

$$dN/dt = rN. \tag{2.1}$$

This has the familiar solution

$$N(t) = N(0) \exp(rt). \tag{2.2}$$

There is unbounded exponential growth if $r > 0$, and exponential decrease if $r < 0$. In either event, the characteristic time scale for the 'compound interest' growth process is of the order of $1/r$.

2.2.2 *Density dependent growth*

Such unbounded growth is not to be found in nature. A simple model which captures the essential features of a finite environment is the logistic equation:

$$dN/dt = rN(1 - N/K). \tag{2.3}$$

Here the effective per capita growth rate has the density dependent form $r(1-N/K)$: this is positive if $N < K$, negative if $N > K$, and thus leads to a globally stable equilibrium population value at $N^* = K$. K may be thought of as the carrying capacity of the environment, as determined by food, space, predators, or other things; r is the 'intrinsic' growth rate, free from environmental constraints.

In any such dynamical system, it is useful to christen a 'characteristic return time', T_R, which gives an order-of-magnitude estimate of the time the population takes to return to equilibrium, following a disturbance (for a more formal discussion, see May *et al.*, 1974, and Beddington *et al.*, 1976a). In eq. (2.3), this characteristic time scale remains $T_R = 1/r$. To elaborate this point, we rewrite eq. (2.3) in dimensionless form by introducing the rescaled variables $N' = N/K$ and $t' = rt = t/T_R$. This gives the parameter-free equation

$$dN'/dt' = N'(1-N'). \tag{2.3a}$$

Such rescaling arguments are of general usefulness in disentangling those factors which influence the *magnitude* of equilibrium populations from those factors which bear upon the *stability* of the equilibrium. In this particular example, it is clear that the magnitude of the equilibrium population depends only on K, whereas the dynamics—the response to disturbance—depends only on r. This fact underlies the metaphor of r and K selection, developed in the next chapter.

It must be emphasised that the specific form of eq. (2.3) is not to be taken seriously. Rather it is representative of a wide class of population equations with regulatory mechanisms which biologists call density dependent, and mathematicians call nonlinear. A plethora of other such models, taken from the ecological literature, is catalogued in May (1975a, pp. 80–81). All share with eq. (2.3) the essential property of a stable equilibrium point, $N^* = K$, with any disturbance tending to fade away monotonically. One way of justifying eq. (2.3) is to regard it as the first term in the Taylor series expansion of these more general density dependent models.

2.2.3 *Time-delayed regulation*

In eq. (2.3), the density dependent regulatory mechanism, as represented by the factor $(1-N/K)$, operates instantaneously. In most real life situations, these regulatory effects are likely to operate with some built-in time lag, whose characteristic magnitude may be denoted by T. Such time lags may, for example, derive from vegetation recovery times or other environmental effects, or from the time of approximately one generation which elapses before the depression in birth rates at high densities shows up as a decrease in the adult population. A rough way of incorporating such time delays is to rewrite eq. (2.3) as

$$dN/dt = rN[1 - N(t-T)/K]. \tag{2.4}$$

This delay-differential equation was first introduced into ecology by Hutchinson (1948) and Wangersky and Cunningham (1957), and by now it enjoys an extensive literature (for a brief guided tour, see, e.g., May 1975a, pp. 95–98). One way of deriving it is as a crude approximation to a fully age-structured description of a single population, in which case T is the generation time. As before, the detailed form of this equation is not to be taken literally, and in more realistic treatments the regulatory term is likely to depend not on the population at a time exactly T earlier, but rather on some smooth average over past populations. (For a more mathematical discussion, see May, 1973a.) Nonetheless, the general properties of eq. (2.4) are representative of this wider class of models, and will be discussed in this spirit.

The qualitative nature of the solutions of eq. (2.4) follow from precepts familiar to engineers. If the time delay in the feedback mechanism (namely, T) is long compared to the natural response time of the system (namely, T_R or $1/r$), there will be a tendency to overshoot and to overcompensate. For modest values of the time delay this overcompensation produces an oscillatory, rather than a monotonic, return to the equilibrium point at $N^* = K$. As the time delay becomes longer (as T/T_R or rT exceeds some number of order unity), there is a so-called Hopf bifurcation, and the stable point gives way to stable limit cycles. These stable cycles are an explicitly nonlinear phenomenon, in which the population density, $N(t)$, oscillates up and down in a cycle whose amplitude and period is determined uniquely by the parameters in the equation. Just as in the case of a stable equilibrium point, if the system is perturbed it will tend to return to this stable cyclic trajectory. Such stable limit cycle solutions are a pervasive feature of nonlinear systems, for which conventional mathematics courses (with their focus on linear systems) give little intuitive appreciation.

Specifically, eq. (2.4) has a monotonically damped stable point if $0 < rT < e^{-1}$, and an oscillatorilly damped stable point if $e^{-1} < rT < \frac{1}{2}\pi$. For $rT > \frac{1}{2}\pi$, the population exhibits stable limit cycles, the period and amplitude of which are indicated in Table 2.1. These numerical details (e^{-1} and $\frac{1}{2}\pi$) are peculiar to eq. (2.4), but the character of the solution, with a stable equilibrium point giving way to stable cycles once T/T_R exceeds some number of order unity, is generic to a much wider class of models with time-delayed regulatory mechanisms.

Table 2.1. Properties of limit cycle solutions of eq. (2.4).

rT	$N(\max)/N(\min)$	Cycle period, T
1·57, or less	1·00	—
1·6	2·56	4·03
1·7	5·76	4·09
1·8	11·6	4·18
1·9	22·2	4·29
2·0	42·3	4·40
2·1	84·1	4·54
2·2	178	4·71
2·3	408	4·90
2·4	1,040	5·11
2·5	2,930	5·36

In particular, it is worth noting that once stable limit cycles arise in equations of the general form of eq. (2.4), their period is roughly equal to $4T$. A qualitative explanation of this fact is as follows: In the first phase of the cycle, the population continues to grow ($dN/dt > 0$) until the earlier population value in the time-delayed regulatory factor attains the potential equilibrium value ($N(t-T) = K$); at this point, population growth ceases ($dN/dt = 0$), and the population begins an accelerating decline from its peak value. Thus the first phase, where the population grows from around K to the cycle maximum, takes a time T. Similar arguments applied to the subsequent phases of the cycle suggest an overall period of roughly $4T$. The exact results in Table 2.1 show that this rough rule remains true, even as the amplitude of the cycle (population maximum/population minimum) increases over several orders of magnitude.

In short, equations such as (2.4) constitute minimally realistic models for a single population, in which the density dependent regulatory effects (derived from food supply limitations, or crowding, or whatever) operate with a time delay. The consequent population dynamics can be monotonic damping to an equilibrium point, or damped oscillations, or sustained patterns of stable cycles, depending on the ratio between T and T_R. A variety of population data can be surveyed in this light, and this is done in section 2.4 and in chapter 3.

2.3 Discrete growth (difference equations)

At the opposite extreme from section 2.2, many populations are effectively made up of a single generation, with no overlap between successive generations, so that population growth occurs in discrete steps. Examples are provided by many temperate zone arthropod species, with one short-lived adult generation each year. Periodical cicadas, with adults emerging once every 7 or 13 or 17 years, are an extreme example.

In these circumstances, the appropriate models are difference equations relating the population in generation $t+1$, N_{t+1}, to that in generation t, N_t. In contrast to section 2.2, time is now a discrete variable.

2.3.1 *Density independent growth*

The difference equation analogue of eq. (2.1) is the simple linear equation

$$N_{t+1} = \lambda N_t. \tag{2.5}$$

Here λ (conventionally misnamed the 'finite rate of increase') is the multiplicative growth factor per generation; the 'compound interest' growth rate is* $r = \ln \lambda$. Equation (2.5) describes unbounded exponential growth for $\lambda > 1$ ($r > 0$), exponential decline to extinction if $\lambda < 1$ ($r < 0$).

2.3.2 *Density dependent growth*

More generally, and more realistically, we will have a density dependent relation of the form

$$N_{t+1} = F(N_t), \tag{2.6}$$

where $F(N)$ is some nonlinear function of N. A fairly complete catalogue

* Throughout this volume, we follow the conventional practice of using ln to denote natural logarithms (to the base e), and log to denote logarithms to the base 10.

of the many forms which have been proposed as discrete analogues of the logistic eq. (2.3), along with their biological provenances, has been given by May and Oster (1976). These forms $F(N)$ all share the essential features of a propensity to population growth at low densities, to population decrease at high densities, and a parameter (or parameters) which measures the severity of this nonlinear response.

Table 2.2 gives a short list of four such expressions for $F(N)$. Each case is complete with the value of the possible equilibrium point, N^*,

$$N^* = F(N^*), \tag{2.7}$$

and of the characteristic return time, T_R, which describes how quickly the system tends to return to equilibrium following a perturbation.

The first of these expressions, form A, is that preferred by mathematicians, because it is *the* simplest nonlinear difference equation. Indeed it is startling that so simple an equation possesses the bizarre range of dynamical behaviour discussed below. However it has the biologically ugly feature that if the population ever exceeds $K(1+r)/r$, it becomes negative (i.e., extinct) in the next generation: the other three forms all have the more attractive property that population fluctuations are bounded above and below, and that the stability properties are global. Form B has an extensive pedigree in the biological literature, and corresponds to modifying the simple eq. (2.5) with a mortality factor, exp $(-aN)$, which becomes exponentially more severe for large N; this is a plausible model for populations which at high density are regulated by epidemics. Form C is included because it is the basis of a study of field and laboratory data, discussed in section 2.4.3 (see Fig. 2.5). Form D is the model which corresponds to the empirical method of analysing data for density dependence by plotting

Table 2.2. Specific formulae for the function $F(N)$ in eq. (2.6).

Label	Form* for $F(N)$	Equilibrium point, N^*, from eq. (2.7)	Characteristic return time, T_R
A	$N[1+r(1-N/K)]$	K	$1/r$
B	$N \exp[r(1-N/K)]$	K	$1/r$
C	$\lambda N(1+aN)^{-\beta}$	$(\lambda^{1/\beta}-1)/a$	$[\beta(1-\lambda^{-1/\beta})]^{-1}$
D	λN^{1-b}; for $N>\epsilon$ λN; for $N<\epsilon$	$\lambda^{1/b}$	$1/b$

* For a catalogue of the original sources for these various forms, see May and Oster, 1976.

log (N_t/N_{t+1}) against log (N_t): the slope of the consequent regression line is b.

Unlike eq. (2.3), the forms A–D and their friends and relatives do not contain any explicit time lags. However, there is a time delay implicit in the structure of eq. (2.5), in the one generation time step between the expression of the density dependent regulatory effects in generation t, and their manifestation in the census data in generation $t+1$. As above, we then expect exponential damping, or oscillatory damping, or sustained oscillations, depending on the ratio between the time delay (which is now 'T' = 1) and the natural response time (T_R, as catalogued in Table 2.2).

This analogy between delay-differential equations and ordinary difference equations is developed in more detail by May *et al.* (1974). It sheds light on the behaviour of the general eq. (2.6) and the specific examples in Table 2.2, namely a monotonically damped stable point for $T_R > 1$ ($1 > r > 0$ in forms A and B), an oscillatorily damped stable point for $1 > T_R > 0 \cdot 5$ ($2 > r > 1$ in forms A and B), and sustained but bounded oscillations for $T_R < 0 \cdot 5$ ($r > 2$ in A and B), but it gives no hint of the bewildering richness of the spectrum of dynamical behaviour which has recently been uncovered in this oscillatory regime ($T_R < 0 \cdot 5$, $r > 2$).

2.3.3 *Stable points, stable cycles, chaos*

A full mathematical and biological account of this spectrum of behaviour for nonlinear difference equations has been given by May and Oster (1976, see also May, 1974a), with emphasis on the generic character of the phenomenon. What follows is a brief summary, emphasizing the implications for population biology. To be definite, we will refer to the forms A and B of Table 2.2 (see also Fig. 2.5), with numerical illustrations drawn from form B.

So long as the nonlinearities are not too severe, the time delay built into the structure of the difference equation tends to be short compared to the natural response time of the system, and there is simply a stable equilibrium point at N^* [determined by eq. (2.7)].

What happens when this equilibrium point ($N^* = K$ in forms A and B) becomes unstable, as it does for $r > 2$? As this point becomes unstable, it bifurcates to produce two new and locally stable fixed points of period 2, between which the population oscillates stably in a 2-point cycle. With increasing r, these two points in turn become unstable, and bifurcate to give four locally stable fixed points of period 4. In this way there arises, by successive bifurcations, an infinite hierarchy of stable cycles of period 2^n, as illustrated in Fig. 2.1.

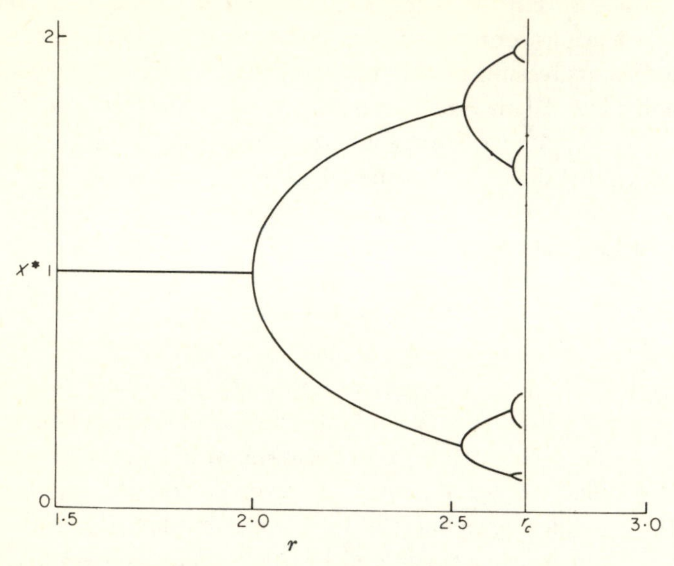

Fig. 2.1. The hierarchy of stable fixed points (X^*) of periods 1, 2, 4, 8, ..., 2^n (corresponding to stable cycles of periods 2^n), which are produced from eq. (2.6) with the form B of Table 2.2 as the parameter r increases. Each pair of points arises by bifurcation as a previous point becomes unstable. The sequence of stable cycles of period 2^n is bounded by the parameter value r_C; beyond this lies the chaotic region.

This sequence of stable cycles with period 2^n converges on a limiting parameter value, r_C say, beyond which the system enters a regime which has aptly been termed 'chaotic' (Li and Yorke, 1975). For any parameter value in this domain there are an infinite number of different periodic orbits, as well as an uncountable number of initial points for which the population trajectory, although bounded, does not settle into any cycle. In particular, Li and Yorke have proved the surprising theorem that, for *any* first order difference equation, once there exists a 3-point cycle (which happens for $r > \sqrt{8} = 2 \cdot 828$ in form A, and for $r > 3 \cdot 102$ in form B) then there necessarily also exist cycles of every integer period, along with an uncountable number of asymptotically aperiodic trajectories.

Another way of characterizing the chaotic regime is to observe that although the equation is rigidly deterministic, with all parameters exactly specified, arbitrarily close initial conditions can lead, after sufficient time, to subsequent population trajectories that diverge widely.

Figure 2.2 aims to illustrate this range of dynamical behaviour for various values of r in the form B of Table 2.2. Table 2.3 summarizes the various regimes for both the forms A and B, and Fig. 2.5 illustrates these regimes, as functions of the two pertinent parameters λ and β, for the form C. The form D is interesting in that (by virtue of the non-

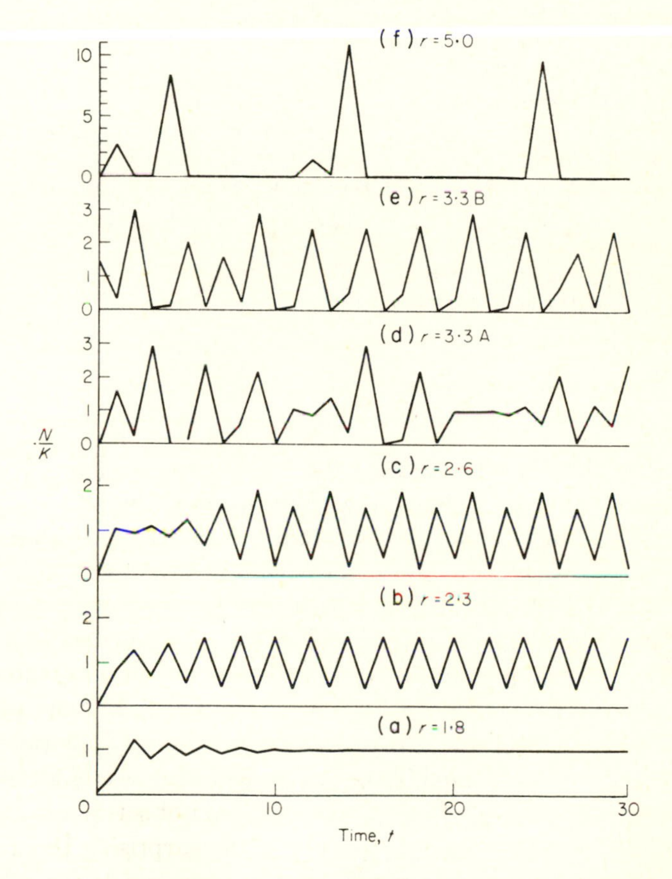

Fig. 2.2. The spectrum of dynamical behaviour of the population density, N_t/K, as a function of time, t, as described by the difference equation (2.6) with the form B of Table 2.2, for various values of r. Specifically: (a) $r = 1 \cdot 8$, stable equilibrium point; (b) $r = 2 \cdot 3$, stable 2-point cycle; (c) $r = 2 \cdot 6$, stable 4-point cycle; (d to f) in the chaotic regime, where the detailed character of the solution depends on the initial population value, with (d) $r = 3 \cdot 3$ ($N_o/K = 0 \cdot 075$), (e) $r = 3 \cdot 3$ ($N_o/K = 1 \cdot 5$), (f) $r = 5 \cdot 0$ ($N_o/K = 0 \cdot 02$).

analyticity of this function) there is an abrupt transition from a stable point if $b < 2$ directly into chaos for $b > 2$, with no intermediate regime of stable cycles.

Table 2.3. Dynamics of a population described by the difference eq. (2.6) with the forms A and B of Table 2.2.

Dynamical behaviour	Value of the growth rate, r	
	Form A	Form B
stable equilibrium point	$2{\cdot}000 > r > 0$	$2{\cdot}000 > r > 0$
stable cycles of period 2^n		
2-point cycle	$2{\cdot}449 > r > 2{\cdot}000$	$2{\cdot}526 > r > 2{\cdot}000$
4-point cycle	$2{\cdot}544 > r > 2{\cdot}449$	$2{\cdot}656 > r > 2{\cdot}526$
8-point cycle	$2{\cdot}564 > r > 2{\cdot}544$	$2{\cdot}685 > r > 2{\cdot}656$
16, 32, 64, etc.	$2{\cdot}570 > r > 2{\cdot}564$	$2{\cdot}692 > r > 2{\cdot}685$
chaotic behaviour		
[Cycles of arbitrary period, or aperiodic behaviour, depending on initial condition]	$r > 2{\cdot}570$	$r > 2{\cdot}692$

Once in the chaotic regime, the population dynamics of these deterministic models is best described in probabilistic terms. This has been done for the form B of Table 2.2, where it is seen that for largish r the trajectories are almost periodic, with an approximate period $[\exp(r-1)]/r$, as illustrated by the top trajectory in Fig. 2.2 (May, 1975b). The task of finding general methods for probabilistic description of the population dynamics in the chaotic regime has received some attention, but much remains to be done before this is in a form where it is a useful tool in the analysis of population data.

Quite apart from their intrinsic mathematical interest, the above results raise very awkward biological questions. They show that simple and fully deterministic models, in which all the biological parameters are exactly known, can nonetheless (if the nonlinearities are sufficiently severe) lead to population dynamics which are in effect indistinguishable from the sample function of a random process. Apparently chaotic population fluctuations need not necessarily be due to random environmental fluctuations, or sampling errors, but may reflect the workings of some deterministic, but strongly density dependent, population model. This question is pursued further in section 2.4.3.

As a postscript, I cannot resist remarking that production of such apparently random dynamics by simple deterministic models is of interest to mathematicians (as a relatively tractable example of a multiple bifurcation process) and to physicists (where it provides a metaphor for the phenomenon of turbulence). To see mathematical ecology informing theoretical physics is a pleasing inversion of the usual order of things.

In all, the message emerging from this section reinforces that of section 2.2. Depending on the ratio between the characteristic return time, T_R, and the time delay in the regulatory mechanism (here 'T' = 1), single populations exhibit either a stable equilibrium point (damped monotonically or oscillatorily) or patterns of self-sustained fluctuations (which can be periodic or chaotic).

2.4 Applications to field and laboratory data

2.4.1 *Nicholson's blowflies*

The first half of this century saw several classic experiments, in which laboratory populations in a constant and limited environment exhibited growth in accord with the logistic eq. (2.3). These experiments, on the protozoan *Paramecium*, on yeast cells, on bacteria such as *E. coli*, on *Drosophila*, and on various flour beetles, are reviewed, for example, by Krebs (1972, pp. 190–200).

Other similar single species experiments have shown sustained patterns of large-amplitude oscillations. The most notable of these is Nicholson's study, illustrated in Fig. 2.3, of the sheep-blowfly *Lucilia cuprina*.

An accurate model for this population would need to include both the details of the reproductive physiology of blowflies, and the fact that there are several distinct but overlapping age-classes. However, the essential dynamics of the system may be captured by the naive eq. (2.4), which includes the three major features: an intrinsic rate of increase, r; a resource limitation, K, set by the amount of the constant supply of ground liver; and a time delay, T, in the action of this limitation, roughly equal to the time for a larva to mature into an adult. After the x and y axes in Fig. 2.3 are scaled, the model gives a 1-parameter fit to the data. This single parameter, namely rT, is chosen to fit the observed ratio between minimum and maximum populations (see Table 2.1): the result is the theoretical curve illustrated in Fig. 2.3. From this particular choice of rT, one can infer an egg-to-adult development time of 9 days, in fair agreement with the actual figure of 11 days.

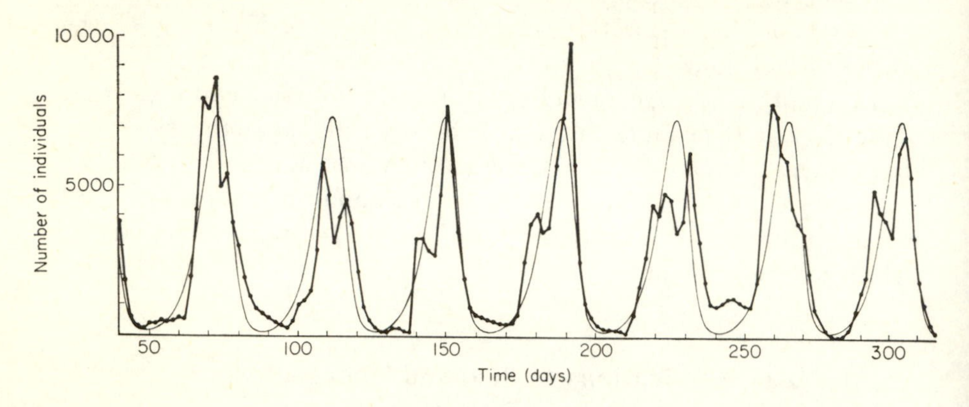

Fig. 2.3. From the one-parameter family of limit cycles generated by the time delayed logistic eq. (2·4), we display that which best fits the oscillations in Nicholson's blowfly populations (namely $rT = 2 \cdot 1$: see Table 2.1). After May (1975a).

Although this crude fit to Nicholson's data has been superseded by more detailed and age-structured models (Oster *et al.*, 1976), it is encouraging to see how the simple ideas developed in sections 2.2 and 2.3 can give a good caricature of the dynamics.

2.4.2 *Responses to temperature change*

Other sets of field and laboratory data show populations with monotonic damping at one temperature, and oscillations (either damped or persistent) at another temperature. These include, *inter alia*, McNeil's field studies of the grassland bug *Leptoterna dolobrata* (see May *et al.*, 1974); Pratt's (1943) laboratory observations on population growth in the water flea *Daphnia magna*, which appears to show a stable point at 18°C and sustained oscillations at 25°C; and the laboratory studies by Beddington (1974; also Beddington and May, 1975) on the Collembola *Folsomia candida*.

Such qualitative changes in the population dynamics may plausibly be associated with changes in the ratio between T and T_R (i.e., with changes in rT) produced by the interplay between external temperature and metabolic rates.

A particularly elegant illustration is provided by Fujii's (1968) laboratory data on three different strains of the stored-product beetle *Callosobruchus*. As one goes from the population illustrated in Fig. 2.4(a) to that in Fig. 2.4(c), the intrinsic growth rate r increases and T_R decreases. Consequently the ratio T/T_R (or rT) increases, and the dynamics goes from the monotonically damped stable point of Fig. 2.4(a), to the damped oscillations of Fig. 2.4(b), to the stable cycle shown in Fig. 2.4(c).

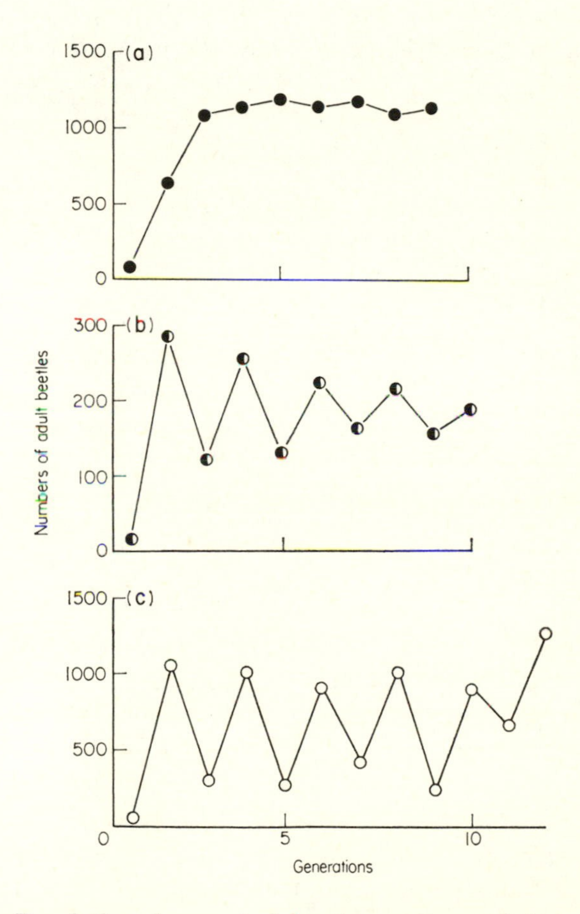

Fig. 2.4. Population changes in laboratory cultures of three different strains of stored-product beetles, displayed as number of adult beetles in successive generations: (a) *C. chinensis* (after Fujii, 1968); (b) *C. maculatus* (after Utida, 1967); (c) *C. maculatus* (after Fujii, 1968).

2.4.3 *Insect population parameters*

From the available literature, Hassell *et al.* (1976a) have culled data on 28 populations of seasonally breeding insects in which generations do not overlap. In each ease, the data have been fitted to the difference eq. (2.6) with the particular form C of Table 2.2 for $F(N)$, and estimates made of the parameters λ, β and a. (The intrinsic growth factor, λ, is first estimated independently, and then β and a are found by fitting the census data: the above reference should be consulted for details).

Figure 2.5 shows the theoretical domains of stability behaviour for this difference equation: depending on the values of λ and β, there is a stable point, or stable cycles in which the population alternates up and down, or chaos. The points show the parameter values for the

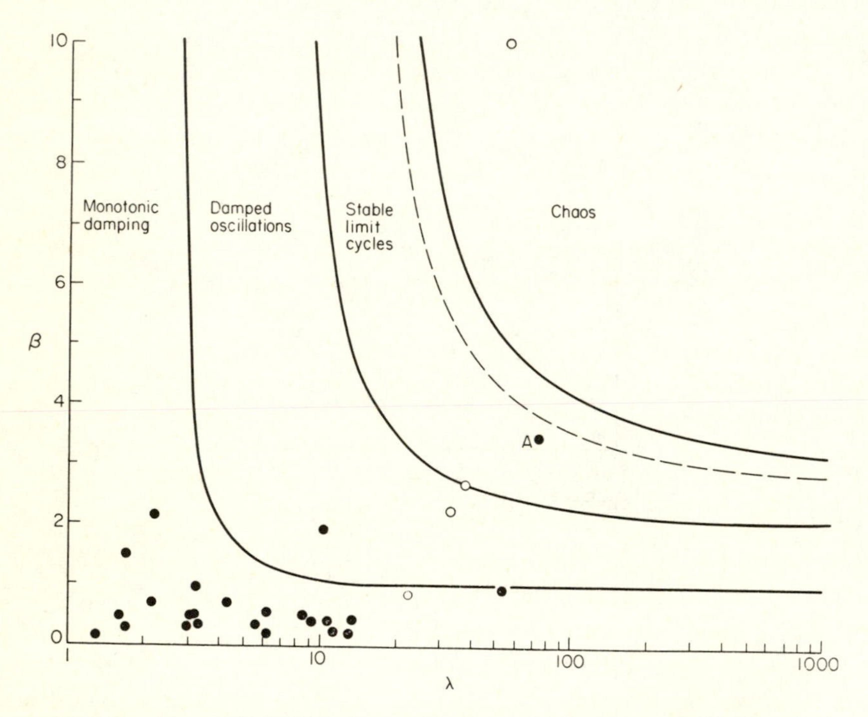

Fig. 2.5. Dynamical behaviour of eq. (2.6) with the form C of Table 2.2. The solid curves separate the regions of monotonic and oscillatory damping to a stable point, stable limit cycles, and chaos; the broken line indicates where 2-point cycles give way to higher order cycles. The solid circles come from the analyses of life table data on field populations, and the open circles from laboratory populations. After Hassell *et al.* (1976a), where details are given.

actual populations, with solid circles denoting field populations and open circles laboratory populations.

Note that there is a tendency for laboratory populations to exhibit cyclic or chaotic behaviour, whereas natural populations tend to have a stable equilibrium point. The laboratory populations are maintained in a homogeneous environment, and are free from predators and many other natural mortality factors, which may well make for exaggeratedly nonlinear behaviour, and give a misleading impression as to their population dynamics in the outside world.

In short, Fig. 2.5 can be interpreted as indicating a tendency for natural populations to exhibit stable equilibrium point behaviour. It is perhaps suggestive that the most oscillatory natural population (labelled A in Fig. 2.5) is the Colorado potato beetle, *Leptinotarsa*, whose contemporary role in agroecosystems lacks an evolutionary pedigree.

These sweeping generalizations should, however, be approached with extreme caution. Quite apart from the biases in data selection, and inadequacies in data reduction, which may be inherent in Fig. 2.5, it must be kept in mind that there are *no* truly single population situations in the real world. To subsume the population's biological interactions with the world around it in passive parameters such as λ and β may do violence to the multi-species reality. Add to this the fact that it requires less severely nonlinear behaviour to take multi-species models into the chaotic regime (May and Oster, 1976; Guckenheimer *et al.*, 1976), and Fig. 2.5 falls far short of a proof that natural populations have stable point behaviour.

2.4.4 '*Wildlife's 4-year cycle*'

The natural world exhibits many instances of 3-to-4 year population cycles, particularly among small mammals in boreal regions. These have attracted much attention, from the time of Elton (1942) onward (see, e.g., Chitty, 1950; Pitelka, 1967; Dajoz, 1974).

Without attempting to explain the biological mechanisms responsible for density dependent regulation in such populations, it is plausible to assume that in strongly seasonal environments such mechanisms operate with built-in time lags of slightly less than, or of the order of, one year. For beasts with a relatively large intrinsic capacity for population growth, r (so that rT significantly exceeds unity), the upshot will tend to be stable limit cycles. As observed in section 2.2.3, regardless of the amplitude, these cycles will have periods of the order of $4T$, which is to say approximately 4 years.

Here is a detail-independent explanation of 'wildlife's 4-year cycle'.

Figure 2.6 shows Shelford's (1943) data for a population of the almost legendary lemming, along with a theoretical curve obtained from eq. (2.4) with $rT = 2.4$ (chosen from Table 2.1 to fit Shelford's numerical data on the amplitude of the cycle) and $T = 9$ months (very roughly the time from the end of one summer to the beginning of the next). Equation (2.4) is, of course, a very crude model. In particular, it necessarily gives uniformly spaced cycles with a period intermediate between 3 and 4 years, whereas the seasonal northerly environment compels the real cycles to be lock-stepped to an integral number of years, which is sometimes 3, sometimes 4 years.

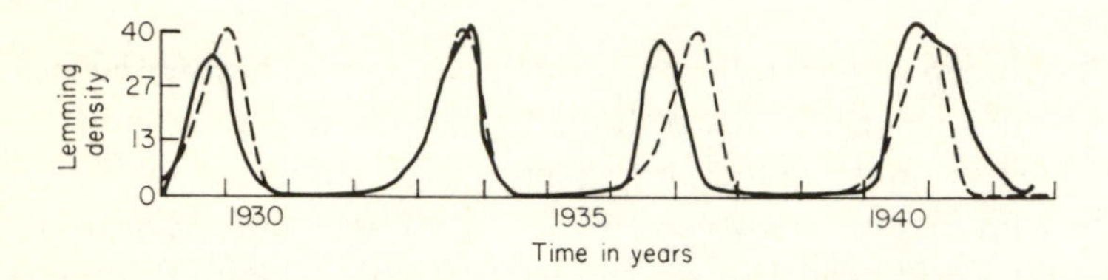

Fig. 2.6. Shelford's (1943) data on the lemming population in the Churchill area in Canada (expressed as numbers of individuals per hectare), compared with a naive theoretical curve (dashed line) obtained from the simple time delayed logistic eq. (2.4); the time delay T is taken to be a little under one year ($T = 0{\cdot}72$ yr.). For further details, see the text.

This explanation is independent of the biological mechanism(s) producing the time delay. Needless to say, elucidation of these biological details remains a fascinating and relevant task.

2.5 Time-varying environments

In all the models discussed so far, the biological and environmental parameters (such as r and K) were taken to be constants. We now examine some of the effects of letting r and K vary in time, in either periodic or random fashion.

There is no difficulty in solving the logistic differential equation (2.3) for arbitrarily time-dependent $r(t)$ and $K(t)$: see, e.g., Poluektov

(1974), Kiester and Barakat (1974), May (1976a). In the particular case when r is constant, the population $N(t)$ tends, after a sufficiently long time, to the asymptotic solution

$$N(t) = \left\{ r \int_0^t [1/K(t')] \exp [r(t-t')] \, dt' \right\}^{-1}. \qquad (2.8)$$

The population's characteristic response time is again $T_R = 1/r$. Equation (2.8) says that $N(t)$ is the inverse of some weighted harmonic average over past values of the carrying capacity, with this average reaching a typical distance T_R into past time.

Consider the situation when the carrying capacity $K(t)$ varies periodically:

$$K(t) = K_0 + K_1 \cos (2\pi t/\tau). \qquad (2.9)$$

Here we require $K_0 > K_1$ if $K(t)$ is not to go nonsensically negative during its cycle. The behaviour of the population, $N(t)$, will now depend on whether its characteristic response time ($T_R = 1/r$) is long or short compared to the period (τ) of the environmental oscillations.

In the limit when T_R greatly exceeds τ ($r\tau \ll 1$), it may be shown that

$$N(t) = \sqrt{K_0^2 - K_1^2} \, [1 + o(r\tau)]. \qquad (2.10)$$

The expression in square brackets specifies that the correction terms are of relative order $r\tau$. In this limit the population averages out the environmental variations: note, however, that this average population value is not simply the average value of $K(t)$, namely K_0, but is less than it. This approximately constant population trajectory in the limit $T_R \gg \tau$ is illustrated in Fig. 2.7(a).

Conversely for T_R short compared with τ ($r\tau \gg 1$), we have approximately

$$N(t) = K_0 + K_1 \cos (2\pi t/\tau) \, [1 + o(1/r\tau)]. \qquad (2.11)$$

The population now tends to 'track' the environmental variations, as illustrated by Fig. 2.7(b).

Similar considerations apply to situations where $K(t)$ has a more general time dependence, with stochastic components (Roughgarden, 1975a). Reechoing the themes which pervade this chapter, we find that populations with relatively large r (short response time T_R) are condemned to track environmental fluctuations, whereas those with relatively small r may average over essentially all fluctuations. More specifically,

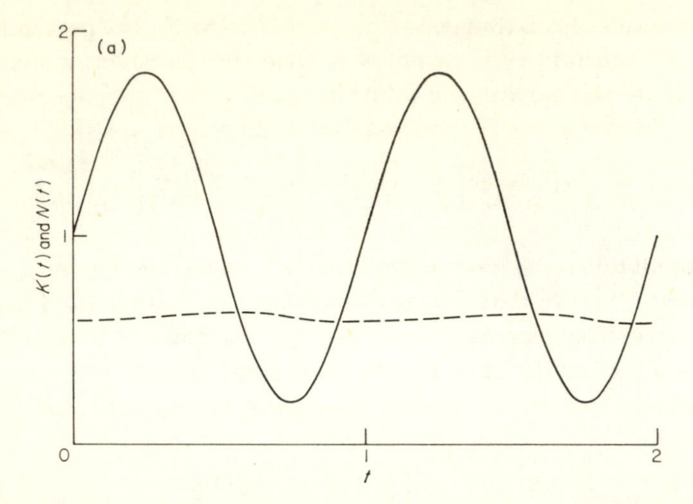

Fig. 2.7a. Illustrating the way a population [N(t), dashed line] varies if it obeys a logistic equation with a time dependent carrying capacity [K(t), solid line]. This figure is for the case when the population's natural response time ($1/r$) is long compared to the periodicity in the environment (τ), i.e., for relatively small r. Specifically, the figure is for eq. (2.3) with eq. (2.9), and $r = 0{\cdot}2$, $\tau = 1$, $K_o = 1$, $K_1 = 0{\cdot}8$; i.e., $r\tau = 0{\cdot}2$.

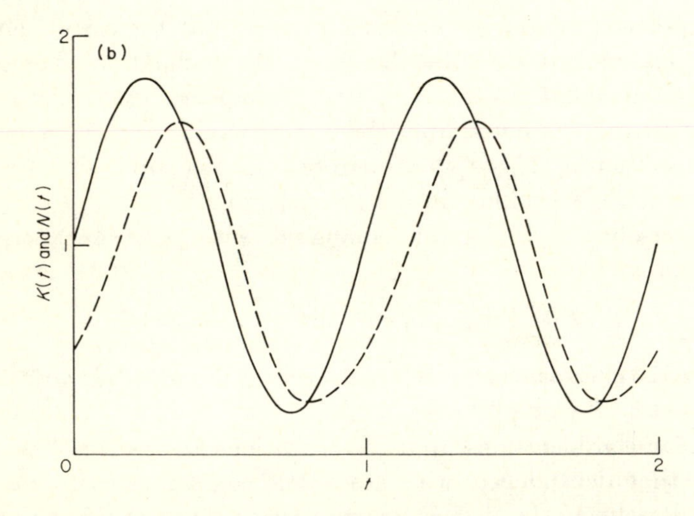

Fig. 2.7b. As for Fig. 2.7a, but in the opposite limit when the natural response time is short (relatively large r) so that the population tracks environmental variations. The details are as for Fig. 2.7a, except that $r = 10$, whence $r\tau = 10$.

populations will tend to average over high frequency (short time scale) components of the environmental noise spectrum, and to track low frequency (long time scale) components; the transition zone occurs for frequency components $\omega \sim r$ (time scales of the order of T_R).

An investigation of the many other aspects of population behaviour in a stochastic environment is largely beyond the scope of this book. Even though the population may average over environmental variations, to maintain an approximately steady value, results such as eq. (2.10) or Fig. 2.7(a) make it plain that this steady value will depend on the relative magnitude of the variations (i.e., on the ratio of K_1 to K_0), and that larger environmental fluctuations will make for lower average population values. Studies of the relation between the magnitude of environmental fluctuations and the probability of the population becoming extinct were initiated by Lewontin and Cohen (1969) and Levins (1969); recent developments are reviewed, e.g., by May (1975a, ch. 5 and pp. 229–231). As a one sentence summary, it may be said that temporal variations in the environment are a destabilizing influence.

2.6 Populations with age structure

The models considered in sections 2.2 and 2.3 pertain to the two opposite circumstances where generations overlap in a continuous manner (leading to first order differential equations) or do not overlap at all (leading to first order difference equations). But many populations have the intermediate structure of several distinct but overlapping age classes. Such is the case, for example, for most commercial fish populations, and most mammals in temperate zones. The consequent description in terms of m interlinked equations for the m age classes has many of the complicating features of multi-species situations, particularly once density dependent effects are included [i.e., the multiple age class analogues of eqs. (2.3), (2.4) and (2.6)]. The hope is that basic elements of such systems are laid bare by considering the opposite extremes of complete and of no overlap, the broad dynamical similarities between which have been the main theme of this chapter.

One result for density independent multiple age class models should, however, be made explicit. This result concerns the relation between the population's actuarial 'life tables' and its intrinsic rate of increase, r. For a single age class model [i.e., eq. (2.5)] this relation is immediate: $r = \ln \lambda$, where λ, the 'finite rate of increase', is the nett number of surviving offspring produced each year, per capita. More generally, in the density independent multiple age class case [the multiple age class analogue of eq. (2.1) or (2.5)], we need to know the effective value of r in terms of the fecundity (m_x) and survival probability (l_x) of the various age classes (x).

The exact form of this relation is well known (see, e.g., Krebs, 1972):

$$\sum_x e^{-rx} l_x m_x = 1. \qquad (2.12)$$

Although this implicit equation for r may be readily solved on a computer, its biological meaning is not at all transparent, and it is ill suited to the sort of general discussion undertaken in chapter 3. An excellent approximation, the biological significance of which is much clearer, is

$$r \simeq (\ln R_o)/T_c. \qquad (2.13)$$

Here R_o is the expected number of offspring (or females) produced by the average individual (or female) over its lifetime,

$$R_o = \sum_x l_x m_x, \qquad (2.14)$$

and T_c is an average generation time, defined in straightforward statistical fashion as

$$T_c = \sum_x x\, l_x m_x / \sum_x l_x m_x. \qquad (2.15)$$

The error introduced by using eq. (2.13) to approximate the exact solution of eq. (2.12) is of relative magnitude $\frac{1}{2}(\ln R_o)\,(\sigma^2/T_c^2)$, where σ^2 is the variance in the generation time. The approximation is good either if $R_o \simeq 1$ (as in human populations), or if the coefficient of variation of the generation time is not too large; a more detailed justification for the use of eq. (2.13) is given by May (1976b).

2.7 Summary

Regardless of whether population growth is a continuous or a discrete process, simple density dependent models show that single populations may exhibit either a stable equilibrium point (damped in a monotonic or an oscillatory manner), or sustained fluctuations (as stable cycles or as chaos), depending on whether the natural periodicities or time-delays in the regulatory mechanisms are short or long compared with the characteristic response time of the system.

Since a population's characteristic response time is often of the order of $1/r$, these considerations introduce elements of self-consistency into the evolution of population parameters. In an unpredictable environment, there will be an advantage in having a largish r, to recover from bad times and to exploit the good times. But large r (short T_R) condemns the population to track environmental fluctuations, and makes for the sort of overcompensation which is inimical to population regulation, thus exacerbating the perceived unpredictability of the environment. Conversely, relatively small r implies a long response time, with the advantage that the population may maintain steady values and may average over environmental variations, but with the disadvantage of slow recovery from traumatic disturbances. Chapter 3 expands and applies these ideas.

This chapter has neglected the effects of spatial heterogeneity, which often has an important stabilizing effect. Such effects are discussed in chapters 3, 5, 9 and 10.

3

Bionomic Strategies and Population Parameters

T. R. E. SOUTHWOOD

In the preceding chapter it has been shown that many of the essential features of the growth of populations of single species in limited environments may be described by simple models. When population growth is a continuous process, an approximate model is

$$dN/dt = rN[1-N(t-T)/K]. \tag{3.1}$$

In the discrete case one has the general eq. (2.6), with, for example, the form B of Table 2.2,

$$N_{t+1} = N_t \exp[r(1-N_t/K)], \tag{3.2}$$

or the form D,

$$N_{t+1} = (\lambda N_t^{-b})N_t. \tag{3.3}$$

The infinite variety of nature may therefore be capable of being caricatured by the few parameters contained in these expressions. Each organism will have a bionomic strategy (i.e., size, longevity, fecundity, range and migration habit) that is summarized by the parameters of these models; this strategy will evolve to maximise the fitness of the organism in its environment. Hence the organism's habitat may be viewed as a templet against which evolutionary forces fashion its bionomic or ecological strategy. This chapter establishes the general form of the relationship between habitat type and bionomic features, as expressed by the impact of habitat on the parameters in the above expressions.

3.1 The parameters

Three parameters may be distinguished:

(1) N^*, the population at equilibrium. Normally in a single species situation this will be $N^* = K$, the carrying capacity.

(2) Time delays, T (the time delay in the response due to some

environmental lag) and τ (the generation time). These time delays have equivalent effects (May *et al.*, 1974), but only τ is directly responsive to evolutionary pressures on the species itself.

(3) The finite rate of increase of the population, $\lambda = \exp r$, itself the difference between gains and losses. This determines the system's return time or natural response time, T_R, which is the time it takes to return to equilibrium, following a disturbance (May *et al.*, 1974). It is a property of the system: for eqs. (3.1) and (3.2), $T_R = 1/r$, and for eq. (3.3), $T_R = 1/b$ (see Table 2.2).

3.2 Habitats

Habitats may be classified according to a number of characteristics:

(1) *Duration stability*, the length of time the particular habitat type remains in a particular geographical location. Thus, for example, a large tree in a climax forest may last for hundreds of years, a herb in successional vegetation for only a handful of seasons, a dung pat for a few weeks. Clearly the significance of this duration stability depends on the relationship between the organism's generation time (τ) and the length of time the habitat remains favourable (H).

(2) *Temporal variability*, the extent to which the carrying capacity (K) of a habitat varies during the time that site is tenable by the organism (i.e., temporal heterogeneity). Variations in K may be predictable, like the spring flush of plant growth in northern deciduous forests, or *un*predictable, like the current destruction of the common elm in Western Europe by Dutch Elm disease.

(3) *Spatial heterogeneity*, continuity versus patchiness. In a tropical rainforest any particular tree species will have a very patchy, scattered distribution; but in northern coniferous forests the same one or two species may continuously cover thousands of square miles.

For any particular animal, the habitat may be defined as that area accessible to the trivial movements of the food-harvesting stages. The range of the animal's movements will therefore determine the scale of the habitat; for the *Drosophila* larva the ripe fruit of a tropical forest tree is a temporary habitat and if it survives to adulthood it will migrate to another site, but for the orang-utan the whole forest is a stable and permanent habitat. Great longevity will tend to reduce the significance of temporal variability and increase the degree of predictability of a given location.

An initial discussion of habitat characteristics may be simplified by confining attention to duration stability; that is, by limiting attention to the value of the ratio τ/H for the habitat.

In those species where τ/H approaches unity, one generation cannot affect the resources of the next; there will be no evolutionary penalty for overshooting the carrying capacity of the habitat. These species are then exploiters, opportunists and (using the terminology of MacArthur and Wilson, 1967) may be referred to as r-strategists.

Conversely, for those animals that occupy long-lived habitats where the carrying capacity (K) is fairly constant, significant overshooting will lower K, and will adversely affect subsequent generations. Many other species will have colonised such stable habitats, and hence interspecific competition in all its forms, including predation, is likely to be intense. Such species are referred to as K-strategists. They are selected for harvesting food efficiently in a crowded environment (MacArthur and Wilson, 1967).

3.3 Size and its implications

A number of extremely significant bionomic characters are closely linked with the size of an organism. In responding to the evolutionary pressures arising from its habitat, a modification of the organism's size will inevitably move it one way or another on the r-K continuum of strategies.

A key parameter is clearly r, the per capita rate of increase. As discussed in chapter 2, r is effectively dependent on the nett reproduction rate, R_o, and the generation time, T_c:

$$r \simeq (\ln R_o)/T_c. \tag{3.4}$$

There is a strong positive correlation between size and generation time in organisms ranging from bacteria to whales and redwoods (Bonner, 1965; see Fig. 3.1). This relationship is probably due to longevity being inversely proportional to total metabolic activity per unit of body weight, and to the fact that the smaller the organism the greater the level of this activity.

A consideration of eq. (3.4) immediately shows that r tends to be more sensitive to changes in the generation time than to changes in R_o. Halving the generation time will double r; but doubling R_o (e.g., by doubling fecundity) will only increase r by the difference of the values

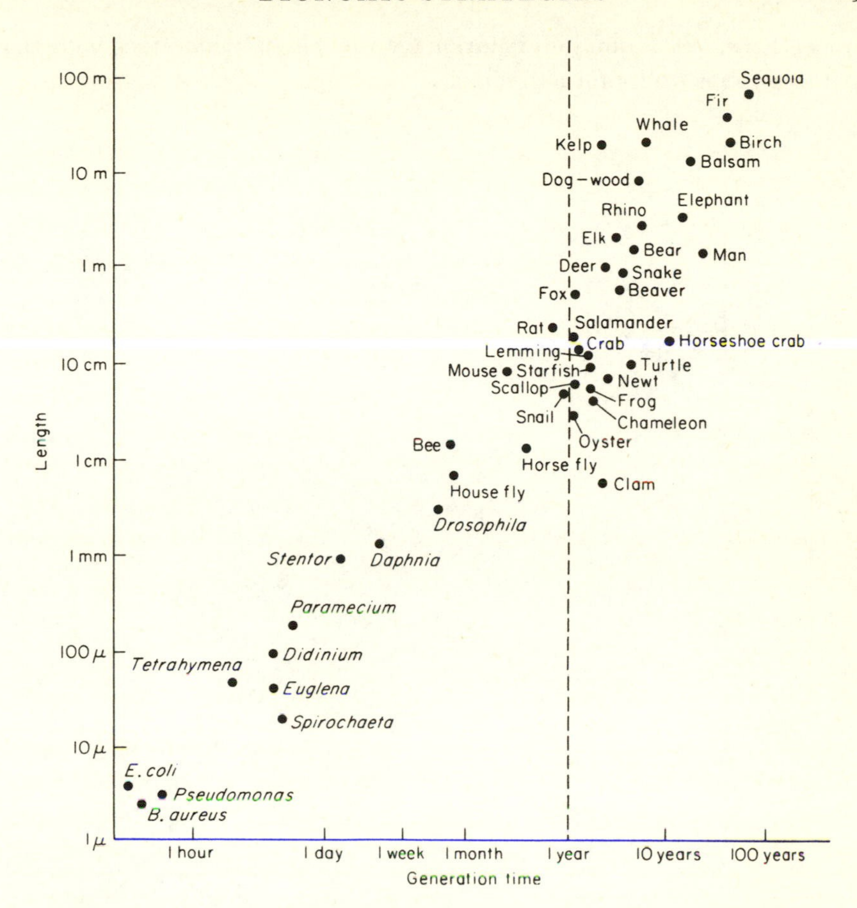

Fig. 3.1. The relationship of length and generation time, on a log-log scale, for a wide variety of organisms (from Bonner, 1965).

of the natural logarithms (i.e., R_0 from 5 to 10 would lead to changes in r from 1.6 to 2.3). It is therefore not surprising that the relationship of r to T_c for a wide range of organisms falls into a narrow straight band, with slope -1, when plotted on a log-log scale (Heron, 1972; see Fig. 3.2). This train of consequences of size change may be expressed as

$$r \propto \frac{1}{T_c} \propto \frac{1}{\text{size}} \propto \text{metabolic rate per unit weight.}$$

Indeed Fenchel (1974) has shown that the relationship of r, and of metabolic rate/unit weight, to organism size are strikingly parallel (Fig. 3.3), both falling to three groups; unicellar, poikilotherms and homeotherms. Each major evolutionary step slightly increases both the metabolic rate and r for a given size.

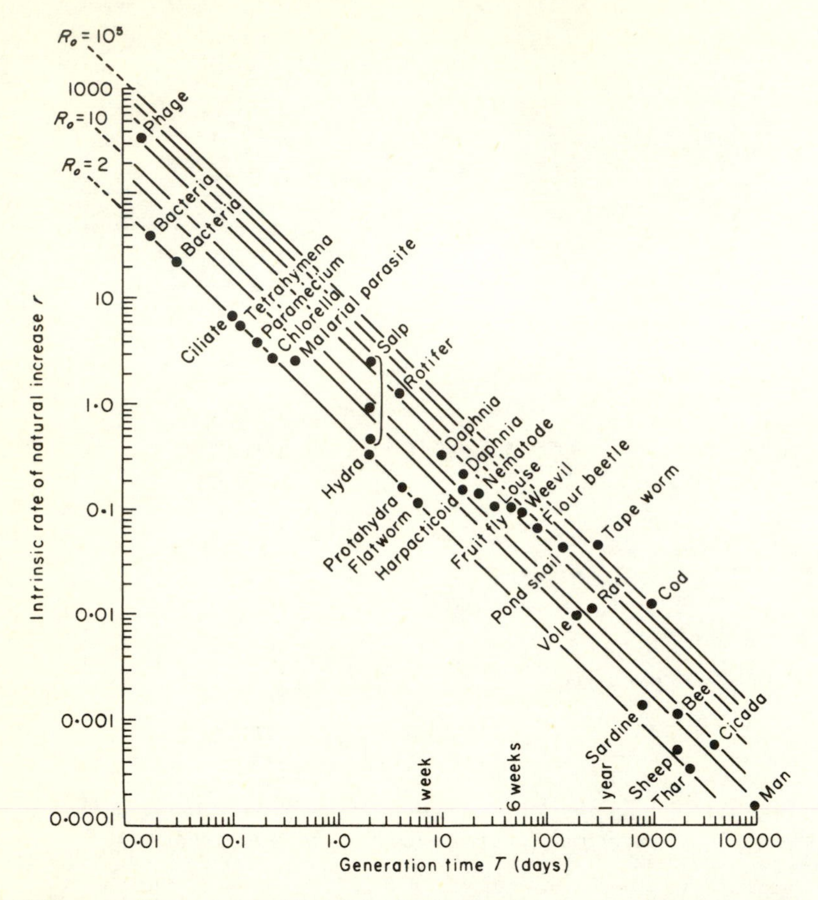

Fig. 3.2. The relationship of the intrinsic rate of natural increase and generation time, with diagonal lines representing values of R_o from 2 to 10^5, for a variety of organisms (from Heron, 1972).

Size change has other implications for the ecology of the species. Larger species are often at an advantage in interspecific competition (lions may drive hyaenas from their kill) or in defence from predators. The allometric growth of offensive or defensive appendages will enhance this ability and, combined with longevity, allows the possibility of a high level of parental care and protection.

The size of an animal will influence the scale of its habitat. To return to the example of two species living on tropical fruits, the size of the orang-utan means that the range of its trivial movements encompasses many types of tree, so that the habitat is predictable and stable. The *Drosophila* larva, on the other hand, is limited to a few centimetres around the oviposition site. The adult stage of *Drosophila*

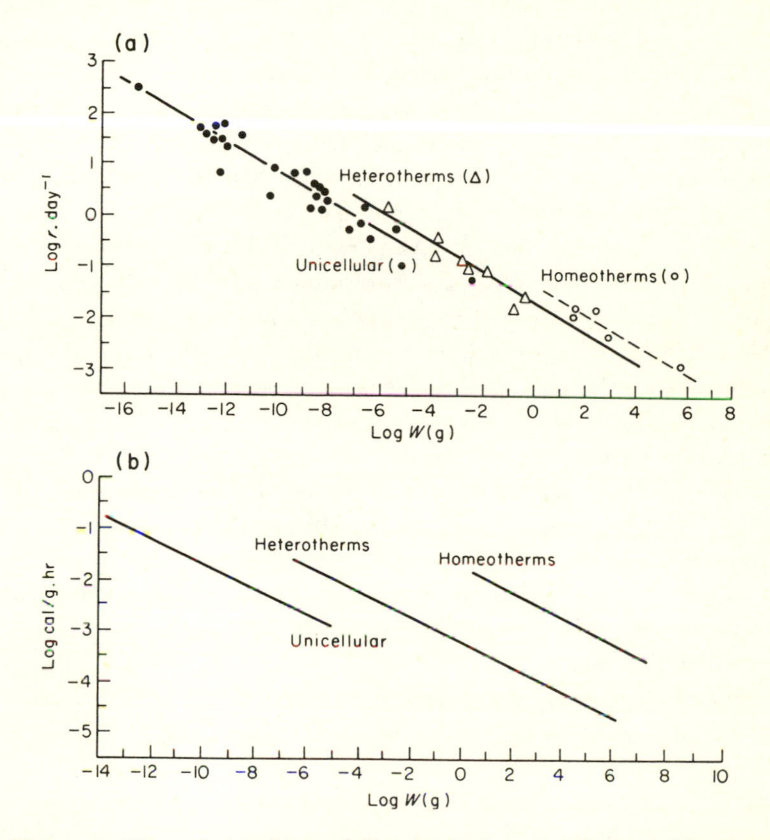

Fig. 3.3. The relationships of the intrinsic rate of natural increase (**a**) and metabolic rate (**b**) to weight for various animals (from Fenchel, 1974).

migrates to new food sources, and indeed many small insects can travel great distances on air currents (Johnson, 1969). But small wingless animals are perforce restricted in the scale of the migrations they can undertake on land. The regular long-distance migrations of the wildebeest on the savanna cannot be emulated as an ecological strategy by a small rodent.

3.4 Fecundity versus longevity

The energy available to an organism capable of reproduction may be directed towards the survival and growth of that organism, or towards the production of offspring, or towards some partitioning between the two. As with size, any change in the allocation of energy will influence the species' bionomic strategy, and hence the population parameters.

The problem can be illustrated by reference to hypothetical and simplified organisms, the parthenogenic 'block-fish' (Fig. 3.4). Each summer the productivity of each block fish is two blocks. These may be added to itself or used in reproduction. Any fish that does not add to itself dies. Every winter each 'block-fish' gains one block; breeding does not occur at this season. As shown in Fig. 3.4, there are two mutations. Mutation 1 puts all its summer productivity into reproduction, and at the end of three years is represented by eight 1-block-fish. Mutation 2 only puts half its summer productivity into reproduction, and at the end of three seasons is also represented by eight individuals; but these show an age and size range from a 7-block-fish to four 1-block-fishes.

From this information we cannot tell which mutation will be successful. Let us consider various possible conditions imposed by the environment.

(1) Carrying capacity limited to a total of 8 blocks. Mutation 1 would remain as shown, but in mutation 2 only the large 7-block-fish and its small 1-block daughter would survive. The population of mutation 2 is thus more vulnerable to extinction, whilst the population of mutation 1 could easily 'bounce back'; even the loss of half its individuals would not affect its size next year.

(2) Half the juveniles (1-block-fish) are killed by predators each year. If this happens, mutation 1 remains as one 1-block-fish, whereas mutation 2 has a population of three (7-, 3- and 1-block).

Various other conditions can easily be envisaged and their outcome determined. For example: (a) block-fish over a certain size are able to exploit additional resources (*K* measured in standing crop of blocks is increased); (b) larger block-fish become less efficient, and so their productivity falls below three blocks per year; (c) the productivity is a proportion of size; (d) mortality is higher on larger block-fish; (e) the block-fish has a strict size limit (which is realistic for the majority of organisms other than fish and plants). In this last case (e), surviving individuals simply become older; often, as in most higher vertebrates,

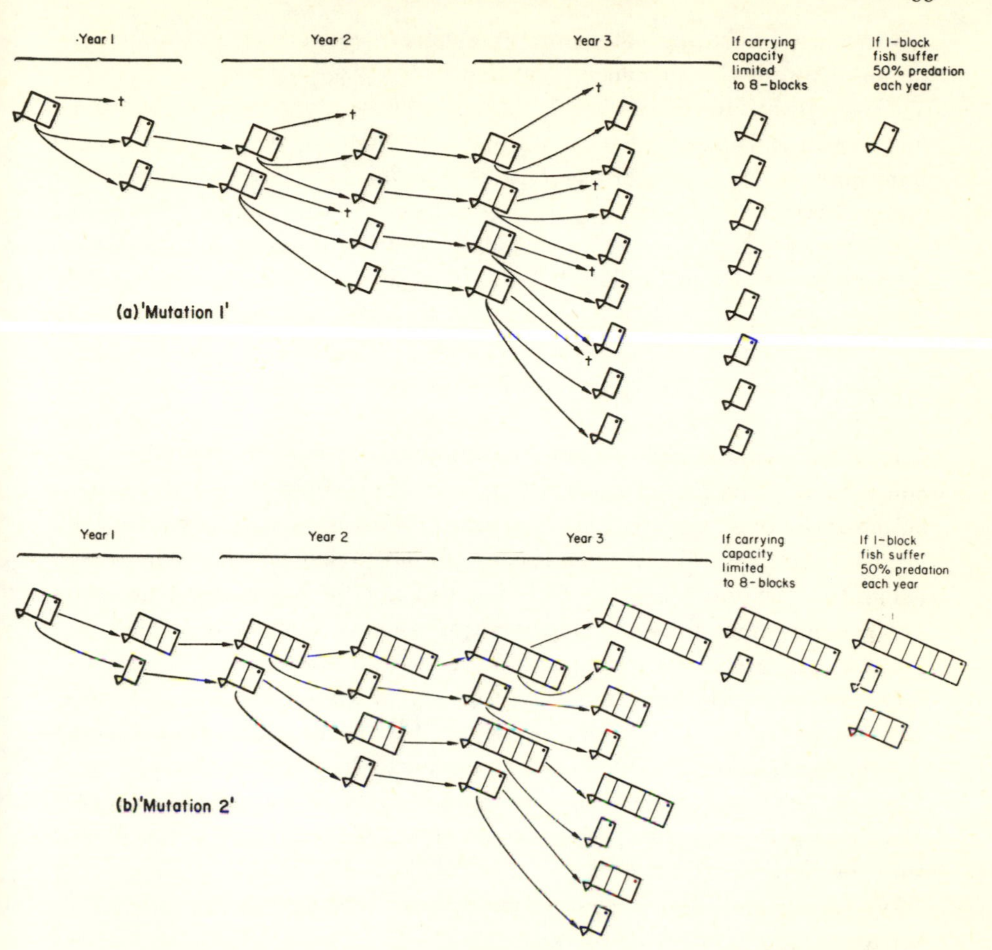

Fig. 3.4. The outcome of two different strategies for allocating resources in a hypothetical animal, the block-fish: (**a**) Mutation 1 in which summer productivity (2 blocks) is entirely allocated to reproduction, (**b**) Mutation 2 in which one block of summer productivity is allocated to reproduction and one block to growth and, hence, to adult survival.

this is accompanied by increased competitive advantage (for example in holding territory).

The optimum allocation of resources between reproduction and maintenance, illustrated above, has been explored algebraically by several workers (e.g., Schaffer, 1974; Smith and Fretwell, 1974). Their conclusions are essentially those that are intuitive from the above: the optimum strategy depends on the different competitive advantages (including, of course, ecological efficiency) and survival probabilities of the various ages and sizes.

The allocation of resources to territory holding may be viewed in the same way. In general it relates to condition (e) above: but sometimes, as in the lizard *Sceloparus jarrovi*, larger individuals hold larger territories (Simon, 1975). An analysis of the costs of territory defence in relation to food availability has been made in the sunbird, *Nectarinia reichenowi* (Gill and Wolf, 1975a).

3.5 The *K*-strategy

The *K*-strategists have a stable habitat (τ/H is very small), and consequently they evolve towards maintaining their population at its equilibrium level, and towards increasing their interspecific competitive ability. Thus they will often be selected for large size, which will increase generation size and lower r. But this will not be a disadvantage: the evolutionary pressures are for the population to remain at or close to K, but not above it—this could lead to habitat degradation. High levels of fecundity are thus not essential if the reduction in births can be matched by increased survival. Plants and animals that are *K*-strategists make a significant investment in defence mechanisms. Parental care is facilitated by low fecundity (small litters or clutches but large size offspring), longevity and size. This reduction in mortality may be considered to lead to more efficient use of energy resources (Cody, 1966): indeed extreme *K*-strategists like the Andean condor, the albatross and large tropical butterflies (e.g., *Morpho*) are noteworthy for their low energy movement, namely gliding.

However, when *K*-strategists do suffer mortality and perturbations, their populations need to return quickly to equilibrium levels or competitors may seize the resources. This implies that b in eq. (3.3) should approximate to unity, and, because there is little mortality, this will tend to be accomplished through the birth rate. In other words the birth rate will be very sensitive to population density, and will rise rapidly if density falls (Southwood *et al.*, 1974). In many vertebrates this is achieved by increased litter and clutch size and by 'bringing' non-breeders into breeding, either by a shortening of the pre-reproductive period or by a modification of the social structure that had previously excluded them from territories (Wynne-Edwards, 1962; Southwood, 1970). The large reserves of tree propagules that rest in forest soils, germinating only when a clearing arises, provide another strategy for the same ends.

K-strategists can therefore be recognised by large size, longevity, low recruitment and mortality rates, high competitive ability and a large investment in each offspring (Pianka, 1970; Southwood *et al.*, 1974; Rabinovitch, 1974). Their population levels will stay close to the equilibrium level and their mating tactics will be geared to this density. The communal nuptial displays observed in bowerbirds, manakins, birds of paradise, certain humming birds, the ruff and the blackcock provide examples (Wynne-Edwards, 1962). Harems are another behavioral adaptation of this type, with their size being that which maximises the longevity of the individual female (Elliott, 1975).

K-strategists are unlikely to be well adapted to recover from population densities significantly below their equilibrium level, and if depressed to such low levels they may become extinct. These organisms, rather than r-selected species, need the concern of the conservationist. The fossil record shows that many lines of animals increase continually in size until extinction: this has been called Cope's rule. These lines have become progressively more K-selected; more and more closely adapted to a specialised and hitherto stable habitat (Bretsky and Lorenz, 1970). Thus the extinction of the dinosaurs was probably due to their inability, because they were extreme K-strategists, to respond to the changes in climate at the end of the Cretaceous (Axelrod and Bailey, 1968; Southwood *et al.*, 1974).

3.6 The *r*-strategy

The r-strategists are continually colonizing habitats of a temporary nature (τ/H is not small), and they are exposed to selection at all population densities. Their strategy is basically opportunistic, 'boom and bust'. Migration will be a major component of their population process, and may even occur every generation (Southwood, 1962; Dingle, 1974; Southwood *et al.*, 1974; Kennedy, 1975).

Selection will favour a high r, arrived at by a large fecundity (large R_o) and short generation time (T_c). As the habitats they colonize are often virtual ecological vacuums, high competitive ability is not required, and they will typically be small in size. Mortality rates may be high; migration, which is an essential component of their fugitive existence, is invariably wasteful. Their main defences against predators, other than a high fecundity, are often a measure of synchrony (and hence temporary satiation of the predators) and their mobility—a hide and seek strategy. With plants this relative lack of defence is very neatly shown by Cates and Orians' (1975) studies on the palatability to slugs of plants from various successional stages (Table 3.1). Only those plants later in the succession make a significant investment in chemical or physical defence.

Table 3.1. Average palatability to slugs, *Arion* and *Ariolimax*, of herbs from various successional stages (after Cates and Orians, 1975).

Plant community	Number of plant species tested	Palatability index* for *Arion*	Palatability index* for *Ariolimax*
Early successional annuals and biennials	18	0·99	0·96
Early successional perennials	45	0·69	0·77
Later successional and climax plants	17	0·40	0·46

* Palatability index is defined as log (amount of test material eaten)/log (amount of control eaten).

Very high values of r lead to instability (Levins, 1968; May, 1974a, 1975a), but the extreme r-strategist is by definition not likely to be affected (Barclay, 1975). A population will not spend many generations, perhaps only one or two, in that particular habitat, and migration is such a capricious process that only a small fraction of mortality will be density dependent. Thus unrealistically high values of r would be necessary if they were, of themself, to produce instability (Southwood, 1975). In other words, at the extreme end of the r-spectrum populations come into being in a new (or newly discovered) habitat with a few colonies; they pass through at the most a handful of generations when the models expounded in chapter 2 apply; and then as the habitat deteriorates (perhaps overcrowded with strong density dependence acting), migrants depart and establish new populations, freed of density restraints.

Because r-strategists occur at such varying densities, mate-finding tactics are likely to be efficient at low numbers and, although extinction will regularly be the fate of individual populations (as their habitats change), the species as a whole will be very resilient. Furthermore, their high mortalities, wide mobility and continuous exposure to new situations are likely to make them fertile sources of speciation.

3.7 Population dynamics and habitat characters

Given the differences in population parameters outlined above, the population growth curves of the extreme *r*- and *K*-strategists will be of very different forms (Fig. 3.5). The *K*-strategist will tend to have a stable equilibrium point (*S* in Fig. 3.5), to which the system returns after moderate disturbances; but if the population declines below some lower threshold (in Fig. 3.5) it cannot recover, and it decreases

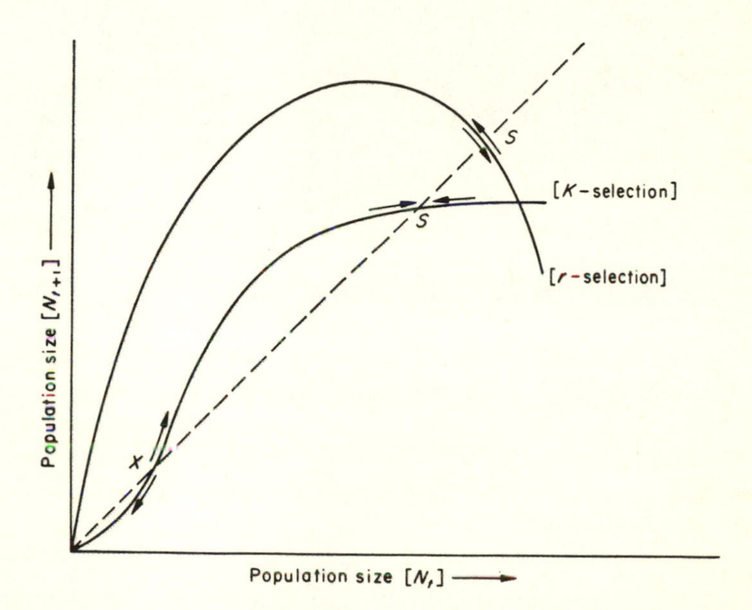

Fig. 3.5. The population growth curves of *r*- and *K*-strategists, showing stable (*S*) equilibrium points and an unstable point, the extinction point (*x*) (from Southwood *et al.*, 1974).

ineluctably to extinction. Conversely, the *r*-strategist's population grows rapidly at low densities, has an equilibrium point (*S* in Fig. 3.5) about which it is liable to oscillate, and crashes down from high densities. Natural enemies are unlikely to be important for these extreme strategies: for the *r*-strategist because enemies will be unlikely to colonize in sufficient numbers sufficiently quickly; for the *K*-strategist because of its large size and high competitive ability. Most organisms are, of course, in an intermediate position and here natural enemies will have a significant role, and may provide at least one further equilibrium point part-way up the population growth curve. This concept is developed in chapter 5.

On the basis of the accumulating evidence from natural populations,

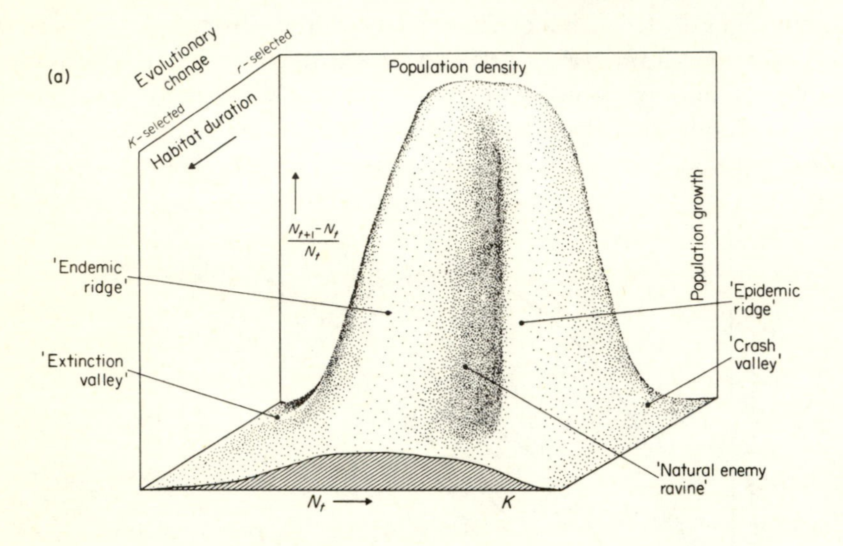

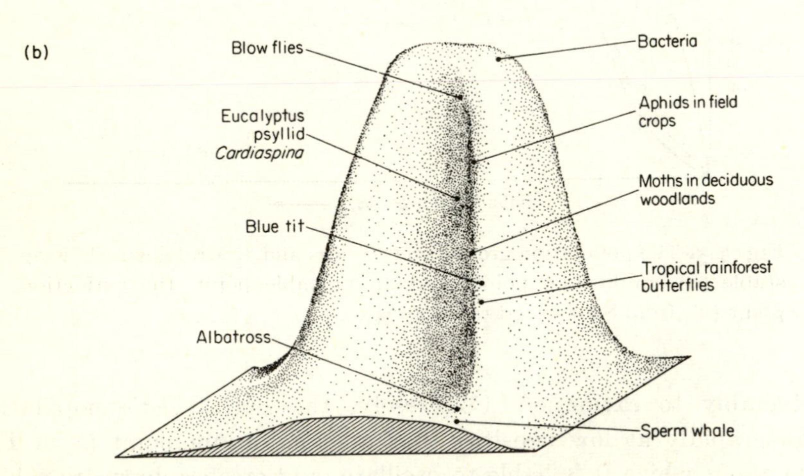

Fig. 3.6. The synoptic model of population dynamics showing (**a**) the general features of population growth in relation to population density and habitat duration and (**b**) the positions of some organisms on the landscape. (**a**) modified from Southwood (1975).

it is possible to construct a three-dimensional synoptic model of population dynamics (Fig. 3.6, after Southwood, 1975). Besides illustrating the points made above as to the role of habitat, this model stresses that *r*- and *K*- are a convenient shorthand for the ends of a continuum and not the branches of a dichotomy.

In addition to the variations shown by the synoptic model (which is, of course, deterministic) other patterns will arise in species population dynamics due to spatial and temporal heterogeneities in the habitat. As already stressed, for an animal the extent of the spatial heterogeneity of a given area will be influenced by the size and mobility of the animal. In general, spatial heterogeneity tends to be stabilizing. Variations in many environmental factors are unlikely to occur synchronously in all the patches, and thus the species 'spreads the risk' through its occupancy of many small habitats (den Boer, 1968; May, 1974b). However, if the numbers in these populations drop to very low levels, so that population size can no longer be regarded as a continuous variable, then a new factor, the stochastic variation of small integer numbers, enters. This increases instability and the chances of random extinction. The

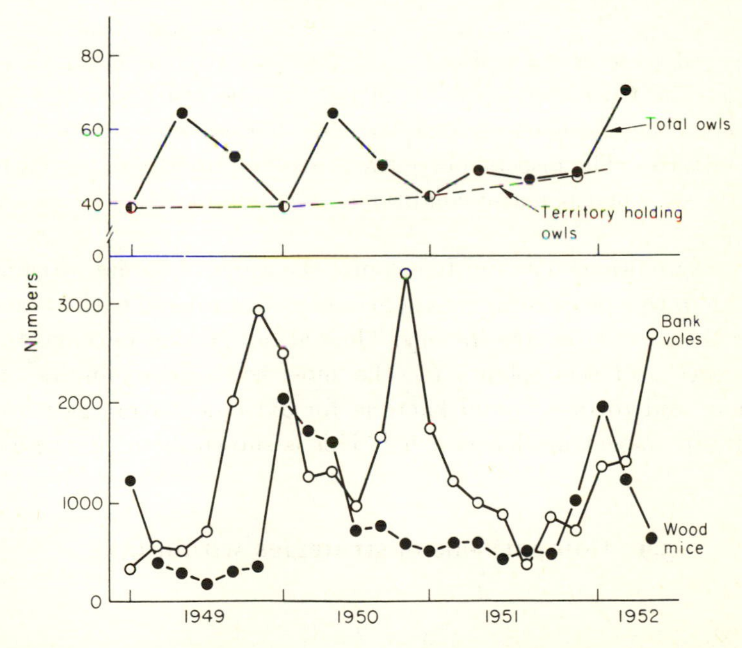

Fig. 3.7. Fluctuations in the numbers of animals with different longevities in the same habitat. (Tawny owls, Bank voles, and woodmice in Wytham Wood, 1948–52: data from Southern, 1970.)

extinction of a Californian colony of the checkerspot butterfly, *Euphydryas editha*, (Ehrlich *et al.*, 1975) and the many British colonies of the Large Blue Butterfly (*Maculinea arion*) (Muggleton and Benham, 1975) may be examples of this.

Temporal heterogeneity in the habitat will increase instability (Blackith, 1974; May, 1974b, 1976a), but again large, long-lived animals with low r-values and other strategies will average out the variations: species with short generation times will track the variation of the environment. The different population traces for tawny owls, voles and mice living in the same woodland illustrate the different scale and amplitude of fluctuations in different animals with different strategies; see Fig. 3.7.

3.8 Comparison of strategies between major taxa

As large organisms are basically K-selected and small ones r-selected, it is not unreasonable to consider vertebrates as K-selected and insects as r-selected (Pianka, 1970). It is interesting to note that the overall conditions prevailing at different periods of geological time probably encouraged the evolution of these different groups. The main bursts of evolution in the vertebrates occurred in the Jurassic, the lower Cretaceous, the Eocene and the Oligocene, periods of warm, moist and stable climates. The insects, in contrast, evolved rapidly in the Permian and Triassic when climatic conditions are thought to have been very variable.

All taxa however attempt to colonize the range of habitats available, although in this process they are, of course, encumbered with attributes due to their evolutionary history. Thus although one must go to the vertebrates and seed plants for the most extreme examples of K-selection, and to insects and bacteria for extreme r-strategies, within each group a spectrum has evolved. This is shown in the next section.

3.9 Comparisons of strategies within taxa

3.9.1 *Birds*

A measure of the reproductive effort of birds is easily obtained from the clutch-size, and there has already been considerable controversy over its significance (Wynne-Edwards, 1962; Lack, 1966, 1968; Mountford, 1973). Much of the disagreement seems to arise because of the different mechanisms operating in the *K*- and *r*-selected species (Cody, 1966; Southwood *et al.*, 1974).

Examples of some of the most extreme *r*-selection in birds are provided by the zebra finch, *Taeniopygia castanotis* and the budgerigar, *Melopsittacus undulatus*, which are nomadic in Central Australia and have some of the shortest times to breeding. Another species with a short generation time (small T_c) is the Japanese quail, *Coturnix coturnix japonica*, a migrant. The largest clutch size (R_o) recorded by Lack (1968) for a passerine is that of the blue tit, *Parus caeruleus*. This bird may also be regarded as opportunistic: its breeding season food, moth larvae in deciduous woodlands, fluctuates greatly and its own population can show a two-fold change over a single season. All these birds are small, or small for their orders: the budgerigar is among the smallest of the parrots and the Japanese quail is tiny for a game bird.

At the other end of the spectrum are some excellent examples of *K*-selection. The large condors, *Gymnogyps* and *Vultur*, lay only a single egg every other season. The conservation of both is a matter of concern. *Vultur vultur*, with a wing span of over ten feet, is the largest of birds of prey, and this family (Cathartidae) contained *Teratornis*, the largest flying bird known from the fossil record. Less endangered, probably only because its breeding sites are relatively free from human disturbance, is the wandering albatross, *Diomedea exulans*. Again it only breeds in alternate years (when successful) and has a clutch size of one and a wing span of over ten feet; its period of immaturity (9 to 11 years) is the longest for any bird. One of its breeding populations, on Gough Island, is considered to have remained almost constant at 4,000 since 1889 (Elliott, 1971). All the Procellariiformes, the order to which the albatross belongs, and the related Sulidae are noteworthy for their low *r* values, arising from long pre-reproductive periods (large T_c), and low clutch size (small R_o). They select breeding sites free of predators and forage over great areas of ocean (Lack, 1968). This *K*-strategy can be accounted for in terms of individual Darwinian fitness (Goodman, 1974). Foraging for squid and fish may be considered to require experience, and therefore if young birds bred they would be unable to support their own offspring: indeed there is evidence that they often cannot support themselves (Jarvis, 1974). Recent studies on the South African gannet, *Sula capensis*, have shown that although birds may rear two young these are lighter than chicks from single clutches, and their chances of survival through the post-fledgling period, the time of heaviest mortality, are significantly reduced (Jarvis, 1974). In other words the populations of these species are close to the effective carrying capacity of the habitats around the breeding grounds; a pair of birds cannot easily double the amount of food harvested. With a small clutch size there will be strong selection against any activity that lowers adult survival. The position is similar to strain 2 of the 'blockfish', with heavy mortality (>50 per cent) of the juvenile 1-block-fish. The population will depend on survival of the older block-fish to reproduce in several years, and it will be seen that doubling their annual reproduction would not be successful if it halved the survival of the adults.

Goodman (1974) calculated from a model for *Sula sula* that fledgling success needed to be increased at least 19 times more than the fractional amount by which this increase was at the expense of decreased parental survival: this model underestimated the function because it did not allow for the high post-fledgling mortality, since shown by Jarvis (1974). High fledgling loss is also a feature of the tawny owl, *Strix aluco*, another *K*-strategist (Southern, 1970); see Fig. 3.7. Therefore any tendency for parent birds to decrease their survival by foraging for too large a clutch size, or by breeding away from the 'protected colonies', will be selected against. There is thus no conflict between this *K*-strategy and natural selection as it is normally understood. These birds are exploiting habitats that, on the scale on which they forage, are very stable; they cannot greatly increase the rate at which they harvest resources. The long life span of adults makes the population very resistant to variations in recruitment, but recovery from abnormal adult mortality is difficult. These are characteristics of *K*-strategies.

The more the habitat varies, the larger the clutch size; the species can exploit the additional food by faster population increase. This

Table 3.2. Clutch size in passerine birds in relation to habitat (from Southwood *et al.*, 1974).

Habitat	Number of species	Average clutch size
Tropical forests	82	2·3
Tropical savanna, etc.	260	2·7
Tropical arid areas	21	3·9
Middle Europe	88	5·6

general tendency is illustrated in Table 3.2, which shows the relations of clutch size to habitat for passerine birds.

3.9.2 *Insects*

In common with a number of other invertebrate groups, insects have a complex metamorphosis which allows the possibility of any individual having more than one habitat during its life. However,the major part of the food harvesting normally occurs in the larval stage, and thus it is normally the character of this habitat that dominates the species strategy.

Lepidoptera

Various tropical butterflies have been noted for their longevity and/or the stability of their populations: *Morpho* (Young and Muyshondt, 1972); *Heliconius* (Ehrlich and Gilbert, 1973); *Charaxes* (Owen and Chanter, 1972); *Hamadryas* (Young, 1974). These butterflies are often large and territorial. Some *Morpho* may take up to 10 months to reach maturity. These characters place them towards the *K*-selected end of the spectrum. Adult survival is relatively high and, like the sea birds discussed above, the number of eggs laid at any one time is small (compared with most *Lepidoptera*), although the total for the whole adult life span is not dissimilar to that of related species. The wide distribution of eggs in space and time allows these species to compete successfully in habitats where there is heavy competition from predators and other herbivores, and where resources are scattered and limited. Again like the sea birds, their populations are resilient to fluctuations in recruitment.

Relatively migrant pest species exhibit *r*-strategies. The army worm, *Spodoptera exempta*, may lay up to 600 eggs: generation time is just over 3 weeks and large populations (outbreaks) occur on the young growth of graminaceous plants (Brown, 1962).

Parasitic Hymenoptera

As larvae, members of this group are mostly internal parasites of other insects. In the Ichneumonidae the females place the eggs in or on the host's body: if the host is inaccessible the ovipositor will be very long, as with *Rhyssa* that bores through timber to reach its host, wood-boring sawflies. Generally, the longer the ovipositor the more effort the female has to expend ovipositing each egg and, although less precisely correlated, the better protected the host (and hence the parasite larva) from predation. Within Ichneumonidae, Price (1972, 1973) has found a close relationship between a long ovipositor, lowered fecundity and increased egg size. More specifically, he has shown that with the same host, the sawfly *Neodiprion*, the fecundity (expressed as number of ovarioles per ovary) falls the later in the host's development the attack is made; see Fig. 3.8a. When this curve is compared with a survival curve

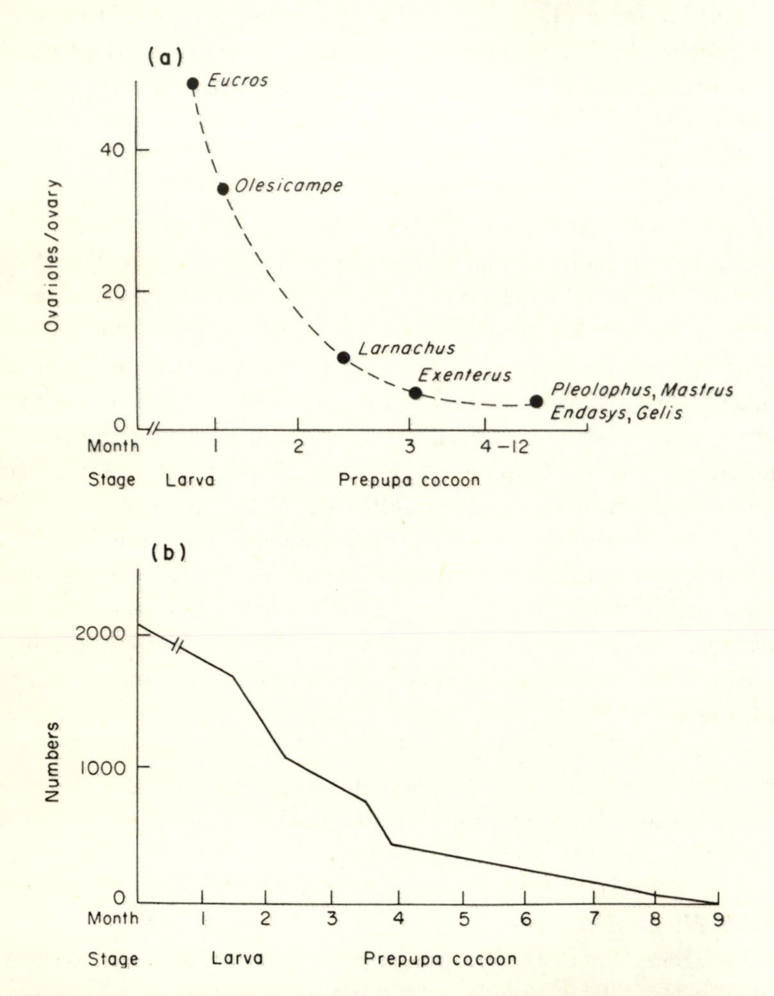

Fig. 3.8. The fecundity of various parasitoids in relation to the survival of their host, the sawfly *Neodiprion*, at the time of parasitism: (**a**) the number of ovarioles per ovary in various parasitoids against the chronology and stage of attack (after Price, 1972); (**b**) survivorship curve for a species of *Neodiprion* in California (after Dahlsten, 1967).

for the host (Dahlsten, 1967, see Fig. 3.8b), the correlations among high host availability, low host life expectency, and high parasite fecundity are apparent.

Sarcophagidae

This family of dipterous flies are commonly known as flesh flies and the viviparously produced larvae of most species are carrion feeders—strongly opportunistic and *r*-selected. However, *Blaseoxiphia fletcheri* larvae live in the fluid in the tubular leaves of pitcher plants. This is a much more stable habitat. Forsyth and Robertson (1975) found that the larval populations of this fly remained close to the carrying capacity of the habitat, the larvae were territorial, the fecundity was lower than normal (11 larvae per female compared with 50–170) and the individual first stage larve much larger (6·9–7·0 mm compared with a range of 1·2–5·4 mm for other species). This provides an excellent example of how the stability of the habitat influences the range of characters, covered by the shorthand '*r*-*K* continuum', in a relatively narrow taxon.

3.9.3 *Plants*

One of the characters of a *r*-strategy is the high allocation of energy to reproduction, whilst *K*-strategists allocate more to mortality avoidance and competitive ability. Following the lead of Harper and Ogden (1970), several recent studies have assessed the resources plants allocate to reproduction. For example, the dandelion, *Taraxacum officinale*, exists in various biotypes; the more disturbed and transient the habitat, the larger the proportion of those biotypes that devote more of their biomass to reproduction (Gadgil and Solbrig, 1972).

Another plant with a range of habitats is *Polygonum cascadense*. Hickman (1975) found that the allocation of resources to reproduction was highest in those habitats where species diversity was lowest, vegetative cover was least and sap potential low. This is illustrated in Fig. 3.9.

A number of different species of goldenrod, *Solidago*, occur in eastern USA in sites ranging from woodlands to open dry disturbed sites, which are early successional stages. The biomass allocation to flowers and stems in these two extremes are compared in Fig. 3.10 (Abrahamson and Gadgil, 1973). This clearly shows the greater allocation to reproduction in the more unstable habitat, a distinction that is greater where

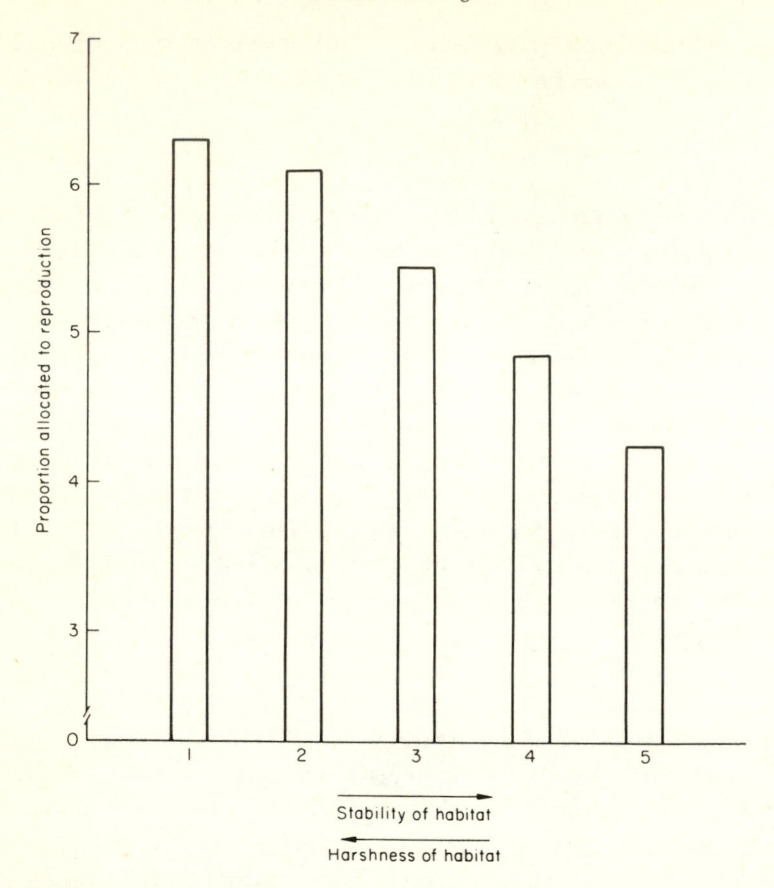

Fig. 3.9. The proportion of total weight allocated to reproductive organs in five populations of the plant, *Polygonum cascadense* in Oregon (data from Hickman, 1975).

two different species (*S. rugosa* and *S. nemoralis*) are involved than between different forms of the same species (*S. speciosa*).

Reedmace or catstails (*Typha* spp.) are plants of damp habitats, and seeds will only colonize new habitats, because *Typha* litter inhibits germination (McNaughton, 1975). Thus an r-K selection continuum might be expected in relation to the extent of colonizing opportunites, i.e., habitat duration (τ/H). McNaughton found strong evidence for such a continuum. In the most unstable habitats *Typha* species had genotypes with greater developmental speed (small τ), with higher fecundity (high R_o), but with less energy in each offspring (i.e., smaller size). These genotypes had high colonizing ability, but low competitive

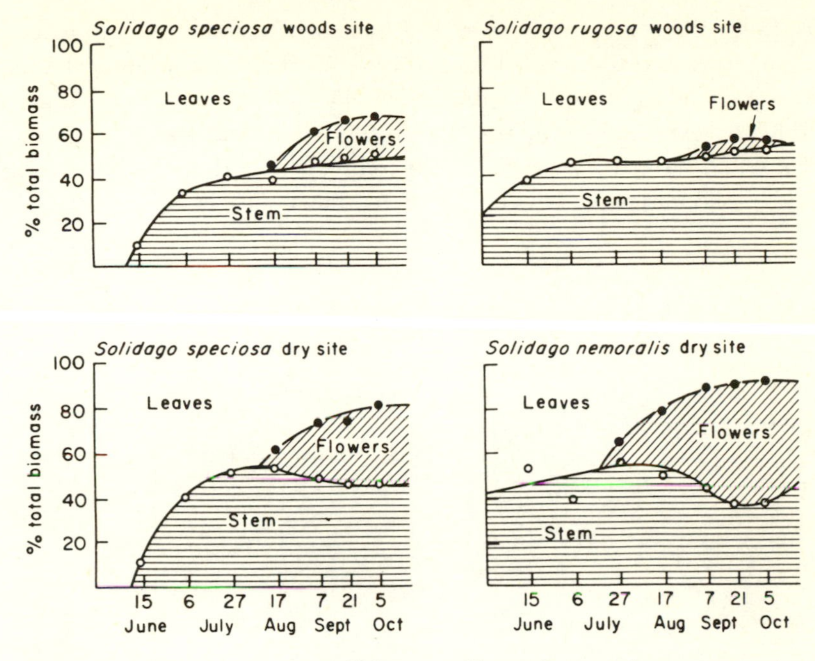

Fig. 3.10. The percentage of biomass allocated at various seasons, to stems and flowers in four goldenrod populations; the species (*Solidago speciosa*) that colonizes both types of habitat shows the differences, but to a less marked extent, that are found between the two species with more limited habitat range (from Abrahamson and Gadgil, 1973).

ability: the genotypes in more stable habitats showed a reversal of these traits.

3.9.4 *Conclusion*

Thus one can see how the interrelated bionomic characters of an organism, such as size, longevity and fecundity, have evolved to give a pattern of population dynamics that is adapted to the features of its habitat. The organism's 'preception' of the scale of its habitat's heterogeneity in both space and time is itself influenced by these bionomic characters. All these variables interact to produce a broad continuum of strategies from small opportunists to large dominants, conveniently referred to as an *r*-*K* continuum. Within any taxa a portion of this continuum will be exposed; in addition to those referred to above it has been recognized in other groups from kangeroos (Richardson, 1975) to corals (Loya, 1976). However, every species is to some extent a prisoner of its evolutionary history; the processes of mutation and selection govern the rate with which it can respond to changes in the features of its habitat. As the environment changes, often particularly rapidly under man's influence, the extent of adaptation may be incomplete. The interlocking features of bionomics, ecology and habitat as we see them today are, as it were, but a single frame in the film of the evolution of the biosphere.

4
Models for Two Interacting Populations

ROBERT M. MAY*

4.1 Introduction

This chapter outlines some of the general dynamical features of models for two interacting populations.

Such pairwise interactions may have one or other of three basic forms: if the effects are such that the growth rate of one population is decreased, and of the other increased, we have the prey-predator situation (the '– +' case); competition occurs if both growth rates are depressed by the co-occurrence of the two species (the '– –' case); and reciprocal enhancement of growth rates corresponds to mutualism (the '+ +' case). These three cases are considered in sections 4.2, 4.3 and 4.4, respectively.

The general discussion in chapter 4 sets the stage for the next three chapters, which describe some particular instances where these theoretical insights can be put more or less closely beside field and laboratory data.

4.2 Two populations: prey-predator

4.2.1 *Continuous growth (differential equations)*

Most elementary ecology texts contain an account of the simplest model for a prey-predator system, namely the classical Lotka-Volterra differential equations, which were first studied around 1920:

$$dN/dt = aN - \alpha NP \tag{4.1}$$

$$dP/dt = -bP + \beta NP. \tag{4.2}$$

* Some of this work was supported in part by the U.S. National Science Foundation, under grant number 13MS 75–10646.

Here the prey population, $N(t)$, has a propensity for unbounded exponential growth, aN, which is limited by predation: the effect of the predators upon the prey population is measured by the 'functional response' term, αNP. The predator population, $P(t)$, has an intrinsic death rate, $-bP$, and a growth rate or 'numerical response' which depends on the prey abundance as βNP.

This system has pathological dynamical properties, namely the neutral stability of the frictionless pendulum. The system oscillates with a period determined largely by the parameters of the model (the period is approximately $2\pi/\sqrt{ab}$, but does depend weakly on the amplitude of the oscillation), but with an amplitude determined solely, and forever, by the initial conditions. If the system is disturbed, it will then oscillate in some similar neutrally stable cycle, with a similar period but with a new amplitude determined by the disturbance. This pathological neutrally stable behaviour depends sensitively on the model possessing the exact structure of eqs. (4.1) and (4.2); the slightest alteration in the mathematical expression given to the various terms in these equations will tip the dynamics towards a stable point, or towards a stable limit cycle, as discussed below. Such a model is called 'structurally unstable'. Structurally unstable models have no place in biology (although they do have legitimacy in physics, where they derive from deep and special symmetries, such as translational invariance, in the physical world: for fuller discussion, see May, 1974b and 1975a, pp. 50–53).

These severe reservations having been expressed, it remains true that the classic Lotka-Volterra model does lay bare one of the general properties of prey-predator models, namely a propensity to oscillations. The mechanism is clear: high prey densities tend to produce high predator densities, which tend to depress prey densities, which makes for lower predator densities, which leads to higher prey densities, and so on. Whether or not this oscillatory tendency is damped depends on the details.

Relatively realistic models may be obtained by modifying the predator-free prey growth term (aN) to include density dependence, and by modifying the crude 'binary collision' functional and numerical responses (αNP and βNP) to allow for effects such as saturation in the predator's capacity to respond to increasing prey densities.

More specifically, the prey growth rate may be represented by the logistic eq. (2.3), or by one of the broadly equivalent forms referred to in section 2.2.2. A detailed discussion of various forms of predator functional response is given by Hassell in section 5.2. If the basic modification is the inclusion of saturation effects, we have a 'Holling type II' or 'invertebrate' functional response (see Fig. 5.1b); this has a destabilizing effect, because the predators are relatively less effective at high prey densities, just when they are most needed as a regulatory mechanism. More complicated modifications allow for predators to become increasingly efficient as prey numbers increase, until saturation eventually sets in (see Fig. 5.1c); such a 'Holling type III' or 'vertebrate' functional response is stabilizing at low prey densities, destabilizing at high densities. In short, a more realistic prey growth equation is

$$dN/dt = rN(1-N/K)-PF(N,P). \tag{4.3}$$

Some forms for the function $F(N,P)$ are catalogued in Table 4.1, and illustrated in Fig. 5.1.

In a like fashion, the predator growth equation can be written more realistically as

$$dP/dt = P\,G(N,P). \tag{4.4}$$

Table 4.1. Explicit forms for the functional and numerical responses per predator, $F(N, P)$ and $G(N, P)$, in eqs. (4.3) and (4.4). (For references, and other forms, see May, 1975a, pp. 81–84).

F or G?	Label	Formula	Remarks
F	(i)	αN	unsaturated (Lotka-Volterra)
F	(ii)	k	constant attack rate
F	(iii)	$kN/(N+D)$	Holling type II, 'invertebrate' (Holling)
F	(iv)	$k[1-\exp(-cN)]$	Holling type II, 'invertebrate' (Ivlev)
F	(v)	$k[1-\exp(-cNP^{1-b})]$	Holling type II, 'invertebrate' (Watt)
F	(vi)	$kN^2/(N^2+D^2)$	Holling type III, 'vertebrate'
F	(vii)	$k[1-\exp(-cN^2P^{1-b})]$	Holling type III, 'vertebrate' (Watt)
G	(viii)	$-b+\beta N$	Lotka-Volterra
G	(ix)	$-b+\beta F(N, P)$	F and G linearly related. [Caughley's 'laissez-faire' case if $F = F(N)$ only].
G	(x)	$s[1-\gamma P/N]$	logistic, with carrying capacity proportional to N.

In chapter 6, there is elaborated a distinction between 'laissez-faire' predators, where the per capita numerical response function G is itself independent of predator numbers, P,

$$dP/dt = P\,G(N), \tag{4.5}$$

and 'interferential' predators, where it is not. Some forms for the function $G(N,P)$ are summarized in Table 4.1. The functional and numerical responses are often linearly related, as in the general form (ix) of Table 4.1: since most of the forms (i) to (vii) for $F(N,P)$ are independent of P, this often leads to the 'laissez-faire' model (4.5).

Some of these modifications, leading to eqs. (4.3) and (4.4), tend to stabilize the system, and some to destabilize it, compared with the razor's edge of neutral stability manifested by the Lotka-Volterra eqs. (4.1) and (4.2).

A powerful mathematical theorem for 2-dimensional systems of differential equations, the Poincaré-Bendixson theorem, may be used to show that essentially all such prey-predator models have *either* a stable point *or* a stable limit cycle (as defined on pages 7, 8). This result was obtained in general form by Kolmogorov (1936), and has recently been more specifically applied to most of the models in the ecological literature (May, 1972a; 1975a, ch. 4). Figure 4.1 illustrates one particular example, constructed from eqs. (4.3) and (4.4) with the forms (iii) and (x) of Table 4.1 for F and G respectively, with parameters so chosen that the system exhibits stable cycles; other parameter choices could give stable equilibrium point solutions.

These prey-predator systems tend to be in tension between the stabilizing prey density dependence, and the often destabilizing predator functional and numerical responses. As one or other of the relevant parameters is tuned, the pair of 'eigenvalues' which characterize the stability of the system can be seen to move from the left half of the complex plane (corresponding to damped oscillations), to cross the imaginary axis into the right half plane (giving rise by a Hopf bifurcation to a stable limit cycle); see May (1975a, ch. 2 and appendix I). The neutral stability of the Lotka-Volterra model is reflected in the eigenvalues always lying exactly *on* the imaginary axis.

Some general correlations between a system's parameter values and its stability behaviour are worth noting, and comparing with observed properties of real populations.

If the environmental carrying capacity for the prey, K, is much larger than the equilibrium prey density in the presence of predators,

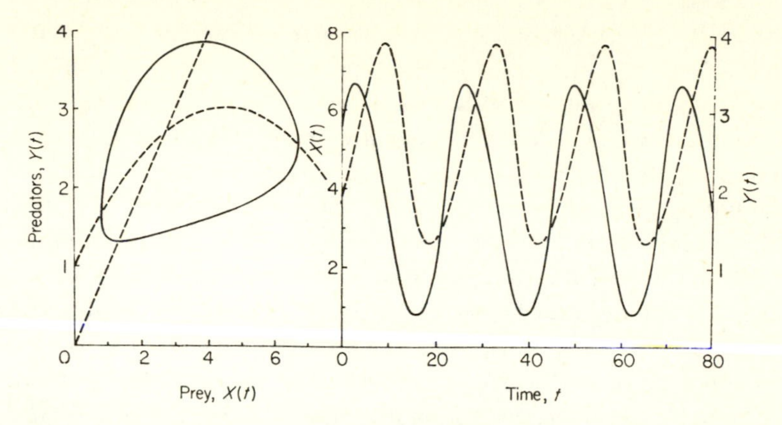

Fig. 4.1. A specific example of a prey-predator limit cycle. For the system obeying eqs. (4.3) and (4.4), with the functional and numerical responses F and G given by the forms (iii) and (x) of Table 4.1, respectively [see also eqs. (6.12) and (6.13)], we form dimensionless variables $X(t)$ and $Y(t)$ from the prey and predator populations: $X = N/D$, $Y = \gamma P/D$. To the left is shown the stable limit cycle endlessly traced out by these population numbers, for the parameter choice $r/s = 6$, $K/D = 10$, $k/\gamma r = 1$; the dashed lines are the isoclines of stationary prey ($dX/dt = 0$) and predator ($dY/dt = 0$) populations, and their intersection is the equilibrium point, here unstable. To the right we display the cyclic population numbers of prey (solid line) and predator (broken line) as functions of time, rt. If displaced from these stable cyclic trajectories, the system tends to return to them.

the stabilizing elements contributed to the dynamics by the prey density dependence will be relatively weak. This underlies the 'paradox of enrichment', whereby increasing K makes for lowered stability, and eventually for stable limit cycle behaviour. The 'paradox' was noted (and christened) by Rosenzweig (1971) on the basis of numerical studies, and is possibly exemplified by natural populations such as the larch bud moth in Switzerland (whose population cycles in the optimal part of its range, between 1700 and 1900 m., but is relatively steady in marginal habitats: Baltenzweiler, 1971). Some recent laboratory studies, such as Luckenbill's (1973) on *Paramecium* and its predator *Didinium*, also tend to support this idea; there is much room for further such exploration of prey-predator dynamics under controlled conditions.

Stability studies of prey-predator models by May (1975a, ch. 4) and Tanner (1975) have further shown a propensity to stable cycles when the intrinsic growth rate of the prey population exceeds that of its predators. The typical situation is indicated by Fig. 4.2, which shows the stability properties of a model constructed from eqs. (4.3) and (4.4) with the forms (iii) and (x) of Table 2.1 (the same system as is illustrated in Fig. 4.1), as a function of the relevant parameter ratios K/D and r/s. The systems which tend to cycle are indeed those where the prey population has a relatively high growth rate (r/s large) in an environment with a relatively large carrying capacity (K/D large). The natural history data reviewed by Tanner is summarized schematically in Table 4.2, and it accords with the above notions. Roughly speaking (and this exercise is necessarily very approximate), the hare-lynx system is the only one of the eight whose population parameters lie in the hatched area of Fig. 4.2, combining relatively large K with an r/s that is not small, and it is also the only one which conspicuously exhibits stable cycles.

Plant-herbivore systems are, of course, special cases of the general prey-predator relationship, with the plants as 'prey' and the herbivores

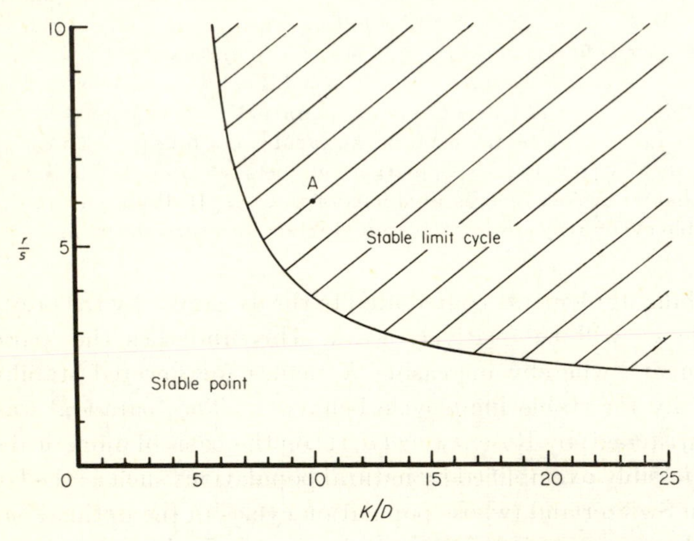

Fig. 4.2. This figure indicates the stability domains of the prey-predator system defined in Fig. 4.1. [see also the text, and eqs. (6.12) and (6.13)], in terms of the parameter ratios r/s (prey/predator intrinsic growth rates) and K/D (appropriately normalized value of the carrying capacity for the prey). In the unhatched region of this parameter space, the equilibrium point is stable; in the hatched region, there are stable limit cycles. The figure is drawn for $k/\gamma r = 1$, but qualitatively similar stability boundaries pertain to other values of this ratio. The point labelled A shows the parameter values corresponding to the explicit example illustrated in Fig. 4.1. After May (1975a, p. 192).

as 'predators'. Chapter 6 puts together theoretical insights and empirical observations for such systems. A more technical elaboration of this theme, applied to natural and managed grazing systems varying from intensive pastures to extensive ranges, is due to Noy-Meir (1975).

Table 4.2. Impressionistic summary of life history data for 8 natural prey-predator systems. (After Tanner, 1975).

Prey-predator	Geographical location	Is K relatively large?	Approximate* ratio r/s	Apparent dynamical behaviour
sparrow—hawk	Europe	no	2	equilibrium point
muskrat-mink	central North America	no	3	equilibrium point
hare—lynx	boreal North America	yes	1	cycles
mule deer—mountain lion	Rocky Mountains	yes	0·5	equilibrium point
white-tailed deer—wolf	Ontario	yes	0·6	equilibrium point
moose—wolf	Isle Royale	yes	0·4	equilibrium point
caribou—wolf	Alaska	yes	0·4	equilibrium point
white sheep—wolf	Alaska	yes	0·2	equilibrium point

* Estimated from Tanner's 'maximum survival and maximum fertility' column, as most appropriate for intrinsic r and s.

4.2.2 *Prey-predator models and single species models*

The relation between these 2-species models and the earlier single species models deserves mention.

Suppose we try to construct a single species equation for just the prey population in a prey-predator model: the population growth rate will depend on the predator population, which itself is derived (in a formal sense) as some integral over past prey populations. The upshot will be an equation involving only the prey population, but where the density dependent regulatory mechanism depends on past values of the prey population.

To be specific, suppose the growth of the predator population is described by an equation of the general form of eq. (4.5), with the numerical response per predator depending only on the prey density. This can be integrated to give

$$P(t) = P(o) \exp \int_0^t G[N(t')]\, dt'. \tag{4.6}$$

The predator population depends on some explicit average over past prey populations. The growth rate of the prey population itself depends nonlinearly on both $N(t)$ and $P(t)$, as in the general eq. (4.3). Use of eq. (4.6) enables this to be reduced to a single equation involving only the prey population:

$$dN/dt = \text{function of} \left\{ N(t) \text{ and } \int_0^t G[N(t')]\, dt' \right\}. \tag{4.7}$$

That is, we have an equation with a time-delayed regulatory mechanism, of the same general character as eq. (2.4). The time delay roughly corresponds to the time for the predator population to respond to prey changes (if our single species equation describes the prey population), or for the vegetation to recover (if our equation describes the herbivores). This point is discussed more fully by Caughley (1976).

As a variation on this theme, consider a plant-herbivore or substrate-plant system in which the 'predator' population, $N(t)$, obeys the logistic eq. (2.3), but where the carrying capacity $K(t)$ (representing the population of the plant or substrate 'prey') has a dynamical dependence on $N(t)$ via an equation of the general form

$$dK/dt = [K_o - K(t)]/\tau - bN(t). \tag{4.8}$$

Here K_o is the saturation level of the resource (plant or substrate), and b the rate of depletion by consumers. The rate at which the resource recovers depends both on its intrinsic regeneration time, τ, and on the degree to which it is depressed below saturation, $K_o - K$. Equation (4.8) can be integrated

$$K(t) = K_o - b \int_{-\infty}^t N(t') \exp[-(t-t')/\tau]\, dt', \tag{4.9}$$

thus again leading to a form of time-delayed logistic equation for the 'predatory' herbivore or plant population. Stability analysis of eq. (2.3) with eq. (4.9) for $K(t)$ shows that (like the ordinary logistic) the equilibrium point at $N^* = K_o/(1+b\tau)$ is always stable, but that (unlike the ordinary logistic) perturbations die away in an oscillatory way once

$$b > (1-r\tau)^2/(4r\tau^2). \tag{4.10}$$

High depletion rates have a tendency to produce oscillations. For a detailed treatment, see McMurtrie (1976a).

In short, the ideas developed in chapter 2 also, in a broad sense, include prey-predator situations.

4.2.3 *Discrete growth (difference equations)*

As discussed for single populations, there will be situations where generations do not overlap, and the prey-predator dynamics are described by difference equations. Arthropod prey-predator systems provide many examples, particularly in seasonal climates. Such systems are discussed in detail in chapter 5.

We content ourselves here with some general remarks about the dynamics of these systems. Although prey-predator systems with continuous growth (differential equations) usually possess either a stable point or a stable limit cycle, a similarly generic discussion of discrete growth models is difficult, and their dynamical behaviour is not typically so simple. Two recent studies by Beddington *et al.* (1975, 1976a), however, are illuminating.

These authors base their work on the classic Nicholson-Bailey model, generalized to include density dependence in the host population growth (for further details see section 5.7):

$$N_{t+1} = N_t \exp\,[r(1-N_t/K)-aP_t], \tag{4.11}$$

$$P_{t+1} = \alpha N_t[1-\exp\,(-aP_t)]. \tag{4.12}$$

Numerical studies show regimes of stable points, of stable cycles (which, unlike the single species models in section 2.3, need no longer have an integer period, and in general do not), and of chaos. The onset of chaotic behaviour, where the population trajectories are effectively indistinguishable from the sample function of a random process, takes place at lower r-values than for the corresponding single species model. The paper by Beddington *et al.* (1975) is rich in fascinating detail, illustrated by photographs of oscilloscope trajectories. More recently, the analytic understanding which underpins section 2.3 has been generalized to multi-species situations by Guckenheimer *et al.* (1976). This elegant mathematical work paves the way for a generic discussion of the dynamic complexities of such models, and, as one special case, illuminates the numerical results mentioned above.

In the continuous prey-predator models, the stability properties tend to be global, in the sense that the system returns to its stable configuration regardless of the magnitude of the perturbations. A separate complication in discrete (difference equation) models is that such global stability behaviour is not typical: given the system possesses a stable point (or a stable cycle), the system will tend to return to the stable point following a small disturbance, but not following a large one. Large disturbances typically lead to the extinction of the predator population, if not of both populations. The term 'resilience' has been introduced (Holling, 1973; see also Orians, 1975 and May, 1975c) to characterize the magnitude of the population perturbations the system will tolerate before collasping into some qualitatively different dynamical regime. Beddington *et al.* (1976a) have used a slight generalization of eqs. (4.11) and (4.12) to study general biological features of this resilience in arthropod prey-predator systems. To date, the concept of 'resilience' has been mainly a useful metaphor; the above work opens the door to experimental and theoretical studies of a quantitative kind, albeit on relatively simple systems.

4.3 Two populations: competition

Simple models for two competing populations, $N_1(t)$ and $N_2(t)$, were also studied in the 1920's by Lotka and Volterra. These models are the direct extension of the single species logistic equation:

$$dN_1/dt = r_1 N_1[1-(N_1+\alpha_{12} N_2)/K_1] \qquad (4.13)$$

$$dN_2/dt = r_2 N_2[1-(N_2+\alpha_{21} N_1)/K_2]. \qquad (4.14)$$

Here K_1 and K_2 are the carrying capacities of the environment, as seen through the eyes of species 1 and 2, respectively; r_1 and r_2 are the intrinsic growth rates; α_{12} is a competition coefficient which measures the extent to which species 2 presses upon the resources used by species 1; and α_{21} is the corresponding coefficient for the effect of species 1 on species 2. The multi-species generalization of these competition equations is discussed in chapter 7 [eq. (7.1)].

The stability character of these equations is well known. If intraspecific competition is stronger than interspecific (corresponding to $\alpha_{12}\alpha_{21} < 1$), there can be a stable equilibrium point with both species coexisting (provided also that the carrying capacity ratio obeys $1/\alpha_{21} > K_1/K_2 > \alpha_{12}$, which will easily be true if both α_{12} and α_{21} are small); if interspecific competition is stronger than intraspecific ($\alpha_{12}\,\alpha_{21} > 1$), no stable coexistence is possible. Particularly interesting is the special case when the two species use the resources in identical fashion (whence $\alpha_{12} = \alpha_{21} = 1$ and $K_1 = K_2$); again the species cannot persist together. Other models, which generalize other single species equations of the same generic character as the logistic, lead to similar conclusions (see, e.g., Gilpin and Justice, 1972).

This early theoretical work led to the enunciation of the 'competitive exclusion principle', that is, the notion that species which make their livings in identical ways cannot stably coexist, which in turn stimulated many interesting experiments. This material is reviewed in chapter 7 (see also Hutchinson, 1975). The concept of 'niche' is used as a shorthand for the constellation of ecological factors which specify just how a species does make its living in the world; the competitive exclusion principle then says that no two species can occupy the same niche. Empirical and theoretical aspects of the niche are developed in detail in chapter 7, which emphasizes the difficulties involved in quantifying a species' niche once several ecological dimensions are relevant. A good definition of both the niche and the competitive exclusion principle have been given by Dr Seuss (Geisel, 1955), and put into the literature by Levin (1970):

> 'And NUH is the letter I use to spell Nutches
> Who live in small caves, known as Nitches, for hutches.
> These Nutches have troubles, the biggest of which is
> The fact there are many more Nutches than Nitches.
> Each Nutch in a Nitch knows that some other Nutch
> Would like to move into his Nitch very much.
> So each Nutch in a Nitch has to watch that small Nitch
> Or Nutches who haven't got Nitches will snitch.'

Although these ideas provided much early impetus, they deserve close scrutiny. At a semantic level, one may ask how identical is 'identical'? On a more pragmatic plane, one may point in nature to many instances where organisms appear to co-occur with, at very least, substantial overlap between niches. MacArthur's guild of five very similar warblers, *Dendroica*, which are found together as insectivores in spruce trees in northern New England, provides one such apparent paradox, made famous by his resolution of it.

Following Hutchinson's (1959) 'Homage to Santa Rosalia, or why are there so many kinds of animals?', and the work of MacArthur and others, contemporary attention is focussed on the more substantial question of how similar can competing species be, yet stably persist together? What are the limits to niche overlap, the limits to similarity, among coexisting competitors?

Theoretical investigations of these questions begin with models for the detailed competition mechanism, which may involve interspecific territoriality, or selection of items from some shared spectrum of resources, or whatever. The overall competition coefficient, α_{ij}, in some population equation such as eqs. (4.13) and (4.14) may then be calculated on this basis. In this way, the macroscopic population dynamics are related to the underlying microscopic mechanisms of competition, and one can ask how much niche overlap is compatible with the 2-species system having a stable equilibrium point.

In particular, consider the admittedly highly special case where the competing species differ *only* in their use of some 1-dimensional spectrum of resources. This situation is illustrated in Fig. 4.3. Here $K(x)$ represents the resource spectrum, which may be amount of food as a function of size or weight, or amount of vertical habitat as a function of height, or in general amount of a resource K as a function of some variable x. The function $f_i(x)$ describes the way the ith species utilizes this resource; each species has its preferred position in the spectrum (with the mean separation between the utilization functions of adjacent competitors being characterized by d), and a degree of variability about this preferred position (typified by a width w). The degree of niche separation clearly may be characterized by the ratio d/w, which will be large for well separated niches, small for highly overlapping ones. A more detailed and realistic version of Fig. 4.3 is given and discussed in chapter 7 (Fig. 7.1).

Following Levins (1968) and MacArthur (e.g., 1972, ch. 2 and 7; see also May, 1975a, pp. 142–147), the competition coefficients in eqs. (4.13) and (4.14) and their multi-species version eq. (7.1) may then be regarded as the overlap integrals of the utilization functions for the species involved:

$$\alpha_{ij} = \int f_i(x)\, f_j(x)\, dx. \tag{4.15}$$

For 2-species competition, we may now choose a particular shape for the utilization functions (e.g., bell-shaped gaussian curves), and then study the macroscopic population dynamics of eqs. (4.13) and (4.14) as a function of the niche separation d/w, and of the parameters K_1 and K_2 which describe the shape of the resource spectrum as perceived by the individual species. We first examine the *statics* of the situation, to see if a 2-species equilibrium exists, and then study the *dynamics*, to see if the equilibrium is stable.

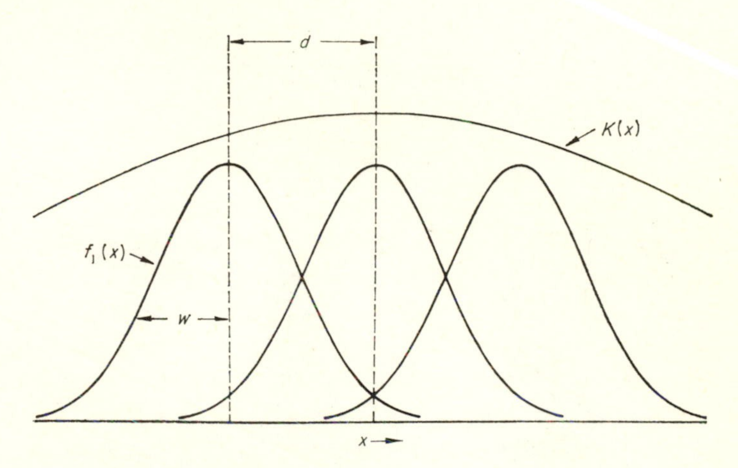

Fig. 4.3. The curve $K(x)$ represents some resource continuum (say amount of food, K, as a function of food size x), which sustains various species whose utilization functions $f(x)$ are, as illustrated, characterized by a width w and a separation d.

For the *statics*, we find the potential equilibrium populations N_1^* and N_2^* by putting $dN_1/dt = dN_2/dt = 0$ in eqs. (4.13) and (4.14). The results are shown in the 'horizontal plane', d/w versus K_2/K_1, in the schematically '3-dimensional' Fig. 4.4. For small niche separation, $d/w \lesssim 1$, a 2-species equilibrium point exists (i.e., both N_1^* and N_2^* are positive) only in a narrow band of K_2/K_1 values; otherwise the equilibrium configuration contains only species 1 or only species 2, depending on whether K_1 is significantly larger or smaller than K_2. Conversely, for large niche separation, $d/w \gg 1$, a 2-species equilibrium point exists for essentially all resource spectrum shapes. Extensions to 3 or more competing species confirms the general tendency for a multi-species equilibrium point to be possible for a wide variety of resource spectrum shapes if d/w is large, but to impose very narrow constraints on the shape if d/w is small. Studies of this kind were begun by MacArthur and Levins (1967), and carried forward by Roughgarden (1974a), May (1974c, 1975a) and others; a good review is by Abrams (1976a). Although not setting any limit to niche overlap in principle, this line of attack suggests that most resource spectrum shapes will require $d/w \gtrsim 1$ if the competitors are to coexist.

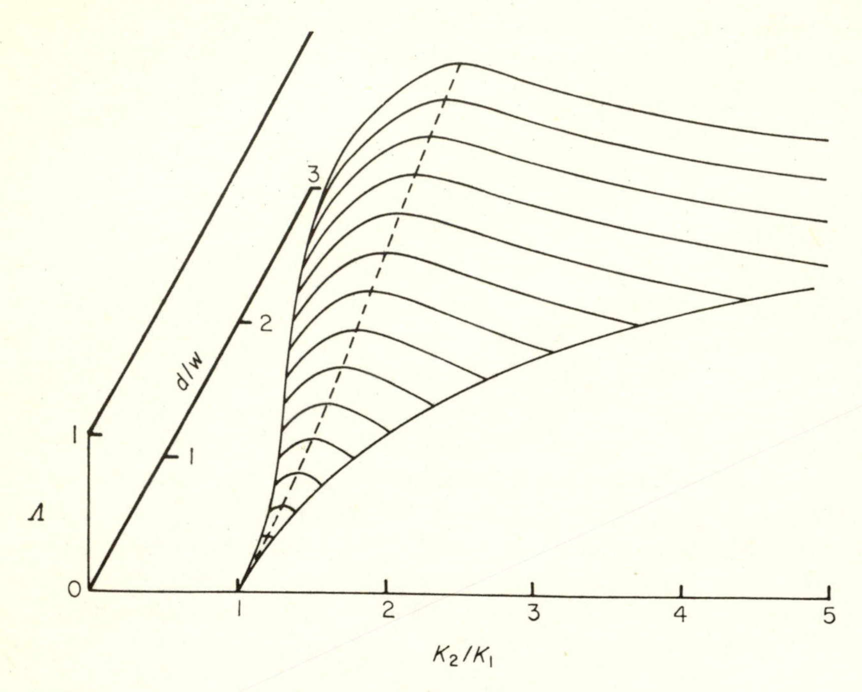

Fig. 4.4. This figure aims to give an impression of the dynamical behaviour of the model for two competing species which is outlined in the text. The quantity Λ represents the rate at which disturbances are damped (i.e., the reciprocal of the characteristic return time for the system); it is plotted as vertical height above the 'horizontal plane' of values of d/w (the degree of niche separation) and K_2/K_1 (the carrying capacity ratio). The value of Λ is shown along contours for fixed values of d/w from 0 to 3, spaced at intervals 0·2 apart. These contours are bounded by the two curved lines in the d/w versus K_2/K_1 plane, and outside these boundaries no 2-species equilibrium point exists. The dashed line indicates the ratio K_2/K_1 which gives the fastest damping rate, for given d/w. The features of this figure are as emphasized in the text.

The *dynamics* of the situation enters Fig. 4.4 via the vertical axis, Λ. This quality Λ characterizes the rate at which the system tends to return to equilibrium, following a disturbance. For simplicity, we have in Fig. 4.4 used $r_1 = r_2 = r$, and expressed Λ as a ratio to r;

thus Λ involves the multi-species generalization of the characteristic return time T_R of section 2.2, and here the explicit normalization is $\Lambda = 1/(rT_R)$. In the absence of competition, $\alpha_{ij} \to 0$, the damping rate tends to r as for the ordinary logistic, and consequently $\Lambda \to 1$. We notice from Fig. 4.4 that *if* a 2-species equilibrium point exists (N_1^* and N_2^* both positive), it is necessarily stable, that is it necessarily has some finite positive damping rate, Λ. But this rate is very small (and the recovery time correspondingly very long) for small d/w, whereas the rate is in general high (and the recovery rapid) for large d/w.

This analysis of the *dynamics* may be pushed one stage further by noting that, in the real world, the environment is likely to exhibit elements of unpredictable variability. Thus the resource spectrum (as measured by quantities such as $K(x)$ or K_1 and K_2) is unlikely to be deterministic, but rather will, to a greater or lesser degree, contain randomly fluctuating components. In place of the deterministic differential eqs. (4.13) and (4.14), which lead to explicit dynamical trajectories for $N_1(t)$ and $N_2(t)$, we will now have stochastic differential equations for the populations, and their behaviour will be described in statistical terms, as fuzzed out probability clouds for the population values. Once such a spectrum of random environmental fluctuations is woven into the fabric of the model, the requirement that the damping rate merely be positive ($\Lambda > 0$) is no longer sufficient to ensure long-term stability, but rather the damping must be strong enough to prevent one or other of the populations from drifting to such low values as to court extinction. (Very crudely and qualitatively, we require $\Lambda > \sigma^2$, where σ^2 characterizes the variance in the environmental noise spectrum.) As is made plausible by the relation between Λ and d/w in Fig. 4.4, the analysis of competition between species in a randomly varying environment suggests the qualitative requirement that d/w should be greater than unity if the species are to persist together. Such studies were initiated by May and MacArthur (1972) and May (1973b). This initial work makes many mathematical approximations of one kind or another, some of which have been refined in subsequent studies by Ludwig (1975), Abrams (1976b), McMurtrie (1976b), May (1974c, d, 1975a) and others. In particular, if the noise spectrum for the random environmental fluctuations has a high degree of correlation (either between different points in time, or between different points on the resource axis), a high degree of niche overlap may be possible (Feldman and Roughgarden, 1975; Abrams, 1976b;

May, 1973b). In the limit of perfect correlations, an effectively deterministic environment is recovered, and we have the situation discussed above where there is in principle no limit to d/w, although exquisitely precise resource spectrum shapes are required if species are to coexist with d/w significantly less than unity.

All the above work is based on the static or dynamic behaviour of models in which the utilization functions are immutably fixed. A different, and much more fundamental, approach has recently been developed by Roughgarden (1975b, 1976): in essence, he allows for evolution in the way the species use the resource spectrum. That is, the shape of the functions $f_i(x)$ in Fig. 4.3, particularly as reflected in their modal position along the resource axis (and thence in the parameter d), are subject to natural selection from generation to generation, and the system evolves to maximize the fitness of the constituent species. The result is, in general, that the model tends to evolve towards a configuration such that d/w is of order unity. This is path-breaking work, lying squarely at the interface between population genetics and population biology; between evolution and ecology (see also section 7.5 below).

No one of the above approaches is decisive. But it is suggestive that three quite distinct lines of attack—static and dynamic behaviour with fixed utilization functions, and models in which utilization functions evolve—give the same answer, namely that there tends to be a limit to niche overlap, such that the average difference *between* species exceeds the typical variability *within* a species. That is, in these special 1-dimensional models,

$$d \gtrsim w. \tag{4.16}$$

It is strongly to be emphasized that these estimates as to the limits to similarity among coexisting competitors are rough, order of magnitude, ones. The sign in the equations is most decidedly $\sim$ not $=$. Thus factor-of-2 differences between calculations based on differently shaped utilization functions, or on different criteria as to how small the population must be before it is highly likely to become extinct, are not really subject to experimental test (at least as matters stand now).

The reason why the different approaches lead to similar conclusions is basically simple. All the models tend to have the feature that the *total* population density, $\sum N_i(t)$, is density dependent in an essentially logistic fashion (with $T_R \sim 1/r$), independent of the value of d/w. The density dependence of the individual populations is, however, sensitively influenced by the degree of niche overlap: for large d/w, each population is well regulated; but for $d/w \ll 1$, the individual populations have very weak density dependence, and they tend to drift up or down (keeping the total population relatively constant), until one or other fluctuates to extinction (see May, 1974b, particularly Figs. 5, 6). That is, individual populations tend to be robustly density dependent for $d/w \gg 1$, and to exhibit fragile density independent behaviour for $d/w \gg 1$. Hence the tendency to infer the limit of eq. (4.16) from various superficially different approaches.

The bulk of this section's discussion of models for two competitors has been devoted to the light they shed on the issues of limiting similarity. This is because I regard it as the central contemporary problem in the area. Chapter 7 goes on from this point, to discuss pertinent observational data. Chapter 7 also discusses other aspects of competition theory, as well as many ways in which the theoretical enquiry into limiting similarity could profitably be extended, to include: direct interference between species (such as territorial squabbles); multiple age classes and phenomena associated with animals with indeterminate growth; competition in many dimensions; time delays in the effects of the interactions; and other things. As Pianka observes at the end of chapter 7, the overriding need is, however, for more experiments and observations rather than for more theory in this area.

4.4 Two populations: mutualism

Most contemporary ecology books have chapters, complete with simple mathematical models, on competition and on prey-predator. Analogous chapters on mutualism are usually absent. I think the reasons for this are partly historical (Lotka and Volterra studied models for competition and prey-predator, but not for mutualism), and partly because mutualism is a relatively inconspicuous feature in temperate zone ecosystems. Be this as it may, mutualism is a conspicuous and ecologically important factor in most tropical communities, and I hope that the next generation of ecology texts will treat all three types of pairwise interaction between species on a roughly equal footing.

One initial difficulty is that the simple, quadratically nonlinear, Lotka-Volterra models that capture some of the essential dynamical features of prey-predator (namely, a propensity to oscillation) or competition (namely, a propensity to exclusion) are inadequate for even a first discussion of mutualism, as they tend to lead to silly solutions in which both populations undergo unbounded exponential growth, in an orgy of mutual benefaction.

Minimally realistic models for two mutualists must allow for saturation in the magnitude of at least one of the reciprocal benefits. This tends to produce a stable equilibrium point, with one (and usually both) of the two equilibrium populations being larger than it would be in the absence of the mutualistic interactions; this, of course, is what mutualism is all about. On the other hand, this equilibrium configuration tends to be less stable, in the sense that perturbations are typically damped more slowly than they are in the corresponding system without mutualistic interactions. Some explicit such models are discussed by Whittaker (1975, pp. 39–41). Qualitative features of the stability properties associated with mutualism are discussed more fully in May (1973c), and a very general formulation of the dynamical character of these systems is in Hirsch and Smale (1974, ch. 12, problem 2).

To illustrate these remarks, consider the unrealistically simple case of two populations $N_1(t)$ and $N_2(t)$ which obey

$$dN_1/dt = rN_1[1 - N_1/(K_1 + \alpha N_2)] \tag{4.17}$$

$$dN_2/dt = rN_2[1 - N_2/(K_2 + \beta N_1)]. \tag{4.18}$$

Here each population obeys a logistic equation, eq. (2.3), with the modification that each species has its carrying capacity increased by the presence of the other, so that $K_1 \rightarrow K_1 + \alpha N_2$ and $K_2 \rightarrow K_2 + \beta N_1$. This overly simple model lacks any saturation effects, and we must limit the amount of mutualistic interaction by demanding $\alpha\beta < 1$, or else the system will 'run away', with both populations growing unboundedly large. As a further simplification, we take the populations to have equal intrinsic growth rates, r, in order not to lose the message of eq. (4.20) in a fog of distracting notation. In this model, the equilibrium populations are made larger by the mutualistic effects:

$$N_1^* = (K_1 + \alpha K_2)/(1 - \alpha\beta) > K_1 \tag{4.19a}$$

$$N_2^* = (K_2 + \beta K_1)/(1 - \alpha\beta) > K_2. \tag{4.19b}$$

On the other hand, the characteristic return time, T_R (see section 2.2), for the system to return to equilibrium following a disturbance is now

$$T_R = \frac{1}{r(1 - \sqrt{\alpha\beta})}, \tag{4.20}$$

which is to be compared with that for each population in the absence of interactions, namely $T_R = 1/r$. The larger the degree of mutualism (the larger $\alpha\beta$), the more pronounced are both these effects.

Many mutualistic systems evolve to the point where at least one of the partners finds the other not merely helpful, but necessary for its existence. This is the limit of 'obligate' mutualism. In this event, models will need to include not only saturation effects, but also threshold effects.

Figure 4.5 attempts to capture some of the dynamical essentials of this situation. To fix ideas, we may think of the figure as pertaining to a plant-pollinator system, in which the plant population, $X(t)$, is an obligate outcrosser and cannot reproduce in the absence of its only pollinator, with population $Y(t)$, which in turn is sustained solely by the nectar of this particular plant; clearly the ideas extend to plants and seed-dispersers, ants and acacias, and other mutualistic systems. The pollinator population, Y, may be thought of as obeying a logistic equation, in which the carrying capacity is proportional to the plant population, X. Then the possible equilibrium values of Y are proportional to X, and lie along a straight line in the X–Y plane (the pollinator 'isocline', where $dY/dt = 0$, as labelled in Fig. 4.5). If the plant density is too low, pollinators may have difficulty finding more than one plant, and plant reproduction may fall below replacement levels; conversely, there will be an upper limit to plant population growth, set by physical limitations regardless of how abundant pollinators may be. The upshot is typically that the equilibrium values of X follow the shape labelled $dX/dt = 0$ (the plant 'isocline') in Fig. 4.5, with a threshold density of pollinators necessary before any equilibrium is possible, and an upward slope as plant population density approaches some saturation value at large Y. The directions in which plant and pollinator population trajectories move in the various domains of X–Y space are as shown by the arrows in Fig. 4.5. The point A is a stable point, and attracts all trajectories originating outside the hatched region abutting the axes. The point B is unstable. All trajectories originating inside the hatched region are attracted to the origin, i.e., the populations become extinct. This model is discussed with more mathematical detail in May (1976a).

If the system whose dynamics is described by Fig. 4.5 is embedded in an environment which is subject to large amplitude perturbations, it is sooner or later liable to be swept into the hatched area in Fig. 4.5, whereupon extinction will probably follow. The long term persistence of mutualistic systems of this kind will be favoured by environments which are stable and predictable.

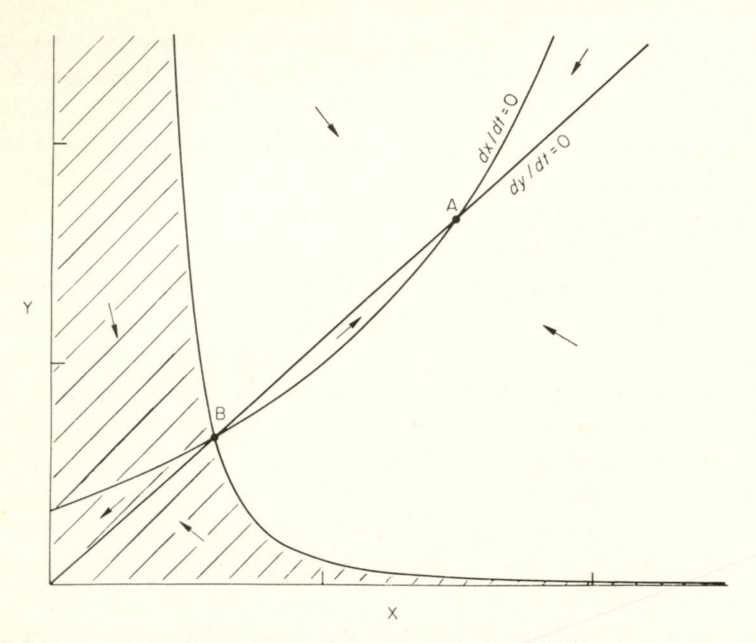

Fig. 4.5. This figure illustrates the dynamical behaviour of two mutualistically interacting populations, X and Y, as discussed in the text. The two labelled curves correspond to possible equilibrium values for X ($dX/dt = 0$) and Y ($dY/dt = 0$), respectively. Population trajectories originating outside the hatched region are attracted to the stable point A, while trajectories originating inside the hatched area are attracted to the origin (that is, extinction).

These dynamical considerations may go some way toward explaining why obligate mutualisms of this sort are less prominent in temperate and boreal ecosystems than in tropical ones. A good survey of empirical evidence bearing on this point is given in Farnworth and Golley (1974, pp. 29–31), where it is observed that not only are there no obligate ant-plant mutualisms north of 24°, no nectarivorous or frugivorous bats north of 33°, no orchid bees north of 24° in America, but also within the tropics mutualistic interactions are more prevalent in the warm, wet evergreen forests than in the cooler and more seasonal habitats.

There are, of course, many instances of obligate mutualism that have gone so far as to blur the distinction between a single organism versus two mutualistic ones. Examples are the fungus-algae associations that constitute lichens; wood-digesting protozoans in the intestines of termites; nitrogen-fixing bacteria in the roots of leguminous plants; mycorrhizal fungal filaments on the roots of many higher land plants. Indeed such associations with various bacteria are essential for the functioning of all higher animals. However, the discussion in this section 4.4 has been conducted in the same spirit as the discussions of prey-predator and competition, in that we have in mind mutualistic interactions between two species, each of which has, as it were, a life of its own.

Following my admonition in the opening paragraph of this section, I would have liked to include a chapter, along the lines of chapters 5, 6, 7, on mutualism. But I think that neither theoretical nor empirical aspects of the subject are yet sufficiently developed to justify this. Many insightful field studies are currently being done, and they should soon provide the basis for such a synthesis. Examples include work on plants pollinated by butterflies (e.g., Gilbert, 1975), by hummingbirds or sunbirds (e.g., Gill and Wolf, 1975b; Stiles, 1975; Colwell, 1973), by bats (e.g., Howell, 1976), and by bees (e.g., Heinrich and Raven, 1972; Janzen, 1971). Levin has recently built upon his empirical studies of the pollination ecology of plants, to explore simple models of the plant-pollinator system (King *et al.*, 1975, and references therein).

In mutualistic models for such systems as plants and pollinators or plants and seed-dispersers, even more than for prey-predator or competition, it will often be necessary to introduce explicit time delays in the interaction processes, or else to work with difference equations. Whereas 2-dimensional systems of ordinary differential equations in general exhibit only stable points or stable limit cycles, these more general systems will display the rich range of dynamical behaviour alluded to earlier, from stable points, to stable cycles (with or without integer periods), to chaotic fluctuations. Such complications have been omitted from the discussion in this section.

4.5 What next?

One logical next step would be to go on to consider *three* interacting populations. For the theoretician's usual case of continuous growth, leading to ordinary differential equations, the step up to three dimensions introduces not only a confusing proliferation of parameters, but also a qualitative change in the dynamical complexity. A 1- or 2-dimensional system of ordinary differential equations is rather tame, exhibiting only stable points or stable cycles (a fact which is intimately connected to the observation that here one can distinguish the inside from the outside of a closed curve); 3-dimensional such systems can display a full and rich dynamical complexity ('strange attractors'), akin to that we first met in section 2.3. Such messy behaviour is manifested even by the simple Lotka-Volterra equation for three competitors, and this example is discussed as a sort of mathematical morality play by May and Leonard (1975; see also Gilpin, 1975a). It is partly for this reason that we count like the Australian Arunta tribe, 'one, two, many', and move on (after chapters 5, 6, 7) directly to multi-species communities.

For some North American bird watchers, beatitude is to join the '600 club'; there are a little over 600 species of birds to be seen in the continent. These, and other broadly similar numbers, are shaped by the underlying limits to similarity among coexisting competitors. Although I believe competition usually to be the basic force, the numbers undoubtedly will be modified to a greater or lesser extent by prey-predator and mutualistic interactions among the multitude of organisms in the community. Other aspects of these general questions are explored in chapters 9 and 12, and in May (1975a, ch. 7). All this work represents only the first tentative steps, but I hope that as our understanding advances it will be possible to explain, in a fundamental way, why the number of bird species is 600 rather than 6 or 60,000, and maybe even why it is 600 rather than 60 or 6,000. The precision of the physical sciences will never be possible; it will never be possible to explain why the typical number is 600 rather than 500.

These are central questions, and ultimately as deep as the origin of species.

5
Arthropod Predator-Prey Systems

M. P. HASSELL

5.1 Introduction

The wealth of examples where animals show cryptic, mimetic or aposematic colouration is itself testimony to the widespread importance of predation as a factor reducing survival. Predation, like competition, is therefore an important factor in the evolution of new species characteristics by natural selection, with selection favouring the efficient predator and the elusive prey. This chapter is largely devoted to a review of some theoretical studies on predation, focusing in particular on the components of predator searching behaviour and how these affect the interaction between predator and prey populations.

The predators we have in mind are arthropods—predominantly insects—for two important reasons.

(1) There is considerable laboratory information available on many of the components of search by arthropod predators. This has suggested the functional form of many of the components included in the models and also indicates realistic ranges of parameter values.

(2) There is amongst the arthropods one very large category of insects whose life cycle has many of the features of the population models to be discussed. These are the insect parasitoids (often loosely called 'insect parasites'). They are very abundant, making up about 10 per cent of the one million or so known insect species, and largely belong to the Diptera (flies) and Hymenoptera (ants, bees, wasps). They differ from true parasites (such as liver flukes), in the strict zoological sense, in that they almost always kill their host, which is most often the egg, larval or pupal stages of some other insect. A significant difference between predator and parasitoid life cycles is that only the adult female parasitoid searches for hosts, and then primarily to oviposit on, in or near the host rather than to consume it. The parasitoid larvae may then feed from the outside or from within the host, but in either

case they cause little serious damage until approaching maturity when they feed on vital organs and the host is killed.

Several features of this life cycle represent important simplifications over that of other predators. In the first place, host mortality depends upon the searching ability of only a single stage in the parasitoid life cycle (the adult female), while males, females and often the immature stages of 'true' predators must all secure prey and will do so with differing searching efficiencies. The second simplification is that parasitoid reproduction is necessarily defined, in a simple, direct way, by the number of hosts parasitised. Such a close correspondence is lacking in predators, whose rate of increase depends upon the survival of each developmental stage and on the fecundity of the resulting adults. It is to insect parasitoids, therefore, that we should look for the most detailed correspondence between the simple population models discussed below and what has actually been observed. However, specific reference to parasitoids will only be made where some distinction from other predators need be made.

Predator-prey models have traditionally been couched in one of two mathematical formats—as differential or difference equations. Each of these are appropriate to quite different kinds of life history. Differential equations apply where generations overlap completely so that birth and death processes are continuous, while difference equations deal with population changes over discrete time intervals. They are, therefore, ideal for systems where generations are quite distinct. In limiting ourselves here to difference equation models for predator-prey interactions, we are restricting ourselves to the kinds of life cycle more typically found in temperate parasitoids and their hosts. Thus, all the models to be discussed below will have the same basic form:

$$N_{t+1} = \lambda N_t f(N_t, P_t) \qquad (5.1)$$

$$P_{t+1} = N_t[1 - f(N_t, P_t)]. \qquad (5.2)$$

Here N_t, N_{t+1} and P_t, P_{t+1} are respectively the prey and predator numbers in successive generations and λ is the finite nett rate of increase of the prey. Predation at time t remains at present an unspecified function of prey and predator population sizes. The use of this format does not necessarily mean, however, that any conclusions to be drawn will not also apply to the many instances of slight overlap in age-classes and generations. This has recently been emphasised by Auslander *et al.* (1974) who have shown that difference equations can remain useful in describing some systems with complex age-class interactions which, by virtue of their internal dynamics, approach a state where the overlap in generations is reduced.

There are other details of predator biology which will shape the structure of population models. For instance, the degree of prey specificity is very important. Most predators are oligophagous or polyphagous, attacking several prey species; completely specific or monophagous species are few. Despite this, and for the obvious sake of simplicity, most theoretical studies have concentrated on single predator-single prey systems. These have the virtue of exposing some fundamental properties of predation, but we should bear in mind that they will be difficult to extrapolate directly to most natural systems.

In this chapter are distinguished two aspects of predation:

(1) The death rate of the prey (due to predation). In particular, we shall consider how predator efficiency is affected by the abundance of prey, the abundance of other predators and the relative prey and predator distributions.

(2) The rate of increase of the predator population. Here the influence of prey density on the survival and developmental rates of immature predators and on the fecundity of adults is to be considered.

This distinction [stressed by Beddington *et al.* (1976b) and Hassell *et al.* (1976b)] parallels in some respects that proposed by Solomon (1949) and Holling (1959a) between functional and numerical responses*. It is, however, a somewhat broader framework since the functional response is restricted to relationships between the number of prey attacked per predator and *prey* density. The separate treatment of factors affecting the prey death rate and the predator rate of increase is much less important for parasitoids than for predators, each host parasitised often leading to one parasite progeny in the next generation. Indeed, this is exactly the assumption made in eq. (5.2) above, and one that is also found in many of the so-called 'predator-prey' models in the literature (e.g. Lotka, 1925; Volterra, 1926).

It is convenient to commence with a brief discussion of the model of Nicholson (1933) and Nicholson and Bailey (1935). This provides an edifice upon which to build and explore the effects of more realistic predator behaviours. Nicholson made three assumptions upon which all his models rest. He assumed (1) that predators search randomly for their prey (i.e. their behaviour is not influenced by the density and

* A *functional response* defines the relationship between the numbers of prey attacked per predator at different prey densities, and the *numerical response* between predator numbers and prey density.

distribution of prey or other predators), (2) that their appetite (or fecundity, if parasitoids) is unlimited and (3) that the area effectively searched (i.e. in which all prey are found) is constant for a given predator population. Nicholson called this searching efficiency, the 'area of discovery' (a). Given these assumptions, the function in eq. (5.1) now becomes

$$f(N_t, P_t) = \exp(-aP_t), \tag{5.3}$$

leading to the population model

$$N_{t+1} = \lambda N_t \exp(-aP_t) \tag{5.4}$$

$$P_{t+1} = N_t[1 - \exp(-aP_t)]. \tag{5.5}$$

[Notice that the overall form of these equations is determined by the first term of the Poisson distribution, which serves to distribute the encounters between predators and prey randomly amongst the N_t prey available].

The properties of this model are well known. For each combination of λ and a there exists a unique equilibrium position, but one that is unstable, since the slightest disturbance leads inevitably to oscillations of increasing amplitude with the predator population lagging behind that of the prey. While such an unstable outcome has been observed in a few laboratory experiments [notably, Gause (1934) and Huffaker (1958)], it is not a feature of the real world where coupled predator-prey interactions have been seen to persist over long periods of time. This disparity in itself, however, does not condemn eq. (5.3) as a description of predation, since the models may be made quite stable by simply allowing some resource limitation or other density dependence to act upon the prey population. The effective rate of increase, λ, would now be a function of prey density as discussed in chapter 2. Having said this, it remains unlikely on *a priori* grounds that the persistence of all coupled predator-prey systems is dependent upon some external density dependent factors, rather than on the internal dynamics of the predator-prey system itself.

We now proceed to consider how eqs. (5.3)–(5.5) can be modified in the light of known predator responses to prey density, predator density and to the distribution of prey.

5.2 The response to prey density

It is implicit in eq. (5.3) that the number of encounters between a predator and a population of prey increases linearly with prey density as shown by the functional response in Fig. 5.1a. This, of course, is not feasible: no predator has an unlimited appetite, nor parasitoid an untold number of eggs. A further difficulty with eq. (5.3) is its requirement that the time available for search (T_s) by a predator is constant at all prey densities. Without this, the area of discovery (a) cannot remain a constant. It was Holling (1959b) who first argued forcibly that this too is impossible, and that searching time *must* depend upon prey density. Let us consider a to be the product of an instantaneous rate of discovering prey (a') and searching time. Thus,

$$a = a' T_s. \tag{5.6}$$

Whenever a prey is encountered, a finite amount of time must be spent in quelling, killing and eating the prey, together with other related time-consuming activities, all of which Holling called the 'handling time' (T_h). There remains, therefore, progressively less time available for search (T_s) as more prey are eaten (parasitized), since

$$T_s = T - T_h N_a \tag{5.7}$$

where T is the initial, total time available to the predators for discovering prey and N_a is the number of prey attacked.

This simple inclusion of handling time has the profound effect of changing the functional response to a negatively accelerating curve as shown by the example in Fig. 5.1b. Encouragingly, there are now numerous examples in the literature of such responses from arthropod predators and parasitoids, at least under laboratory conditions [see Hassell *et al.* (1976b) for a review]. It remains but a straightforward step explicitly to include handling time in eqs. (5.1) and (5.2). The precise form of the function, however, will differ between parasitoids and predators for the simple reason that hosts are not removed on parasitism and therefore remain available for rediscovery and further 'handling'. The different models that result are clearly derived and discussed in Rogers (1972a). Let us consider the case for parasitoids. The functional responses in Fig. 5.1 are described by the equation

$$N_a = a' T_s N_t, \tag{5.8}$$

the form of the response depending on whether T_s is a constant, or varies with N_a. Substituting eq. (5.7) into eq. (5.8), we have

$$N_a = a' N_t(T - T_h N_a),$$

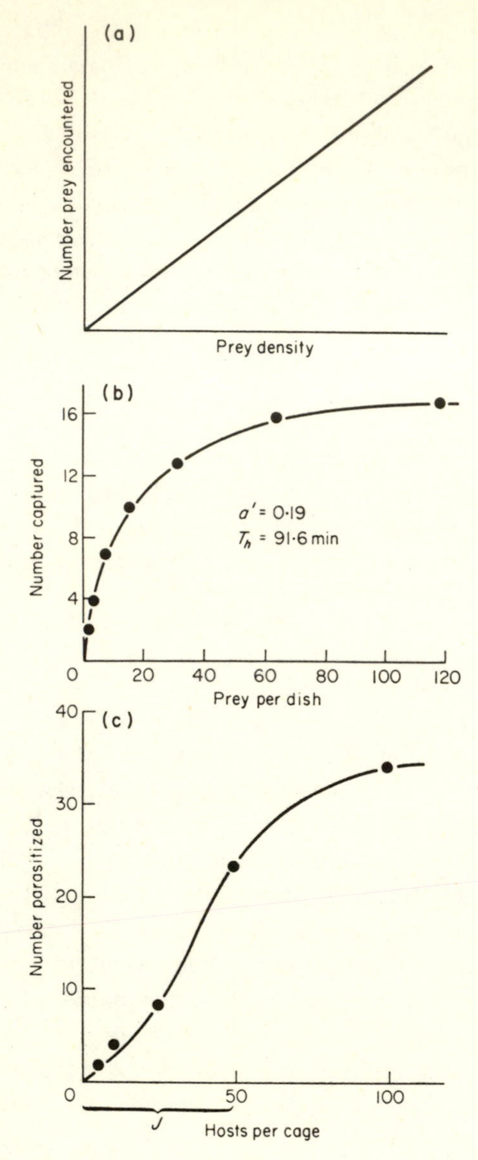

Fig. 5.1. Three types of functional response. (**a**) A linear response implicit in the Nicholson-Bailey model. (**b**) A concave response, obtained from second instar coccinellids (*Harmonia axyridis*) feeding on *Aphis craccivora* (Mogi, 1969). (**c**) A sigmoid response for the braconid parasitoid, *Aphidius uzbeckistanicus*, parasitising the aphid, *Hylopteroides humilis* (Dransfield, 1975). *J* represents the range of prey densities over which the response is density dependent.

that is

$$N_a = \frac{a' N_t T}{1 + a' T_h N_t}. \tag{5.9}$$

The function in eq. (5.1) now becomes

$$f(N_t, P_t) = \exp\left(-\frac{a' T P_t}{1 + a' T_h N_t}\right), \tag{5.10}$$

which clearly reverts to the Nicholson-Bailey model [eq. (5.3)] when $T_h = 0$.

The inclusion of the additional parameter, T_h, is inevitably destabilising in a population model since predation is now proportionately greater at low rather than high prey densities. The extent of this additional instability depends, however, not on the absolute value of

Table 5.1. Estimated values of handling time (T_h) for a selection of parasitoids and predators, using eq. (5.10) [see Rogers (1972a) for details of method] except for *Pleolophus* and *Nemeritis* where T_h was directly observed. The values of T_h/T are based on conservative estimates of longevity.

Parasite or predator species	Host or Prey	Handling time T_h (hrs)	$\frac{T_h}{T}$	Author(s)
Parasitoids				
Nemeritis canescens	*Ephestia cautella*	0·007	<0·0001	Hassell & Rogers (1972)
Chelonus texanus	*Ephestia kühniella*	0·12	<0·001	Ullyett (1949a)
Dahlbominus fuscipennis	*Neodiprion lecontei*	0·24	<0·003	Burnett (1958)
Pleolophus basizonus	*Neodiprion sertifer*	0·72	<0·02	Griffiths (1969)
Dahlbominus fuscipennis	*Neodiprion sertifer*	0·96	<0·01	Burnett (1958)
Cryptus inornatus	*Loxostege sticticalis*	1·44	<0·02	Ullyett (1949b)
Nasonia vitripennis	*Musca domestica*	12·00	<0·1	DeBach & Smith (1941)
Predators				
Anthocoris confusus (5th Instar)	*Aulacorthum circumflexus*	0·38	<0·001	Evans (1973)
Notonecta glauca (1st Instar)	*Daphnia magna*	0·76	<0·005	B. H. McArdle (unpublished)
Ischnura elegans (12th Instar)	*Daphnia magna*	0·82	<0·002	Thompson (1975)
Harmonia axyridis (2nd Instar)	*Aphis craccivora*	1·61	<0·002	Mogi (1969)
Phytoseiulius persimilis (Adult ♀)	*Tetranychus urticae*	1·87	<0·005	Pruszynski (1973)

T_h but on the value relative to the time available (i.e. T_h/T). Table 5.1 gives estimated values for T_h and of T_h/T for a variety of parasitoids and predators. In all cases, handling is a very small fraction of total time and thus will have a very small destabilizing effect on population models. This is argued more formally in Hassell and May (1973).

A population model following from eq. (5.10) would be suitable as a first approximation to describe a parasitoid-host interaction where generations are discrete, search is random and there are no complications of parasitoid mutual interference (see below). It would be much less applicable to true predators, whose searching performance will vary during development. In other words, we may expect the parameters a' and T_h to scale markedly with both the age (= size) of the predator and, similarly, with the size of prey taken. The ways in which they may do so is illustrated in Fig. 5.2 from Thompson (1975). No longer will a single functional response serve as a description for predation as a whole. We now have a matrix of responses and must be able to

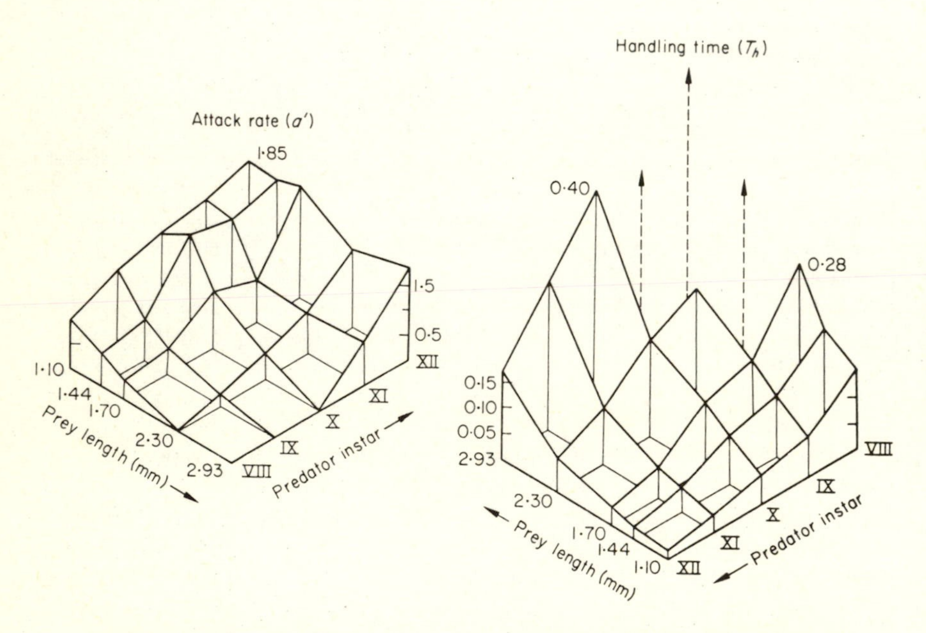

Fig. 5.2. Effect of both predator and prey size on the attack rate (a') and handling time (T_h), for the damselfly, *Ischnura elegans*, feeding on *Daphnia magna*. From Thompson (1975), courtesy of the British Ecological Society.

predict the age structure of both predators and prey in order to model adequately the prey death rate.

Finally, we should note that while many functional responses *are* of the form shown in Fig. 5.1b, several arthropod predators exhibit rather different (e.g. humped, sigmoid) responses which are not adequately described by equations such as eq. (5.9). These are reviewed in Hassell *et al.* (1976b). Of particular interest are the sigmoid responses, since predation is now density dependent, and therefore contributing to stability, over a range of prey densities (J in Fig. 5.1c). The example in Fig. 5.1c results from more active search whenever prey are plentiful rather than scarce. In effect, T or perhaps a' in eq. (5.9) are now themselves a function of prey density. Such examples are probably more widespread amongst invertebrates than previously thought (Hassell *et al.*, 1977).

5.3 The response to predator density

Within the confines of a laboratory cage, predators and parasitoids often react markedly to the immediate presence of another individual of the same species. Thus, coccinellid larvae, predatory mites and the parasitoids, *Nemeritis canescens* and *Diaeretiella rapae* have all been observed to respond to encounters by an increased tendency for local dispersal. The same tendency has been observed in *N. canescens* following the detection by the female that a host has already been parasitized (Rogers, 1972b). In addition, the considerable literature, particularly for egg parasitoids, on the ways that females mark their hosts and show aggressive behaviour to other females, suggests that such mutual interference is widespread.

It is to be expected, therefore, that interference in a simple laboratory system will reduce the searching efficiency per predator as predator density is increased, the underlying cause being a reduction in searching time following each predator encounter. Clear examples of this, for both predators and parasitoids, are shown in Fig. 5.3. Some of the relationships are noticeably curvilinear (Fig. 5.3a, b), as indeed they must be over a sufficient range of predator density. Models that describe such curvilinear relationships have been proposed by Royama (1971), Rogers and Hassell (1974) and Beddington (1975). The discussion is simplified, however, if a more empirical description of the data is adopted, where a linear relationship between log (searching

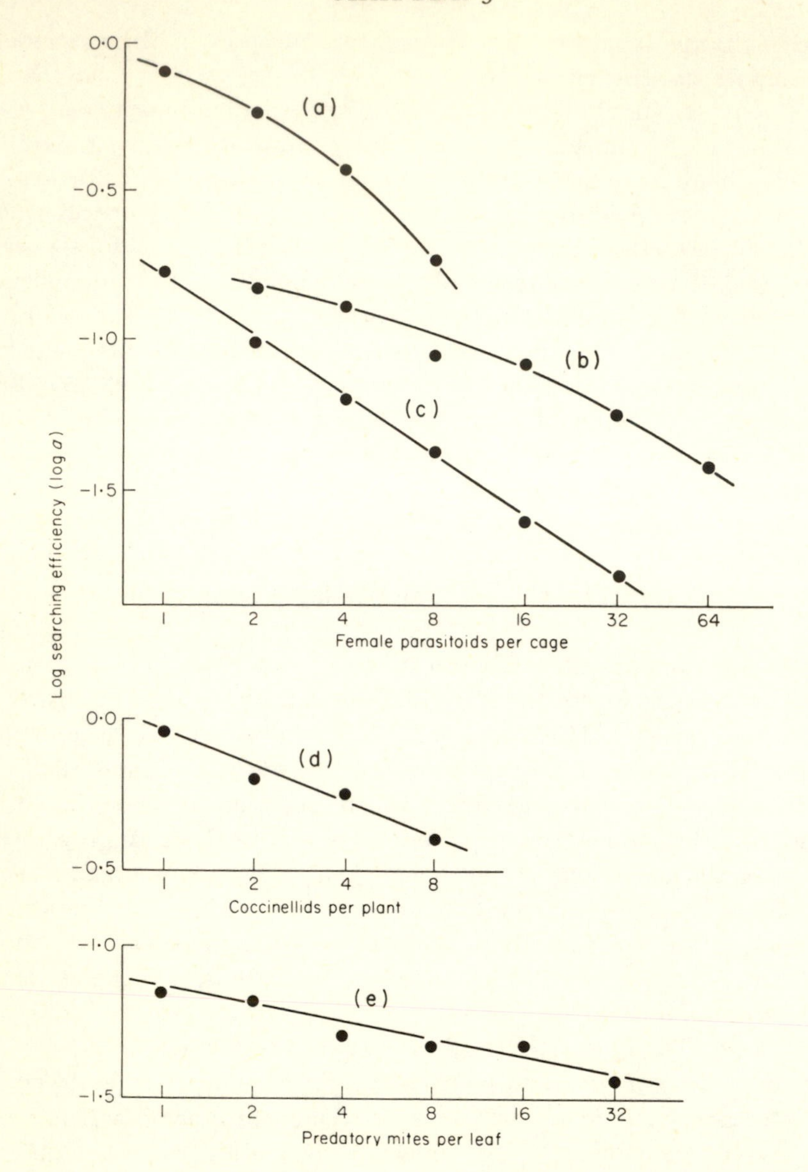

Fig. 5.3. Relationships between searching efficiency (log a) and log density of searching parasitoids or predators. (**a**) *Pseudeucoila bochei* (Bakker *et al.*, 1967); (**b**) *Encarsia formosa* (Burnett, 1958); (**c**) *Nemeritis canescens* (Hassell, 1971); (**d**) *Coccinella septempunctata* (Mickelakis, 1973); (**e**) *Phytoseiulius persimilis* (J. Fernando, unpublished).

efficiency) and log (predator density) is fitted as in Fig. 5.3c, d, e (Hassell and Varley, 1969). Thus,

$$\log a = \log Q - m \log P_t \tag{5.11a}$$

or

$$a = QP_t^{-m} \tag{5.11b}$$

where m is the slope and log Q the intercept of the linear regression. The function for predation in eq. (5.1) now becomes

$$f(N_t, P_t) = \exp(-QP_t^{1-m}) \tag{5.12}$$

[Note that this reverts to the Nicholson-Bailey model, eq. (5.3), when $m = 0$].

We now have a simple population model that includes predator interference and whose stability properties may be readily analysed following the recipe in Hassell and May (1973). This gives the stability boundaries shown in Fig. 5.4 where stability hinges solely on the value of the interference constant m and the prey rate of increase λ. The

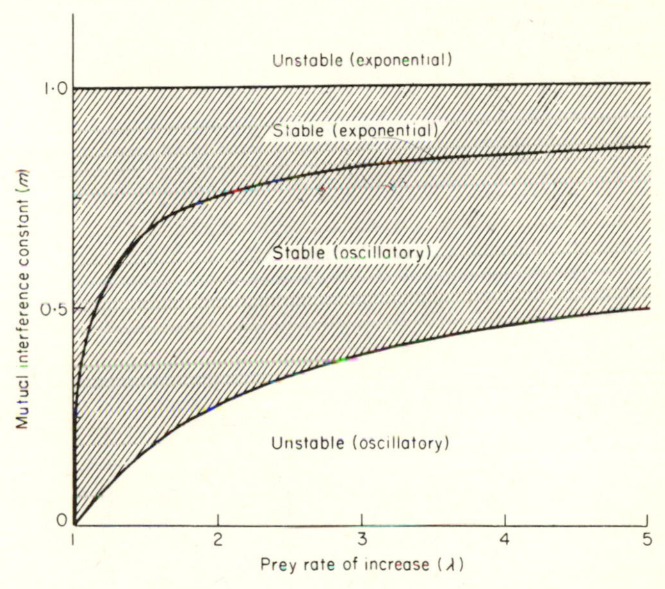

Fig. 5.4. Stability boundaries between the mutual interference constant (m) and the prey rate of increase (λ) [see eqs. (5.1) and (5.12)]. The shaded area denotes the conditions for stability and is divided into two regions: where the equilibrium is approached monotonically and where there are damped oscillations. The line between these regions indicates the conditions for most rapid approach to the equilibria. From Hassell and May (1973), courtesy of the British Ecological Society.

Table 5.2. Interference parameters from laboratory and field studies. Values of Q have been adjusted to give efficiencies over a m^2 per day.

No.	Species	m	Q m^2/day	Field or Lab.	Parasitoid or Predator	Author(s)
1	*Aphytis melinus*	0	0·0003	L	Parasite	D. J. Rogers (pers. comm.)
2	*Phytoseiulius persimilis*	0·18	0·0002	L	Predator	J. H. P. Fernando (pers. comm.)
3	*Dahlbominus fuscipennis*	0·28	0·021	L	Parasite	Burnett (1956)
4	*Aphytis coheni*	0·33	0·0013	L	Parasite	D. J. Rogers (pers. comm.)
5	*Aphidius uzbeckistanicus*	0·35	0·012	L	Parasite	Dransfield (1975)
6	*Coccinella 7-punctata*	0·38	0·033	L	Predator	Michelakis (1973)
7	*Cryptus inornatus*	0·38	0·0054	L	Parasite	Ullyett (1949b)
8	*Encarsia formosa*	0·38	0·024	L	Parasite	Burnett (1958)
9	*Bracon hebetor*	0·44	0·0090	L	Parasite	Benson (1973)
10	*Telenomus nakagawai*	0·48	0·003*	F	Parasite	Nakasuji *et al.* (1966)
11	*Cyzenis albicans*	0·52	0·0038	F	Parasite	Hassell and Varley (1969)
12	*Chelonus texanus*	0·54	0·021	L	Parasite	Ullyett (1949a)
13	*Alloxysta brassicae*	0·55	0·0032	L	Parasite	Chua (1975)
14	*Aptesis abdominator*	0·60	0·073	L	Parasite	von B. Sechser (pers. comm.)
15	*Diaeretiella rapae*	0·65	0·034	L	Parasite	Chua (1975)
16	*Nemeritis canescens*	0·68	1·17	L	Parasite	Hassell (1971)
17	*Pseudeucoila bochei*	0·68	6·31	L	Parasite	Bakker *et al.* (1967)
18	*Cratichneumon culex*	0·86	0·025	F	Parasite	Hassell and Varley (1969)
19	*Apanteles fumiferanae*	0·96	0·020	F	Parasite	Miller (1959)

* Q/day for host egg masses, rather than m^2.

third parameter, Q, only affects the equilibrium levels of the populations without affecting stability. In evaluating the significance of interference, it is a useful step to consider the values of m obtained from laboratory and field studies in the context of this stability diagram. The placing of points on the graph, however, is made difficult by having no real estimates of the prey rate of increase λ which, of course, is not merely the prey fecundity but is the rate of increase per prey after allowance for all prey mortalities other than predation (Hassell *et al.*, 1976a). Known values of m are therefore given in Table 5.2, distinguishing between field and laboratory studies and also between predators and parasitoids. The field values should be treated with caution since any errors in the estimates of the density of searching predators (which will usually be considerable) will tend to produce unrealistically high values for m (Hassell and Varley, 1969). It is clear from this analysis that the degree of interference found from laboratory studies would be a powerful stabilizing mechanism if extended to natural population interactions.

Finally, we should consider the possible evolutionary benefit of interference behaviours, which at first glance just involve a reduction of T_s and hence a reduced searching efficiency, a. Certainly, interference cannot be advantagous for a predator that searches randomly for a homogeneous prey. It is only when we consider the spatial heterogeneity of the prey population and the predators' response to this, that the evolutionary significance of interference becomes clear. A fuller discussion of this is therefore deferred to a later section.

5.4 The response to prey distribution

Most prey populations under natural conditions will exhibit a clumped distribution between units of their habitat, whether they be trees, branches, leaves etc. Within this framework, the random search assumed in eqs. (5.3), (5.10) and (5.12) implies that on average the same number of predators search for the same period of time per unit area. In this way each prey individual has an equal probability of being discovered. While this is a convenient assumption mathematically, it is not in accord with observed, or indeed expected, predator behaviour. Figure 5.5 shows some examples where arthropod predators and parasitoids have clearly spent a disproportionate time in unit areas of high prey or host density. All of these relationships may be sensibly interpreted as lying on different regions of an idealised aggregative response that is sigmoid. There is a lower plateau to the response where the predators are not distinguishing between different low prey density areas, an upper plateau where there is no discrimination between high density areas, all of which yield ample food, and an intermediate region within which the predators react markedly. The possible means by which arthropod predators aggregate in this way are various. They may, for example, respond to a volatile substance emanating from the prey, as clearly shown for some predators or parasitoids of bark beetles (Wood *et al.*, 1968). Alternatively, or in addition, aggregation may be the end result of a tendency merely to remain for longer periods searching the region where a prey has already been encountered (Murdie and Hassell, 1973).

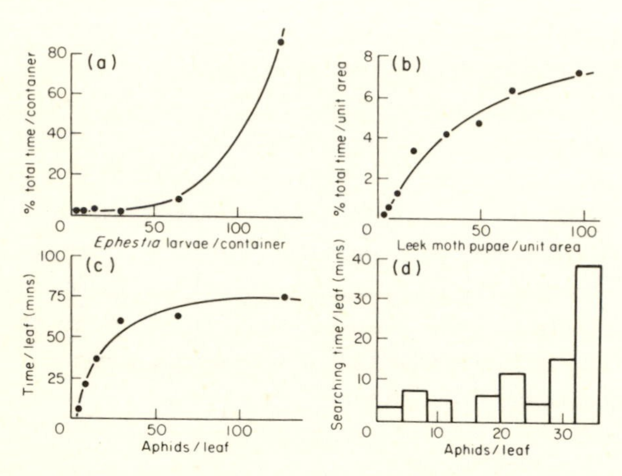

Fig. 5.5. Some examples of aggregative responses in insects. (**a**) the ichneumonid parasitoid, *Nemeritis canescens*, searching for different densities of flour moth larvae (*Ephestia cautella*) per container (Hassell, 1971); (**b**) the ichneumonid parasitoid, *Diadromus pulchellus*, searching for leek moth pupae (*Acrolepiopsis assectella*) per unit area (Noyes, 1974); (**c**) the braconid parasitoid, *Diaeretiella rapae*, searching for aphids (*Brevicoryne brassicae*) per leaf (Akinlosotu, 1973); (**d**) the coccinellid, *Coccinella septempunctata*, searching for aphids (*B. brassicae*) per leaf (M. P. Hassell, unpublished). After Hassell and May (1974).

The importance of such behaviours lies in the way that aggregation can contribute to the stability of the interacting populations. This can be illustrated by a most simple model for predator aggregation. We commence with the Nicholson-Bailey model, eqs. (5.4) and (5.5), but now divide the prey and predator populations between n unit areas such that in the ith unit there is a fraction α_i of the prey and β_i of the predator population. Now the function for predation becomes

$$f(N_t, P_t) = \sum_{i=1}^{n} [\alpha_i \exp(-a \beta_i P_t)], \qquad (5\cdot13)$$

which serves to distribute in each generation P_t predators and N_t prey into the n unit areas in the proportion specified by α_i and β_i. This, of course, will reduce to the Nicholson-Bailey model in the event that the same fraction of the predator population searches in each ith unit area. The extent to which non-random search enhances stability will depend on how clumped are the prey (the α_i set) and how marked is the predator aggregation (the β_i set). A fuller discussion of this, where the prey are distributed according to the negative binomial and the predator response is more in accord with the results in Fig. 5.5, is given in Hassell and May (1974). Murdoch and Oaten (1975) also give a detailed treatment of the effects of such non-random predator behaviour on the stability of predator-prey interactions.

We saw in an earlier section that predators, rather than parasitoids, present the complication of differing functional responses depending upon predator and prey sizes. In a similar way, the level of interference or the degree of aggregation amongst predators is also likely to vary both throughout predator development and also for different prey sizes. Information on such inter-age class effects, however, is at present lacking.

5.5 The predator rate of increase

It has already been stressed that a parasitoid is a rather special type of predator whose rate of increase depends simply on the number of hosts parasitized by the adult female population. Such assumptions are quite inadequate for true predators, whose overall rate of increase depends crucially upon (1) the survival of the immature stages and (2) the fecundity of the surviving adults, both of which have a complex dependence on the number of prey eaten throughout the life cycle. The predator's rate of increase is, therefore, likely to be affected by any factors that have an important effect on searching efficiency; in particular, the predator's response to prey density, predator density and to the prey distribution. Unfortunately, adequate information is only available on one of these—the effect of prey density—which is very briefly examined in this section. A much fuller discussion is to be found in Beddington *et al.* (1976b).

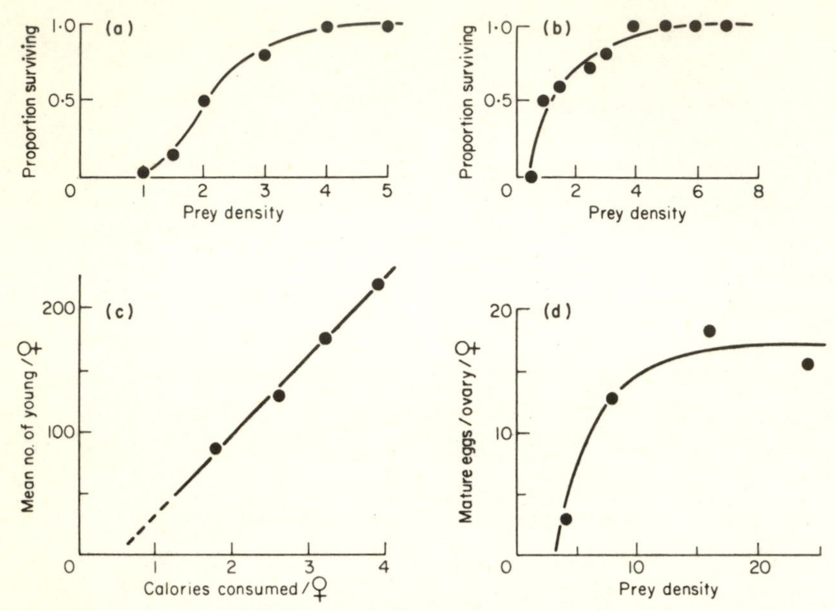

Fig. 5.6. (**a**) and (**b**) Relationships between mean prey density during particular instars and the proportion of individual predators surviving to the end of the instar. (**a**) first instar of the coccinellid, *Adalia bipunctata* (Wratten, 1973); (**b**) survival through first and second instar of the spider, *Linyphia triangularis* (Turnbull, 1962).
(**c**) Reproductive rate of *Daphnia pulex* var. *pulicaria* as a function of feeding rate (Richman, 1958).
(**d**) Fecundity of *Adalia decempunctata* as a function of aphid density available (Dixon, 1959). After Beddington *et al.* (1976b).

The way in which the survival of immature predators to the adult stage can depend upon prey density is shown in Fig. 5.6a, b. Survival here is almost complete at high prey densities, but progressively declines as less and less food is available. In addition to this all-or-none effect (starvation), low prey densities will reduce the predator's growth rate and hence prolong the developmental period. Apart from increasing the likelihood of mortality from other sources, this may have complex effects on the weight and fecundity of the ensuing adult.

Once in the adult stage, there remains an important dependence of fecundity per female on the number of prey eaten. Typically, in arthropods, this takes the form of a linear relationship above some critical level of prey consumption, as shown in Fig. 5.6c; below this, all the energy intake is allocated to maintenance of the predator and none to egg production. Following from this, the relationship between fecundity and prey *density* rather than prey *eaten* (Fig. 5.6d), must reflect the form of the female predator's response to prey density (its functional response), except again for the displacement along the prey axis.

First steps in the mathematical description of these components have been taken by Beddington *et al.* (1976b). There now remains the task of exploring (analytically, if possible) the effects of such more realistic predator reproduction on the dynamics of predator-prey interactions. This is made somewhat more difficult by generation times no longer being absolutely fixed, but dependent on prey density itself. The effects on stability that will emerge are not easy to predict since there are now important time lags associated with larval survival and development rates affecting future adult fecundity.

5.6 Some relationships between parameters

The reproductive success of a predator depends largely on the number of prey eaten (especially so in the case of parasitoids), and hence evolutionary pressures will be towards maximising the prey death rate per predator. It is, therefore, unlikely that the values of the various parameters defining predator performance (i.e. T, T_h, Q, m, μ, where μ is some index of predator aggregation in regions of abundant prey) are simply a pot-pourri mix, bearing no relationship to each other. They should all relate to the details of life history and to the particular searching strategy adopted by the predator. Unfortunately, the discussion must remain speculative at the present since these critical parameters are known for very few parasitoids and even fewer predators.

5.6.1 *Longevity* (T) *and handling time* (T_h)

Table 5.1 shows that the handling time (T_h) is generally a very small fraction of the total time available for search (T) [see eq. (5.9)]. This has the dual benefit to the predator of an increased attack rate and reduced population instability. Selection should therefore act either to increase T or reduce T_h. Two rather different situations can be distinguished. On the one hand, there are those insect parasitoids that are more or less restricted to a single host species whose generations are quite discrete. In these instances, it is to the parasitoids advantage for its life cycle to be closely synchronised with that of its host, which therefore sets the limit on T. To ensure $T_h \ll T$ within this constraint, therefore, requires selection to reduce T_h as far as possible. However, many, if not most, predators and parasitoids attack several species whose generations overlap to some degree. Close synchrony of predator and prey life cycles is no longer possible and selection could now tend both to increase longevity and reduce handling time. This would only be impossible where T_h is unavoidably long as in some pupal and egg parasitoids that must drill through a tough covering of the host before oviposition. To these would accrue considerable advantages in being especially long-lived and so ensuring T_h to remain a very small fraction of T.

5.6.2 *Q, m, μ and the degree of polyphagy*

The level of interference (m) [see eq. (5.11)] is in part a function of the activity of the searching predators and parasitoids. The more active and mobile species—those with higher intrinsic searching efficiencies, a' or Q—are likely to encounter each other more often, so leading to greater interference. That this is so, at least under laboratory conditions, is apparent from Fig. 5.7a. The hollow points come from field examples and have been omitted in fitting the linear relationship to the data. The caveat on page 80, that estimates of m for natural populations will be biased by errors in estimating the searching predator population (P_t), again applies. In addition, we can have little reliance on Q when converted to units of 'm^2 per day'.

A more interesting relationship, involving m and the number of recorded prey species, is given in Rogers and Hubbard (1974), and shown here with additional data in Fig. 5.7b. It implies that there is some selective advantage in enhanced interference as the range of potential prey decreases. It remains, however, unclear if this is the primary relationship. It may well be, for instance, that the basic relationship is between Q and the number of prey species, although there is little sign of this from the available data.

The underlying, causal relationships between these parameters are unlikely to be resolved without considering predator aggregation (μ). Whenever predators aggregate in relatively restricted parts of their habitat, the likelihood of mutual interference is enhanced. It is in this light that the ecological role of interference should be viewed: as a

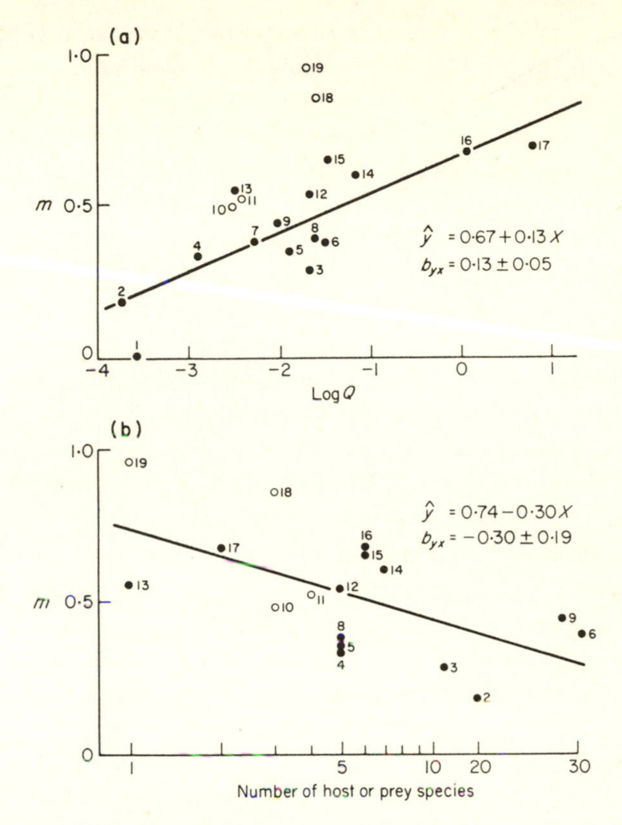

Fig. 5.7. (**a**) The relationship between the mutual interference constant m and searching efficiency parameter Q for a variety of parasitoids and predators. All values of Q are expressed in units of m^2 per predator per day. After Rogers and Hassell (1974) with additional data added.
(**b**) The relationship between m and the number of prey species known to be attacked. The number of prey for points 2 and 6 are guesstimates. After Rogers and Hubbard (1974) with additional data added.
In both graphs, the numbers correspond to those in Table 5.2 and the hollow circles to data from natural rather than laboratory populations.

mechanism ensuring the greater dispersal rate of predators from areas where prey are already, or likely to become, heavily exploited. The dispersing predators then have the chance of locating other less-exploited areas of prey. Whether a given act of interference is beneficial or not, therefore, depends upon whether the reduction in searching time (T_s) due to interference is more than compensated by finding a further area with sufficiently abundant prey. Ideally, the dynamic balance between the two processes of predator aggregation and interference should act to leave an optimum number of predators within a given area of prey. Within this framework, we should expect the level of interference and extent of aggregation to be directly related. It is unfortunate that the aggregative behaviour of most of the species in Fig. 5.7a, b has not been studied, so preventing a similar treatment of m and μ.

With this in mind, some sweepingly general, but hopefully also useful, statements are possible on the relationship of searching parameters to the range of prey species accepted. We start with the observation that predators with a very narrow prey range (i.e. monophagous or oligophagous) are more likely than polyphagous predators to respond to specific cues from a particular prey species. Thus, the parasitoids and predators known to make use of prey pheromones in locating their prey (e.g. Wood *et al.*, 1968; Mitchel and Mau, 1971; Sternlicht, 1973) are restricted to only one or a few prey species. There will be several consequences of such efficient prey location.

(1) Areas of high prey density will be more easily discovered, leading to a higher aggregative index (μ).

(2) The intrinsic searching efficiency (Q), measured for a particular prey species, will be higher than that of a generalist, or polyphagous predator.

(3) More efficient aggregation will lead to more frequent predator encounters and hence more mutual interference (m). This has the benefit of preventing local over-exploitation of the prey. Interference leading to dispersal is more likely where the predator or parasitoid can easily discover other prey areas.

We now have a framework, that is in accord with the results in Figs. 5.7a, b, and in which Q, m and μ are all loosely linked and bear a direct relationship to the range of prey species. The fuller development of such ideas must await the accumulation of further examples where the components of predator search have been properly quantified.

There remains a further field in which the parameters describing predator performance are relevant, namely biological control. Classical biological control has achieved most of its notable successes to date using insect parasitoids in the control of pests of perennial standing crops such as orchard and forest trees, in contrast to many agroecosystems which suffer annual ecological upheavels (see DeBach, 1974). The success of an introduced parasitoid in such relatively stable habitats will depend on many factors (e.g. climate, synchrony of life cycles); but amongst these, the searching characteristics of the parasitoid will always be important. Thus, the ultimate aim of biological control—the establishment of a new equilibrium by the parasitoid at very low host densities—is ultimately dependent on a high searching efficiency (Q). Low searching efficiencies will always tend to result in rather higher equilibrium populations. For the parasitoid to maintain a *stable* equilibrium, requires in addition such behaviours as aggregation (μ) or interference (m) as discussed above. It is striking that these three parameters (Q, m and μ) are just those which seem to be related to the degree of polyphagy. A long-standing tenet amongst biological control workers has been that success is more likely using natural enemies that are fairly specific to the pest species in question [but see Ehler and van den Bosch (1974) for the importance of polyphagous natural enemies in non-perennial crops]. It was supposed that introduced polyphagous natural enemies would not show an adequate numerical relationship with any one of their prey species. While this may well be true, we can now see that the use of specific species has also increased the likelihood of optimum searching characteristics.

5.7 Predation and competition

Predation is only one of several mortalities that affect prey populations under natural conditions. Some of these will depend on the physical environment, others, such as competition, on biological interactions. While the components of predation in these more complex systems will continue to have effects similar to those already described, the overall stability properties of the prey population now depend on the nett effect of a variety of stabilizing and destabilizing factors. This is simply illustrated by combining, within a single model, both predation and intraspecific competition amongst the prey.

Single species competition models and their full range of stability behaviour have been reviewed in chapter 2. In general, they have a stable equilibrium, at least for low rates of population increase, and hence should continue to contribute to stability when included in predator-prey models. This has been well demonstrated by Beddington *et al.* (1975) using an extension of the Nicholson-Bailey model [eqs. (5.4) and (5.5) above] in which they included a version of the discrete logistic model (the form B of Table 2.2). The expression for N_{t+1} in eq. (5.4) now becomes

$$N_{t+1} = N_t \exp\left[r(1 - N_t/K) - aP_t\right] \qquad (5.14)$$

Here K is the 'carrying capacity' of the prey population and $r = \ln \lambda$. In contrast to the Nicholson-Bailey equations, there are now a wide range of conditions in this model in which the equilibrium is stable. This work is discussed more fully in section 4.2.3; see eqs. (4.11) and (4.12).

A stimulating view of the interaction of predation and competition has been presented by Southwood (1975), in which he presents a framework for interpreting several kinds of population fluctuations (see chapter 3 for a fuller discussion of this synoptic approach). The number of prey eaten by the predator population was assumed to be a sigmoid function of prey density. This will occur when:

(1) predation depends upon a sigmoid functional response (Fig. 5.1c) and a predator density that is relatively constant compared to that of the prey (more likely in polyphagous predators);

(2) when *both* the functional and numerical responses are either convex or sigmoid. Such a numerical response is likely, for example, when the predator generation time is very much shorter than that of the prey.

The reproductive rate of the prey (λ) was also assumed to be a function of prey density, but here we can follow previous models in this chapter and assume it to be constant. Finally, competition is included according to the form B of Table 2.2.

The properties of this model are displayed in Fig. 5.8, where the numbers in successive generations are compared (cf. Fig. 3.5). An equilibrium, whether stable or unstable, will occur whenever the curve crosses the 45° line. In this example, there is a lower stable equilibrium (P) maintained by the predator (due to the density dependent part of the predators' combined functional and numerical responses). Should prey density rise above point U, however, the prey population increases, unchecked by the predators which are now limited by handling time, satiation or their slower rate of increase, until finally the prey are regulated at a much higher stable equilibrium (C) by intraspecific competition. Whether the population remains at this level or invariably crashes, to be controlled again by the predator, depends upon the parameters of competition; in particular, on whether or not the prey 'scramble' and overexploit their resources. In this way, the model describes instances where the prey are normally regulated by predators, but occasionally 'escape' and increase to the limit of their resources. Furthermore, it is easily extended, by altering the properties of the predators, to include examples between the extremes of a globally stable predator-prey equilibrium to no such equilibrium at all.

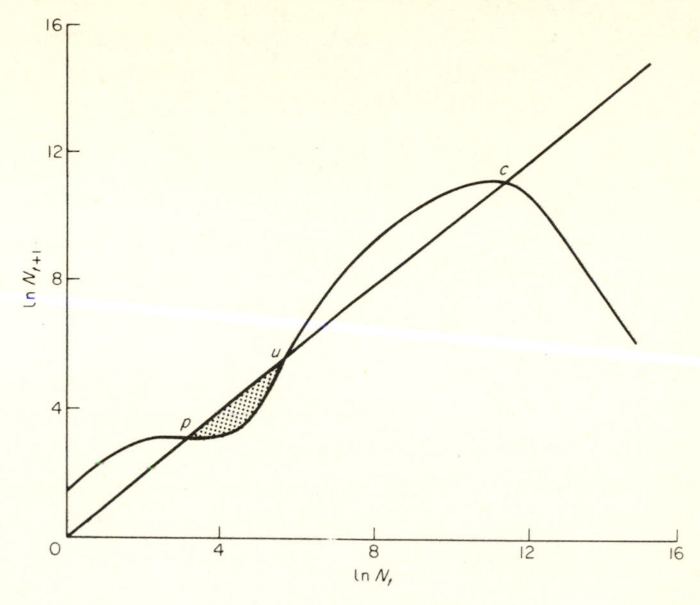

Fig. 5.8. Relationship between population densities in successive generations (ln N_t and ln N_{t+1}) from a predator-prey model with intraspecific competition acting upon the prey. Points P, C and U (where the curve intersects with the 45° line) represent stable equilibria due to predation (P) and competition (C) and an unstable equilibrium (U). After Southwood (1975).

In coupled predator-prey systems where time lags become very important, this elegant picture will be somewhat blurred since predation will now also hinge importantly on the way that the predator population is fluctuating. However, the general picture should remain, provided that predation tends to be more-or-less density dependent up to a critical level of prey abundance.

6
Plant-Herbivore Systems

GRAEME CAUGHLEY

This chapter touches lightly on the congruity of plant-herbivore systems and prey-predator systems, classifies the former according to the interaction between the plants and animals, explores the dynamic behaviour of examples for each category of this classification, and closes with a few speculations on the selective forces operating within these systems.

6.1 Interaction between trophic levels

A stack of trophic levels is a hierarchy in which each level extracts energy from those below and contributes energy to those above. The ultimate source of energy is the sun, the stack of trophic levels acting as a sink. Above the basal level the dynamics of a given layer can be summarised as the rate at which it gains energy from the level below (the functional response) and the rate at which it fixes that energy (the numerical response); this is a more abstract way of expressing the notions developed in sections 4.2 and 5.1.

A population ecologist deals with a tiny sub-set of this system, often restricting his field or laboratory studies to the interaction of two populations, one in each of two contiguous trophic levels. He has willingly parted company with reality, but hopes that the processes he observes within his arbitrarily circumscribed microcosm will be general inter-level processes in the small. Many theoreticians are similarly inclined, and they ask what will happen to two populations if the members of one eat the members of the other.

As observed in section 4.2, an early joust with this windmill was provoked independently by Lotka (1925) and Volterra (1926) who showed that this interaction, if impossibly simplified, would lead to mutual cyclicity. Their model of randomly colliding hunters and hunted, the first with an infinite capacity to eat and the second with an infinite capacity to multiply, is biologically implausible and mathematically naive. That hardly matters. I strongly suspect that, primarily, they sought to demonstrate that such a simplified system could behave in a complex manner; reality could, if necessary, be added later.

But a trivial aspect of their modelling, quite as arbitrary as their assumptions, has survived largely unchallenged. They discussed their model in the context of interaction between the second and third trophic level. Lotka chose the relationship between host and parasite as his paradigm, while Volterra, a close associate of the fisheries dynamicist Umberto D'Ancona, tended, when groping for a suitable example, to think of larger fish eating smaller fish. Subsequent elaborations of the Lotka-Volterra formulation, as discussed in chapter 4, have led to more powerful models that can lay modest claim to counterfeiting reality. With a few exceptions (e.g. Noy-Meir, 1975) and qualifications (e.g. May, 1973a), they have been tagged 'prey-predator' systems and firmly chained conceptually to interaction between the second and third trophic level.

For two good reasons this is a pity. First, the system that Lotka and Volterra modelled, albeit simplistically, was a generalisation of the interaction between any two trophic levels. Second, the subsequent elaborations tend to be better descriptions of what goes on one step down where plants interact with herbivores.

Any model purporting to describe the interaction between two populations in consecutive levels must, as a minimum, have one equation for growth of the population in the higher level and one for growth of the population in the next level down. Since the latter's growth is in turn influenced by the level below *it*, and since it can influence that level itself, ideally the model would include at least one equation for each nether level and one for the sun. Should we wish to truncate this unwieldy list by contracting back to the minimal pair of equations, we are forced to apply a mathematical tourniquet at the point where the two levels of interest are severed from those below. A single assumption serves: we postulate that our lower level cannot influence the rate at which energy flows into it from below, its resources being renewed at a constant rate. Mathematically the assumption is translated into a logistic term, or some such, placed within the equation describing growth at the lower level. As will be argued in section 6.3.1, this simplifying assumption is dubious for a prey-predator system but hovers near the edge of reality for many plant-herbivore systems.

6.2 A classification of grazing systems

At the risk of enraging vertebrate ecologists and range managers who use 'grazing' in the strict sense of 'eating grasses and herbs', the word is used here to mean 'eating any kind of plant or part of a plant'.

Table 6.1. A classification of relationships between vegetation, V, and herbivores, H.

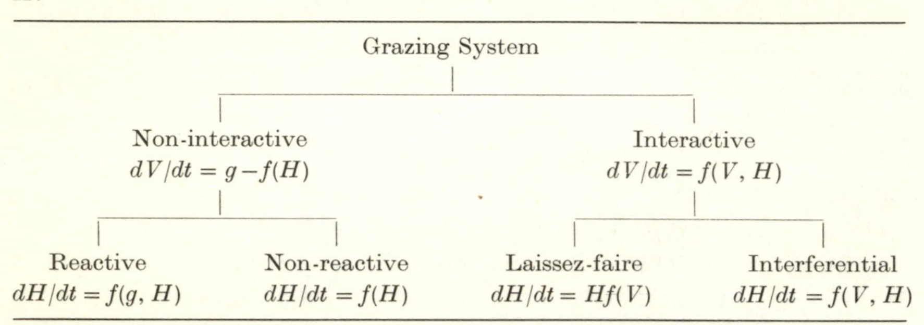

Table 6.1 is a classification of grazing systems in which the primary division is made according to Monro's (1967) dichotomy: between those systems in which the herbivores influence the amount of food available to subsequent generations and those systems where they have no such effect. The first is an 'interactive system' and the second a 'non-interactive system'.

The essence of a *non-interactive system* lies in the herbivore's inability to influence the rate, g, at which its resources are renewed. It subdivides uneasily into 'reactive systems' where rate of increase of herbivores reacts to the rate g at which plant material is renewed (there being no reciprocal reaction), and 'non-reactive systems' in which the herbivores increase at a rate largely independent of the parameters of plant growth.

In an *interactive system* the herbivores do influence rate of renewal of the vegetation, which in turn influences rate of increase of the animals: the two components interact. This class divides again into 'laissez-faire systems' in which the herbivores do not interfere with each other's feeding activities, and 'interferential systems' in which an animal may reduce the ability of another to obtain food that would otherwise be available to it. These two systems have different dynamics.

One or two examples from each of the four terminal divisions of the classification (Table 6.1) will now be examined. In each case we will attempt to summarise the dynamics by a simple model.

6.3 The dynamics of grazing systems: non-interactive systems

6.3.1 *Reactive systems*

When the rate of renewal of plants, g, is independent of the standing crop of plants, V, plant density changes as

$$dV/dt = g. \tag{6.1}$$

When a herbivore population of size H feeds on these plants, with each herbivore eating at a constant rate c, the change in V might become

$$dV/dt = g - cH. \tag{6.2}$$

Alternatively, if each herbivore has a maximum rate of intake, c, which declines as vegetation is thinned out, then

$$dV/dt = g - cH[1 - \exp(-dV)]. \tag{6.3}$$

When the available vegetation, V, comprises falling leaves which are eaten immediately upon impact by one or more insatiable herbivores, we have the extreme limit

$$dV/dt = g\, e^{-\infty H} = 0;\ \text{i.e., } V = \text{constant}. \tag{6.4}$$

That seemingly improbable case repays investigation. If b is defined as the rate of intake sufficient to maintain a herbivore and allow its replacement in the next generation, then the proportion of available food (which, remember, is produced at a rate g) used for maintenance is bH/g, leaving $(1 - bH/g)$ that can be channelled into generating an increase of herbivores. With rising H, the herbivores' rate of increase slows progressively until finally all the food is utilized for maintenance, and the population stabilises at $H = g/b = K$. Hence the herbivore population grows as

$$dH/dt = rH(1 - bH/g). \tag{6.5a}$$

Here r is the herbivore's intrinsic rate of increase, the rate at which it would increase in that environment if no resource were limiting. Substituting K for g/b turns the equation into the common form of that describing logistic growth, eq. (2.3):

$$dH/dt = rH(1 - H/K). \tag{6.5b}$$

The assumptions underlying this model can be retrieved as characteristics a population must possess if it is to grow logistically:

(a) A population's resources are renewed at a constant rate, independent of both the standing crop of the resource and the standing crop of the population, the population having no influence over the amount of resource available to the next generation.
(b) The members of the population have insatiable appetites.
(c) Rate of increase is a linear function of food intake per head.
(d) There is no lag between cause and effect.

Of these four conditions for logistic growth only the first is critical. A slight relaxation of the remaining three has little effect on the trajectory of growth. The replacement of the functional response of eq. (6.4) with that of eq. (6.2) or (6.3), or the introduction of a short time lag, leads to a quasi-logistic pattern of growth.

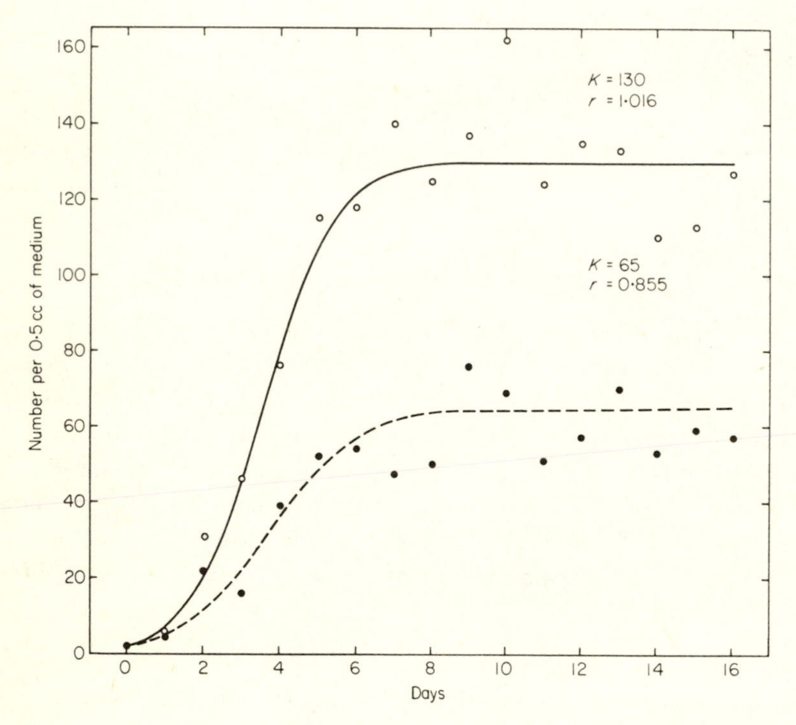

Fig. 6.1. Growth of *Paramecium* populations grazing a vegetation renewed at a constant rate.

Figure 6.1 shows the growth of two herbivore populations in grazing systems conforming closely to all four logistic assumptions. It had to come from the laboratory because such an implausible set of assumptions would seldom be met in the field. The examples are from

Gause's (1934) classic experiments on *Paramecium caudatum* grazing on the bacterium *Bacillus pyocyaneus*. *Paramecium* was cultured in tubes of Osterhaut's medium, a mixture of salts balanced in relation to the physiology of *Paramecium* but which inhibits reproduction of *Bacillus*. At the beginning of each experiment Gause placed twenty *Paramecium* in the tube and added a constant quantity of *Bacillus* to it each day. Figure 6.1 shows how two populations grew, one of which was fed at twice the rate of the other.

These data come from Gause's Appendix Table 4, the fitted logistic curves being calculated from the first seven points and the level of the asymptotes being gauged by eye. Gause (1934), Kostitzin (1939) and Andrewartha and Birch (1954) also fitted logistics to these same data. All differ slightly from one another, but the variation reflects no more than the multitude of ways by which a logistic can be fitted.

The populations acted in a thoroughly logistic manner: the curves track the data closely, K is proportional to rate of renewal of the resource, and r is much the same whether rate of renewal is high or low. These results are entirely predictable, not because *Paramecium* populations have a penchant for growing logistically, but because Gause validated the logistic assumptions by the experimental conditions he imposed upon these grazing systems.

That in itself goes a long way towards making the point of this section: true logistic systems are unlikely to occur in nature. The facts of a few populations (e.g. fruit flies whose larvae feed on the flesh of stone fruit) may be close enough to logistic assumptions to make it a workable model, and other populations whose dynamics are not logistic may have trajectories of growth to which a logistic can be fitted as a pragmatic approximation (see its use by Murphy (1967) for sardines), but as a model of how a population actually works it is seldom likely to teach us much.

There is an exception and that exception is important. Plants cannot influence the rate of renewal of their resources, water and sunlight, and therefore cannot influence the supply of these to the next generation. The growth of many plant populations should therefore be close to logistic, and there is some evidence from the growth of pastures (Brougham, 1955; Davidson and Donald, 1958) that this is so. Theoretically and evidentially the logistic is a first choice to model the growth of a population of plants. It will often be the most appropriate model.

That apparent detour brings us back to the discussion of differences between plant-herbivore systems and prey-predator systems foreshadowed in section 6.1. A logistic term is included in a plant-herbivore model, not as the contrivance introduced into prey-predator models to seal the system mathematically from lower trophic levels, but as a description of the relationship between plants and their resources. That was the basis of the previous suggestion that most prey-predator models are actually plant-herbivore models misnamed.

6.3.2 *Non-reactive systems*

Non-reactive plant-herbivore systems are those in which the rate of change of animal numbers is independent of food supply.

Totally non-reactive systems are exceedingly uncommon, the best known of the handful being *Thrips imaginis* feeding on sap and pollen (Andrewartha and Bircth, 1954). More common are systems in which the herbivore's dynamic reaction to changes in food supply is very weak, plant density accounting for only a small proportion of the variance in animal numbers. Most of these comprise an 'outbreak herbivore' whose food plant is a grass or herb. The grass-hopper, *Austroicetes cruciata* is one such. Its outbreaks are triggered by a rise in the quality of its food (Andrewartha and Birch, 1954). Newsome's (1969) mouse populations living in wheat fields are another example. They erupt in response to a freak sequence of weather conditions that ensures adequate food throughout the summer but cracks the soil early in the season and then softens it enough to allow burrowing towards the end of summer.

The special characteristics of the ecology of each outbreak species rules against a general model for their trajectories of growth. The only common feature is the exponential pattern of growth during the early stages of the eruption.

6.4 The dynamics of grazing systems: interactive systems

One of the commonest grazing systems is that in which rate of change of herbivores is a function of plant density, and rate of change of plants is a function of herbivore density. The two components interact. Formally, the 'plant density' in this context is the quantity of vegetation per unit area which is available both for consumption by animals and for producing plant growth (Noy-Meir, 1975).

A model of this system must include several parameters: usually two for plant growth, two for grazing pressure and two or three for the growth of the herbivore population. Here is a representative set of parameters* for such a model:

r_1 = the intrinsic rate of increase of plants,
K = maximum ungrazed plant density,
c_1 = maximum rate of food intake per herbivore,
d_1 = grazing (searching) efficiency of the herbivore when vegetation is sparse,
a = rate at which herbivores decline when the vegetation is burned out or grazed flat,
c_2 = rate at which this decline is ameliorated at high plant density,
d_2 = demographic efficiency of the herbivore; its ability to multiply when vegetation is sparse.

A number of choices are available for depicting the relationship between plant density and rate of food intake per herbivore (the functional response), and between plant density and rate of increase of the herbivore (the numerical response): see Table 4.1. I have used simple functions that rise with increasing plant density, and saturate when food is abundant. The numerical response used here ignores the effect of 'underpopulation', that is, reduced fecundity at low density reflecting the difficulty of finding a mate when mates are scarce. This is a purposeful dereliction. Underpopulation has not, to my knowledge, been identified for any herbivore, vertebrate or invertebrate. It may occur, but the lack of clear examples suggests that the critical density for underpopulation is lower than the threshold marking a high probability of imminent stochastic extinction.

6.4.1 *Laissez-faire systems*

In a laissez-faire interactive system the herbivores do not interfere with each other's search for food. Non-territorial ungulates provide the type example of this kind of grazing behaviour. Noy-Meir (1975) explored by graphic analysis the theoretical implications of holding

* In particularising from the general prey-predator system of chapter 4, to the plant-herbivore system of this chapter, the symbol for the prey population, N, has already been replaced by V for vegetation, and the symbol for the predator population, P, by H for herbivore. Similarly, in what follows there is no attempt to keep the symbols for the various parameters precisely congruent with those used in chapters 4 and 5 (e.g., those in Table 4.1).

ungulates at an arbitrary constant density, as in farming, and Caughley (1976) looked at the properties of a system containing wild ungulates hunted by man. To keep it short and simple, the present discussion is limited almost entirely to systems in which wild ungulates range free of persecution.

May (1975a, and section 4.2) has summarised the different ways in which this system can be modelled, and he has emphasised that the qualitative behaviour of the model is insensitive to the details of its construction. Depending on the values of its parameters, the system may be characterised by a stable equilibrium point, or by a stable limit cycle, whose amplitude may be so severe as to produce extinction. Of the many available, one model will suffice to sketch in the outlines of this system:

$$dV/dt = r_1 V(1 - V/K) - c_1 H[1 - \exp(-d_1 V)] \qquad (6.6)$$

$$dH/dt = H\{-a + c_2[1 - \exp(-d_2 V)]\}. \qquad (6.7)$$

The symbols are as defined above.

The first equation expresses the rate of change of vegetation by two terms, the first depicting logistic growth and the second the rate of grazing. The functional response of diet to plant density is in Ivlev (1961) form; see form (iv) of Table 4.1. Equation (6.7) summarises the rate of change of herbivores, H, in terms of their intrinsic ability to multiply, as modified by the availability of food. Herbivores can increase at a maximum rate of $\{-a + c_2[1 - \exp(-d_2 K)]\}$, which in most circumstances will equal their intrinsic rate of increase, $r_2 (r_2 = c_2 - a)$, because at high plant density the term inside the square brackets will tend to unity.

Before exploring the applicability of this model to real grazing systems, we will examine in detail the mathematical properties of a system conforming to these rules. Figure 6.2 shows the growth of a population of herbivores, and the resultant changes in plant density, as the two spiral towards their mutual equilibrium point. For this illustration, the parameters were set at

$$r_1 = 0{\cdot}8 \qquad a = 1{\cdot}1$$
$$K = 3000$$
$$c_1 = 1{\cdot}2 \qquad c_2 = 1{\cdot}5$$
$$d_1 = 0{\cdot}001 \qquad d_2 = 0{\cdot}001$$

Although the example is imaginary it can be thought of, without contradicting current knowledge, as white-tailed deer colonising a mosaic of grassland and forest. Wildlife managers will recognise the growth curve as a deer eruption.

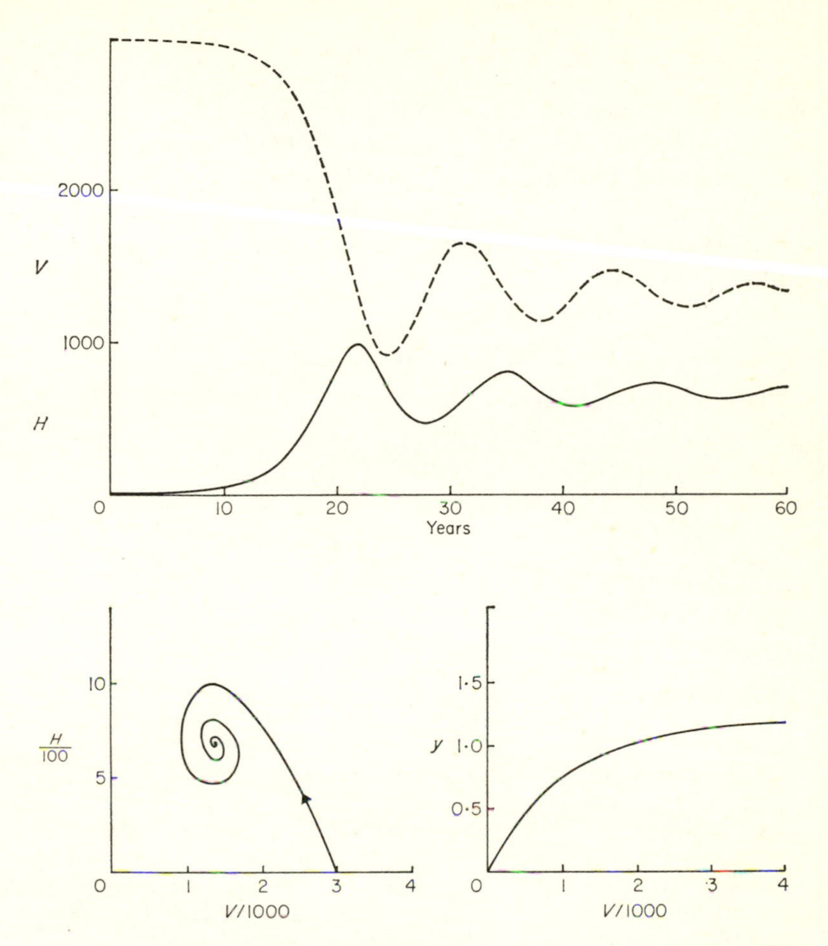

Fig. 6.2. Model of the trend of vegetation (V) and herbivores (H) during an ungulate eruption. Intake (y) of vegetation per herbivore is graphed against density of vegetation.

To gain more general analytic insight into the dynamics of this model, we first observe that the possible equilibrium point (where $dV/dt = dH/dt = 0$) lies at the population values

$$V^* = (1/d_2) \ln [c_2/(c_2 - a)] \tag{6.8}$$

$$H^* = \frac{r_1 V^*(1 - V^*/K)}{c_1[1 - \exp(-d_1 V^*)]}. \tag{6.9}$$

Depending on the specific parameter values, this critical point may represent the locus of a stable equilibrium, or the focus of a stable limit cycle. The former is the case if

$$K < V^*(2e^x - 2 - x)/(e^x - 1 - x), \tag{6.10}$$

and the latter otherwise, with $x = d_1 V^*$. This stability criterion is illustrated in Fig. 6.3. For the parameter values catalogued above, and illustrated in Fig. 6.2, the critical point is given by eqs. (6.8) and (6.9) as $V^* = 1300$ and $H^* = 670$, and as shown by the location of this point in the parameter space of Fig. 6.3, the system has a stable equilibrium point.

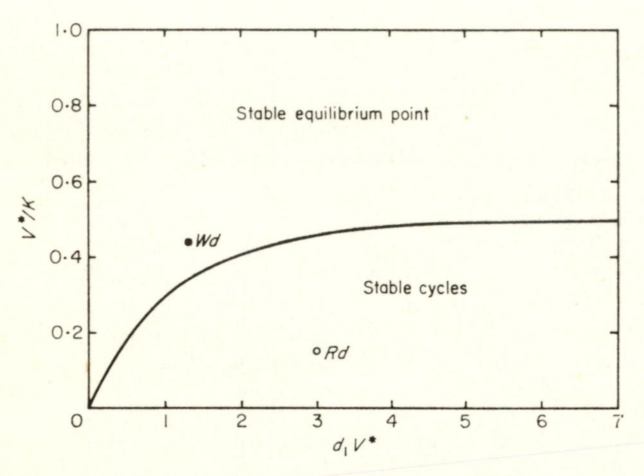

Fig. 6.3. The boundary separating stable equilibrium point from stable limit cycle behaviour in the population described by eqs. (6.6) and (6.7). The parameters are as defined in the text. The dot and the open circle mark, respectively, the critical points for the 'white-tailed deer' model and the 'reindeer' model.

For relatively large values of $d_1 V^*$, the stability condition eq. (6.10) simplifies to

$$K < 2V^*; \tag{6.11}$$

this approximation is valid to better than 2 per cent for $d_1 V^* > 5$. In this form the biological implications are more apparent: a system of a vegetation and a population of efficient herbivores has oscillatory behaviour if grazing pressure tends to hold the standing crop of vegetation below half that of its ungrazed state.

We now use these insights to study the effects of perturbing the parameters in the model. The graphic methods of Rosenzweig and MacArthur (1963) provide the most convenient approach. There will be one or more values of V at which $dV/dt = 0$ when H is held constant at some arbitrary value. The points may represent maxima or minima of cycles, or stable equilibrium. They can be plotted on a graph of H against V. If this procedure were followed for every possible value of H, the resultant points would fall along a curve, the zero 'isocline' of V. The form of this curve, expressing H as a function of V, clearly is given by eq. (6.9). Similarly, there is a set of points at which $dH/dt = 0$ when V is held constant. This isocline of herbivores is an inverted T, the crossbar coinciding with the uninteresting instance of $H = 0$, and the vertical rising from the V axis at the value V^* given by eq. (6.8). Figure 6.4a gives the zero isoclines of V and H for the numerical example above.

The salient feature is whether the position of the isocline of H is to the right of the highest point, indicated by a triangle, of the V isocline. That is a necessary and sufficient condition for point-stability of this model (Rosenzweig, 1971; Gilpin, 1972). The system's equilibrium point [the V^*, H^* of eqs. (6.8) and (6.9)] of course lies at the intersection of the two isoclines.

Figures 6.4b–6.4h show how the isoclines shift as each parameter value is increased in turn by one third, the other parameters in each case being held at the standard values given previously. The dashed isoclines in each graph are the standard set of Fig. 6.4a. Figures 6.4b–6.4e depict the changes wrought by increasing the parameters of plant growth and grazing pressure (r_1, K, c_1, d_1). Only the V isocline changes, the hump being shifted laterally when either K or d_1 is increased. Both changes lower the stability of the system, the increase in K throwing the modelled system out of equilibrium. That effect is by now well known; it is Rosenzweig's (1971) 'paradox of enrichment' (see also the more general discussion in section 4.2).

Figures 6.4f–6.4h show what happens when the parameters of herbivore growth (a, c_2, d_2) are upped by one third. Only the H isocline reacts. A rise in a enhances stability (the H isocline is shifted further to the right of the hump), whereas rising c_2 or d_2 decreases stability. In our example the enhanced stability consequent on an increase of a is of mathematical interest only, since the herbivore has been eliminated. Because a is a rate of decrease, a rise in a reduces the population's intrinsic rate of increase, and actually pushed it below

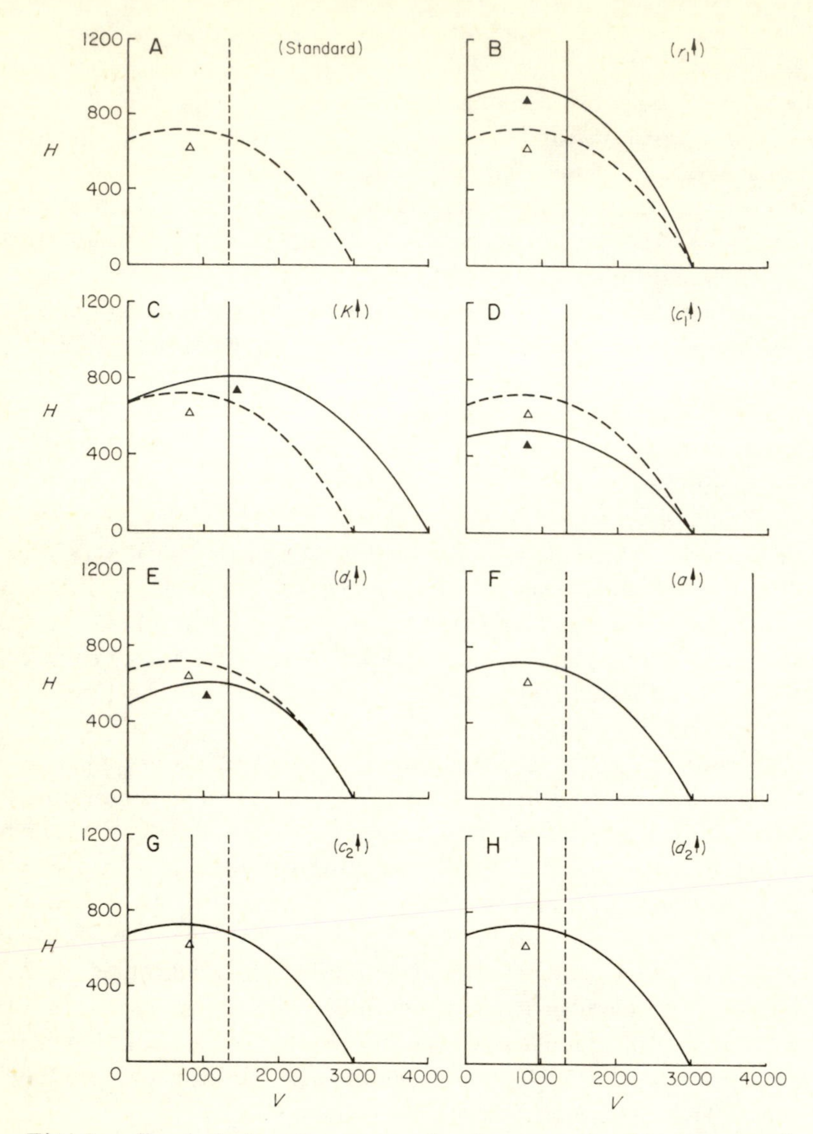

Fig. 6.4. Graph A gives the zero isocline for vegetation (the curve) and the zero isocline for herbivores (vertical line) for the 'white-tailed deer' model. Graphs B–H show how these isoclines shift as each parameter in turn is increased by one third. Triangles indicate the peaks of the V isoclines.

zero in this example, with the system then stabilising at $V^* = K$, $H^* = 0$.

In summary, the system's stability is lowered by an increase in maximum plant cover (K), an increase in grazing efficiency (d_1), an increase in the positive component of the herbivore's intrinsic rate of increase (c_2), and by an increase in its demographic efficiency when food is in short supply (d_2). A large genetic problem rearing up behind those misleadingly simple conclusions is examined in section 6.5.

The laissez-faire model is at its best as a summary of the interaction between a population of ungulates and its food supply. The trajectory of an ungulate population increasing from minimal density while grazing a vegetation at density K follows an eruption to a peak, a crash, and a 'stabilisation' around an equilibrium level well below the peak density (Caughley, 1970). Figure 6.5 shows the eruption of sheep introduced into western New South Wales. The parallel between this trajectory and that simulated in Fig. 6.2 is immediately apparent.

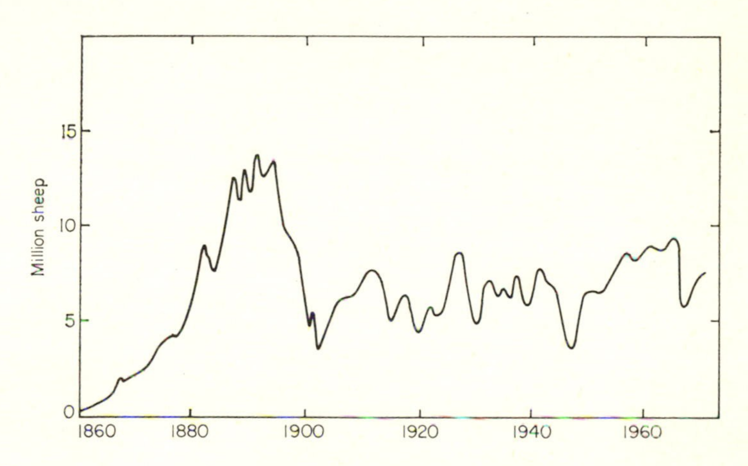

Fig. 6.5. An ungulate eruption: the trend of sheep numbers in the Western Division of New South Wales between 1860 and 1972. Data from Butlin (1962) and N.S.W. Yearbooks (1956–1972).

Ungulate populations are also prone to erupt when the plant-herbivore system is disturbed. Because the herbivores do not start their upswing from minimal density, and because the vegetation is not at its K level, these 'sub-eruptions' are not as spectacular as that of Fig. 6.5. They are, however, much more common. Leopold *et al.* (1947) reported over a hundred for deer populations in the United States between 1900 and 1945.

Probably the most spectacular full-scale eruption on record is that of reindeer introduced onto St Matthew Island in the Bering Sea. Figure

6.6 gives the available estimates of population size (Klein, 1968), to which a trajectory was fitted by eqs. (6.6) and (6.7); see Caughley (1976). The parameter values were guessed rather than estimated; they give the point illustrated in Fig. 6.3. As shown in Fig. 6.6c, the intersection of zero isoclines is to the left of the hump, indicating that the system has no stable equilibrium point, and therefore that the reindeer numbers will oscillate cyclically (in this case violently, and probably to extinction). This is probably an accurate representation of reality.

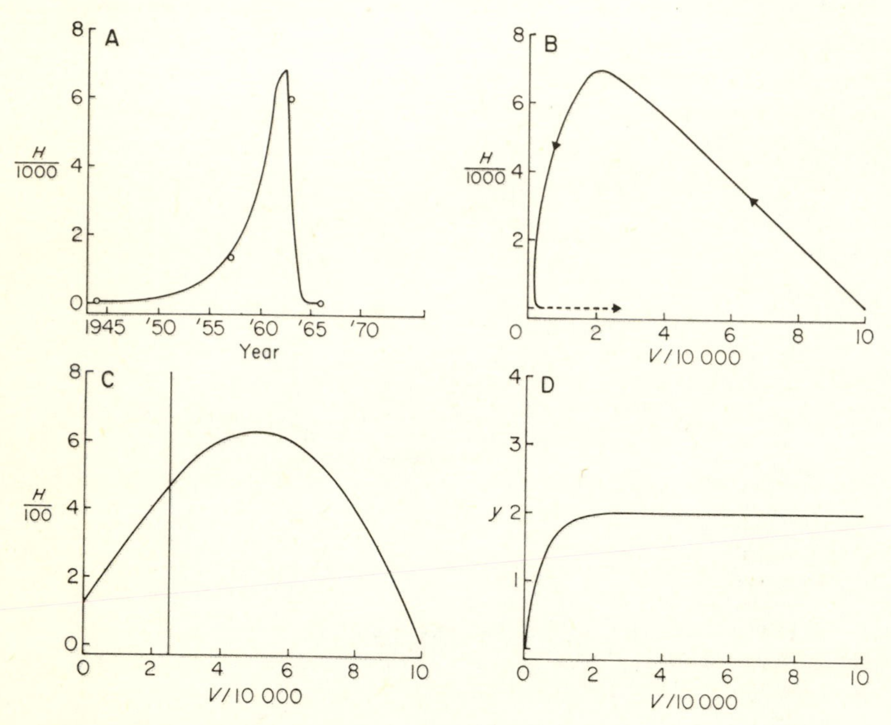

Fig. 6.6. A reindeer population in trouble. Graph A is a trajectory fitted by the laissez-faire model to Klein's (1968) estimates of numbers on St Matthew Island. Graph B shows the same trajectory as a plot of herbivores (H) against vegetation (V); graph C shows the zero isoclines for herbivores (vertical line) and the vegetation (curve), and Graph D is intake (y) per head against density of vegetation.

6.4.2 *Interferential systems*

When herbivores interfere with each other's grazing activities, the previous model is no longer appropriate. In contrast to the laissez-faire systems of section 6.4.1, the per capita rate of increase of herbivores is a function of both plant density and animal density.

From among the many qualitatively similar possible models (see, e.g., Table 4.1), we choose one which has already been discussed in section 4.2 (see Figs. 4.1 and 4.2), and has had detailed attention in the literature (May, 1975a; Tanner, 1975):

$$dV/dt = r_1 V(1 - V/K) - c_1 H[V/(V+D)] \qquad (6.12)$$

$$dH/dt = r_2 H(1 - JH/V). \qquad (6.13)$$

Here D is inversely proportional to grazing efficiency at low plant density (being the characteristic density of vegetation at which the herbivore functional response saturates), and J is a proportionality constant related to the number of plants needed to sustain a herbivore at equilibrium. The other symbols are as previously defined (note that r_1 and r_2 are to be identified with the earlier symbols r and s of chapter 4).

The zero isocline for the vegetation V, along which $dV/dt = 0$ for any given H, follows from eq. (6.12):

$$H = (r_1/c_1)(1 - V/K)(V+D). \qquad (6.14)$$

The zero isocline for H, along which $dH/dt = 0$, is similarly

$$H = V/J. \qquad (6.15)$$

This latter isocline slopes out from the origin, in contrast to the vertical isocline of eq. (6.8). A further contrast lies in the stability properties of this model, which allow the possibility of a stable equilibrium point to either the left or the right of the hump in the V isocline, at $V = (K-D)/2$. The details of the stability properties, as a function of parameters such as K/D and r_1/r_2, are discussed in Fig. 4.2 and the accompanying text; for a fully detailed exposition, see May (1975a, Appendix I).

Our example is provided by the moth *Cactoblastis cactorum*, which was introduced into Australia in 1925 to control the prickly-pear cacti *Opuntia inermis* and *O. stricta*. Its name is honoured in the pantheon of that select company of biological agents that were actually successful in controlling their target species. Extensive stands of dense pear averaging about 500 plants per acre were virtually wiped out within two years of their colonisation by *Cactoblastis* (Dodd, 1940). Dodd's account of the initial spread is worth reading, as are the subsequent reverberations in the ecological literature. The post-crash equilibrium was first interpreted as a game of hide and seek between *Opuntia* and *Cactoblastis* (Nicholson, 1947; Andrewartha and Birch, 1954) and then as a tight grazing equilibrium reinforced by larval interference (Monro, 1967, 1975; Birch, 1971).

The eggs of *Cactoblastis* are not laid at random. Their dispersion is doubly contagious in that they are laid in egg-sticks of around 80 eggs each and the egg-sticks are themselves clumped (Monro, 1967). Hence during the summer some cactus plants receive many more eggs than would be expected by random chance, and others escape infestation completely. A parallel is provided by the large number of invertebrate parasites that distribute their attacks contagiously (Griffiths and Holling, 1969), a behaviour that, theoretically at least, should enhance the system's stability (Hassell and Varley, 1969; Rosenzweig, 1971, 1972; Hassell and Rogers, 1972; Holling, 1973; Hassell and May, 1973, 1974). The contagious distribution of *Cactoblastis* eggs greatly influences the outcome of grazing by the larvae. Since a loading of above about 1·5 sticks per cactus ensures destruction of the plant, much of the larvae's resource is wasted (Monro, 1967).

Dodd (1940) and Monro (1967) present enough data to allow a stab at the values of four of the six parameters of eqs. (6.12) and (6.13). A clue to r_1 is provided by Dodd's observation that *Opuntia* can increase from root stock to 250 tons per acre in two years. Assuming the root stock weighs one ton, the rate of increase is around $r = 2 \cdot 7$. That will be an overestimate of r_1 because the growth is entirely vegetative. My guess for a maximum rate of increase depending partly on sexual reproduction is $r_1 = 2$. The value of K is taken from Dodd's remark that 5000 plants per acre is a fair estimate. Two rough estimates of r_2 can be made. At the beginning of the experiment 2750 eggs were received from South America and hatched in the laboratory. The adults produced 100,605 eggs: $r_2 = 3 \cdot 6$. Dodd indicated that in the field *Cactoblastis* could erupt from 5,000 larvae to 10,000,000 per acre in two years: $r_2 = 3 \cdot 8$. For different reasons both figures are liable to underestimate r_2 which is set at 4. J comes from Monro's measurements of the summer generation at his B1 and B2 sites: $J = 2 \cdot 23$ in units of cactus plants per egg-stick. Parameters c_1 and D cannot be estimated. However c_1 should be large to reflect plants damaged and killed by the larvae as a by-product of their feeding. This structural damage contributes most to c_1, vastly outweighing the contribution of ingested plant tissue. D should be small to reflect the uncanny ability with which female moths search out food plants on which to lay eggs.

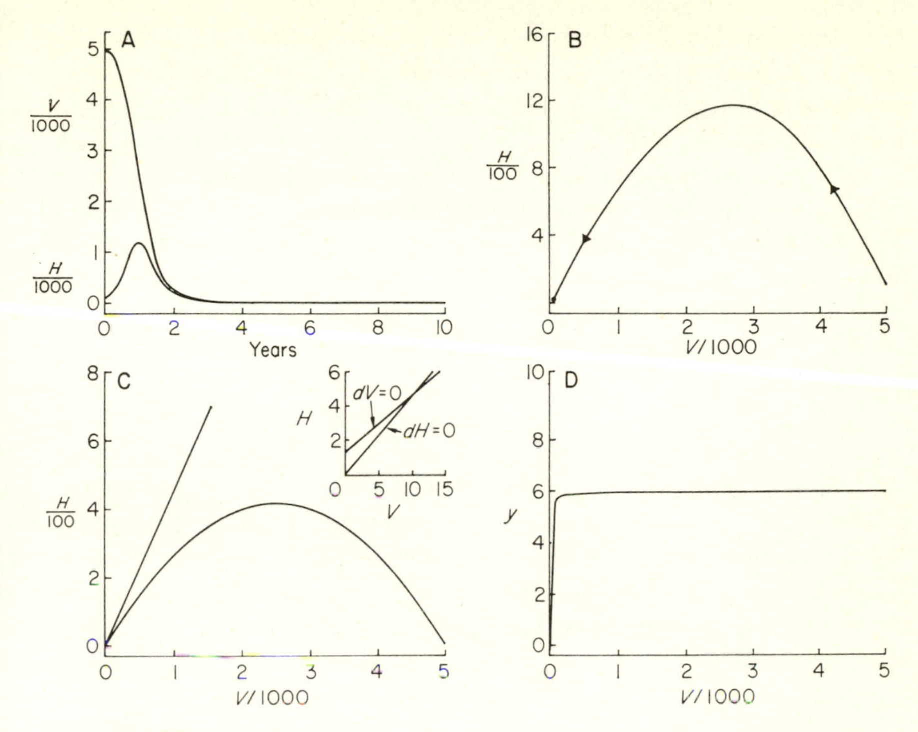

Fig. 6.7. The control of a plant by a herbivore. Graph A gives trajectories by the interferential model for a stand of cactus invaded by the *Cactoblastis* moth. Graph B shows the path to mutual equilibrium in phase space, graph C provides the modelled zero isoclines of cactus (curve) and *Cactoblastis* (line), and graph D is consumption per *Cactoblastis* egg-stick against cactus density.

Three outcomes are demanded of a simulation of this system: *Opuntia* must crash within two years to square with Dodd's observations, it must stabilise at around 11 plants per acre to reflect the summer density at Monro's B sites, and the equilibrium must be highly stable. These requirements are met by setting $c_1 = 6$ and $D = 4$. Figure 6.7 diagrams a simulation of *Cactoblastis* invading a stand of Cactus. *Opuntia* density is expressed as plants per acre and *Cactoblastis* as egg-sticks per acre, the archaic 'acre' being retained to preserve uniformity with Dodd's round figures. *Opuntia* crashes in the required time to a highly stable equilibrium of 11 plants per acre carrying five egg-sticks.

This strategic model containing only six constants has produced a cactus crash looking much like the real thing. A tactical model would contain more detail, particularly on the survival of root stock after the crash which often allowed a temporary resurgence of the cactus; but it is unlikely to produce an outcome differing in kind from that provided by the simple interferential model. We can speculate cautiously that the model has duplicated the major determining forces of the system, a conclusion reinforced by the result from the same data translated into laissez-faire parameters and fed through that model [eqs. (6.6), (6.7)]. It reported that without larval interference a system with these parameters would be spectacularly unstable.

6.5 Selection

Within a non-interactive grazing system the herbivore population's growth may be expressed as two parameters, r and K. If the system is non-reactive, r is the only parameter open to selection (Monro, 1967), whereas both will be selected in a reactive system. A genetic interpretation need not proceed beyond individual selection or kin selection (Birch, 1960; Lack, 1965), unless r is high (May, 1974a, 1975b).

Interactive systems are not so simple. Since at least six parameters are candidates for selection, it is no longer profitable to speculate within the limited framework of an $r-K$ continuum. In terms of the laissez-faire model [eqs. (6.6), (6.7)] such components of fitness as grazing efficiency (d_1), intrinsic rate of increase (c_2-a), and demographic efficiency (d_2) should be increased by individual selection. But the graphic analysis of Fig. 6.4 indicates that such selection would ultimately send the population down the rolling road to extinction. Similarly, individual selection would be expected to reduce the clumping of eggs and hence the interference between herbivores in an interferential system [eqs. (6.12), (6.13)], but that change will also erode the system's stability and increase the probability of extinction. With this selective force and counter-force operating, one on individual herbivores, the other on the population (group) as a whole, the parameters can change only until they have positioned the system's critical point near the watershed dividing stability from cycles and instability. A similar argument is advanced by May (1975a, p. 102) within the context of a delayed logistic model, which in some circumstances (Caughley, 1976; see also section 4.2.2) can serve as a shorthand summary of an interactive model. The extensive overshoot characterising a typical ungulate eruption is precisely what would be predicted for an interactive system with an equilibrium point near the watershed.

The genetical arguments against group selection are well known. They revolve around the difficulty of an 'altruistic' gene spreading through a species when, by definition, it lowers the fitness of each bearer and is therefore selected against (Maynard Smith, 1971a). But that argument can be turned on its head by asking how a 'selfish' gene can spread through a species if each population becomes extinct soon after infection. Such thoughts are pursued in various recent studies: see Gilpin (1975b), D. S. Wilson (1975), and the reviews by E. O. Wilson (1975), May (1975e) and Chapter 11.

A resolution of this dilemma must wait for population geneticists to grow weary of their pivotal assumption that a population has no dynamics, and for population dynamicists to abandon the belief that a population has no genetics.

7
Competition and Niche Theory

ERIC R. PIANKA

7.1 Introduction: definitions and theoretical background

7.1.1 *Competition*

By definition, competition occurs when two or more organisms, or other organismic units such as populations, interfere with or inhibit one another. Organisms concerned typically use some common resource which is in short supply. Moreover, the presence of each organismic unit reduces the fitness and/or equilibrium population size of the other. Competition is sometimes quite direct, as in the case of interspecific territoriality, and is then termed *interference competition*. More indirect competition also occurs, such as that arising through the joint use of the same limited resources, which is termed *exploitation competition*. Because it is always advantageous for either party in a competitive interaction to avoid the other whenever possible, competition presumably promotes the use of different resources and hence generates ecological diversity. The mechanisms by which members of a community of organisms partition resources among themselves and reduce interspecific competition shapes community structure and may often influence species diversity profoundly (see Schoener, 1974, for a recent review).

Ecologists are, however, divided in their attitudes concerning the probable importance of competition in structuring natural communities. Some, myself included, either tacitly or explicitly assume that self-replication in a finite environment must eventually lead to some competition. Other ecologists, particularly those that study small organisms and/or organisms at lower trophic levels, tend to be much more skeptical about the impact of competition upon organisms in nature. While the persistence of this dispute over the strength of competition in natural communities could conceivably reflect a natural dichotomy, it might well be more realistic and more profitable not to view competition as an all-or-none phenomenon. An emerging conceptual framework envisions a gradient in the intensity of competition, varying continuously between the endpoints of a complete competitive vacuum (no competition) to a fully saturated environment with demand equal to supply ('super-saturated' environments are also possible, with demand exceeding supply).

Competition lends itself readily to mathematical models, and an extensive body of theory exists, most of which assumes saturated communities at equilibrium with their resources, sometimes referred to as 'competitive' communities. Much of this theory is built upon the overworked Lotka-Volterra competition equations:

$$\frac{dN_i}{dt} = r_i N_i \left(\frac{K_i - N_i - \sum_{j \neq i}^{n} \alpha_{ij} N_j}{K_i} \right). \qquad (7.1)$$

Here n is number of species (subscripted by i and j), r_i is the intrinsic rate of increase of species i, K_i is its 'carrying capacity,' N_i is its population density, α_{ij} is the 'competition coefficient' which measures the inhibitory effects of an individual of species j upon species i. At equilibrium, all dN_i/dt must be equal to zero, giving the equilibrium population densities:

$$N_i^* = K_i - \sum_{j \neq i}^{n} \alpha_{ij} N_j^* \qquad (7.2)$$

for all i from 1 to n.

The Lotka-Volterra competition equations greatly oversimplify the process of interspecific competition (for examples, see Hairston *et al.*, 1968; Wilbur, 1972; Gilpin and Ayala, 1973; and Neill, 1974). Indeed, the alphas in these equations may well be illusory and may often obscure the real *mechanisms* of competitive interactions. Nevertheless, whatever flaws the Lotka-Volterra equations may have, they have clearly contributed much to current ecological thinking. Not only do they provide a conceptual framework, but they have helped to give rise to many exceedingly useful ecological concepts in addition to competition coefficients, including equilibrial population densities, the community matrix, diffuse competition, r and K selection, as well as non-linear isoclines (see, for example, Ayala *et al.*, 1973a; Schoener, 1974; Gilpin and Ayala, 1973). Some important papers, books, and/or major reviews in the voluminous literature on competition include Crombie

(1947), Birch (1957), Milne (1961), Milthorpe (1961), DeBach (1966), Miller (1967), MacArthur (1972), Grant (1972a), Stern and Roche (1974), Schoener (1974, 1976a) and Connell (1975).

7.1.2 *Niche theory*

Among the first to use the term niche was Grinnell (1917, 1924, 1928), who viewed it as the ultimate distributional unit, thus stressing a spatial concept of the niche. Elton (1927) emphasized more ethological aspects, and defined the ecological niche as the functional role and position of the organism in its community, stressing especially its trophic relationships with other species. Although the term niche has been used in a wide variety of ways by subsequent workers, the idea of a niche gradually became linked with competition. Empirical studies of Gause (1934) and others (see below) showed that ecologically similar species were seldom able to coexist in simple laboratory systems; hence species living together must each have their own unique niche. The one species per niche concept became accepted as ecological dogma, although a few dissenters urged otherwise (Ross, 1957, 1958). Because the term 'niche' has been used in a wide variety of different contexts and is rather vaguely defined, some ecologists prefer not to use the word [see, for example, commentaries on niche in Williamson (1972) and Emlen (1973)]. Recently however, the ecological niche has become increasingly identified with resource utilization spectra (Fig. 7.1), through both theoretical and empirical work of a growing school of population biologists (Levins, 1968; MacArthur, 1968, 1970, 1972; Schoener and Gorman, 1968; Pianka, 1969, 1973, 1974, 1975; Colwell and Futuyma, 1971; Roughgarden, 1972, 1976; Vandermeer, 1972; Pielou, 1972; May and MacArthur, 1972; May, 1974d, 1975d; Cody, 1974; Schoener, 1968, 1975a, 1975b). Such an emphasis upon resource use is operationally tractable, although it largely neglects considerations of reproductive success (some earlier treatments of the niche such as the n-dimensional hypervolume concept of Hutchinson (1957) used fitness to define niche boundaries).

Although niche theory will ultimately have to include aspects of reproductive success as well as resource utilization phenomena, my emphasis here is on the latter. Possibilities abound for significant further work, both theoretical and empirical, on the constraints and interactions between optimal foraging and optimal reproductive tactics (Pianka, 1976).

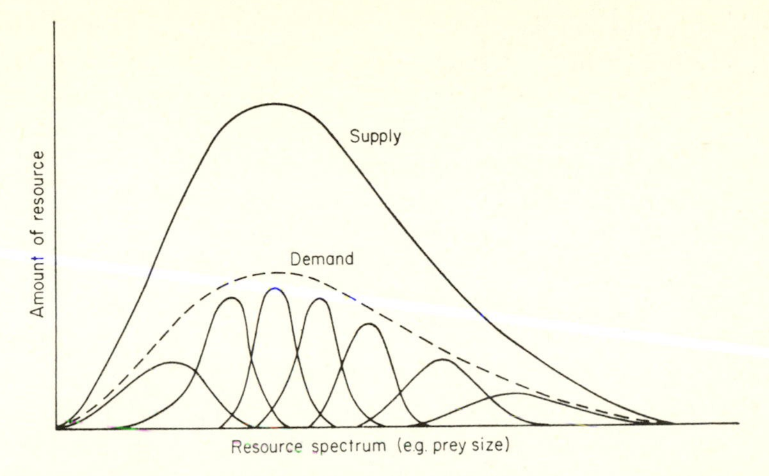

Fig. 7.1. Niche relationships among potentially competing species are often visualized and modelled with bell-shaped resource utilization curves. The uppermost bell-shaped curve represents the supply of resources along a single resource continuum, such as prey size. The vertical axis measures the amount of resource available or used over some time interval. The lower small curves are intended to represent seven hypothetical species in a community, with those species that exploit the 'tails' of the resource spectrum using a broader range of resources (that is, they have broader niches) because their resources are less abundant. The sum of the component species utilization curves, shown by a dashed line, reflects the total use or the overall demand along the resource gradient. Pressures leading to the avoidance of interspecific competition should result in a relatively constant ratio of demand/supply along the resource continuum, as shown.

The concept of niche 'breadth' (niche 'width' or 'size' are frequent synonyms) has proven to be extremely useful. Niche breadth is simply the sum total of the variety of different resources exploited by an organismic unit. In the absence of any competitors or other enemies, the entire set of resources used is referred to as the 'fundamental' niche (Hutchinson, 1957) or the 'pre-interactive,' 'pre-competitive,' or 'virtual' niche. Any real organismic unit presumably does not exploit its entire fundamental niche since its activities are somewhat curtailed by its competitors as well as by its predators; hence its 'realized,' 'post-interactive,' or 'post-competitive' niche is usually a subset of the fundamental niche (Hutchinson, 1957; Vandermeer, 1972). The difference between the fundamental and the realized niche, or the niche change due to competitors, thus reflects the effects of interspecific competition (as well as predation).

Consideration of the variety of factors influencing niche breadth leads into the problem of specialization versus generalization. A fairly substantial body of theory on optimal foraging predicts that niche breadth should generally increase as resource availability decreases (Emlen, 1966, 1968; MacArthur and Pianka, 1966; Schoener, 1971; MacArthur, 1972; Charnov, 1973, 1975). In an environment with a scant food supply, a consumer cannot afford to bypass many inferior prey items because mean search time per item encountered is long and expectation of prey encounter is low. In such an environment, a broad niche maximizes returns per unit expenditure, promoting generalization. In a food-rich environment, however, search time per item is low since a foraging animal encounters numerous potential prey items; under such circumstances, substandard prey items can be bypassed because expectation of finding a superior item in the near future is high. Hence rich food supplies are expected to lead to selective foraging and narrow food niche breadths. A competitor can act either to compress or to expand the realized niche of another species, depending upon whether or not it reduces resource levels uniformly (which leads to niche expansion) or in a patchy manner (which should often result in a niche contraction, especially in the microhabitats used).

Niche theory has only relatively recently begun to distinguish clearly two fundamental components of niche breadth, namely, the so-called 'between-phenotype' versus 'within-phenotype' components (Van Valen, 1965; Roughgarden, 1972, 1974b, 1974c). A population with a niche breadth determined entirely by the between phenotype component would be composed of specialized individuals with no overlap in resources used, whereas a population composed of pure generalists with each member exploiting the entire range of resources used by the total population would have a between-phenotype component of niche breadth of zero and a maximal within-phenotype component. Real populations will clearly lie somewhere between these two extremes, with various mixtures of the two components of niche breadth.

Another central aspect of niche theory concerns the amount of resource sharing, or niche overlap. Ecologists have long been intrigued with the notion that there should be an upper limit to how similar the ecologies of two species can be and still allow coexistence. Concepts that have emerged from such thinking include the so-called 'principle' of competitive exclusion (below), character displacement, limiting similarity, species packing, and maximal tolerable niche overlap (for examples, see Hutchinson, 1959; Schoener, 1965; MacArthur and Levins, 1967; MacArthur, 1969, 1970; May and MacArthur, 1972; Pianka, 1972; Grant, 1972a). A number of models of niche overlap in competitive communities, typically built upon eq. (7.1), have generated several testable predictions (MacArthur and Levins, 1967; MacArthur, 1970, 1972; May and MacArthur, 1972; May, 1974d; Gilpin, 1974; Roughgarden, 1974a, 1975b, 1976). Some of these models suggest that maximal tolerable niche overlap should decrease as the number of competing species increases, with such decreases in overlap approximating a decaying exponential (cf. Fig. 7.13). Indeed, MacArthur (1972) coined the term 'diffuse competition' to describe the total competitive effects of a number of interspecific competitors, implying that a little competitive inhibition by many other species can be equivalent to strong competitive inhibition by fewer competing species.

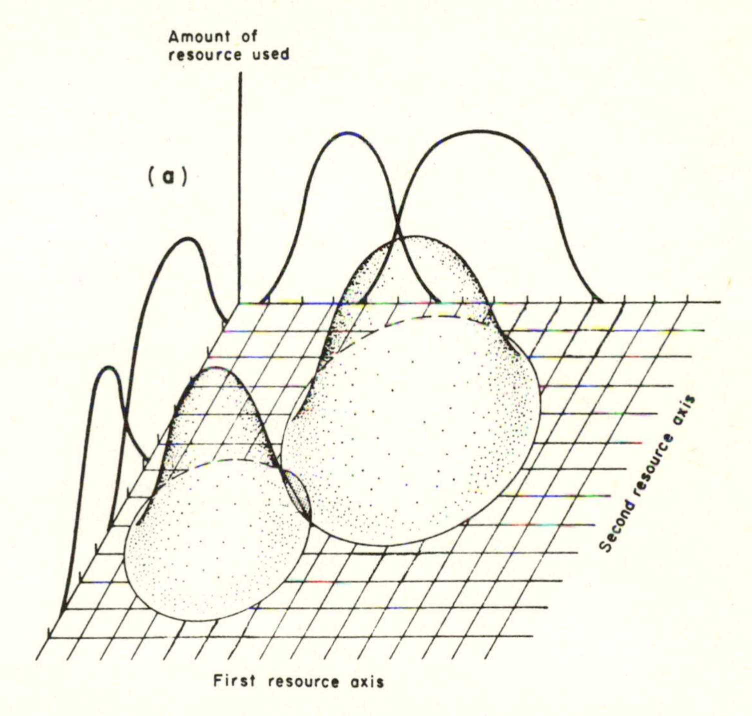

Fig. 7.2. (a) Diagrammatic representation of resource utilization of two hypothetical species along two niche dimensions, such as prey size and height above the ground of foraging. Note that, although the shadows of the three dimensional peaks on each separate niche dimension overlap, the true overlap in both dimensions is very slight. Adapted after Clapham (1973).

One model, that of May and MacArthur (1972), predicts that maximal tolerable overlap should be relatively insensitive to environmental variability; a critical discussion of this and related work is given in section 4.3.

Most existing theory on niche overlap is framed in terms of a single niche dimension (but see recent work by May, 1974d, and Roughgarden, priv. comm.). Real plants and animals differ in their use of just one resource only infrequently, however (for such an example, see Fig. 7.10). Rather, pairs of species frequently show moderate niche overlap along two or more niche dimensions (Fig. 7.2). Ideally, a multidimensional analysis of resource utilization and niche separation along more than a single niche dimension should proceed through estimation of proportional simultaneous utilization of all resources along each separate niche dimension (Pianka, 1974, 1975; May, 1975d). In practice, however, it is extremely difficult to obtain such multidimensional

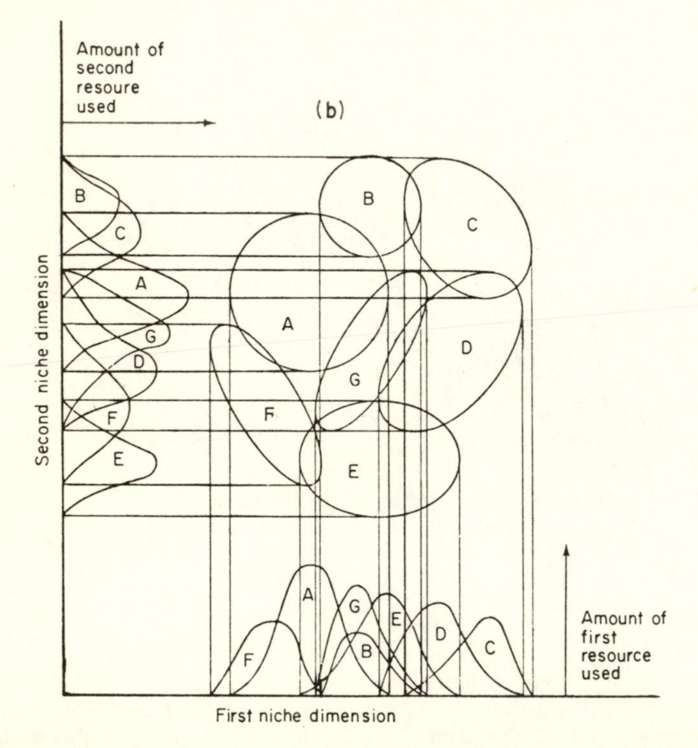

Fig. 7.2. (b) A similar plot for seven species, showing that pairs with substantial or complete overlap along one dimension can avoid competition by niche separation along another dimension.

utilization data because animals move and integrate over both space and time. An individual animal would have to be followed and its use of all resources recorded continually in order to obtain accurate estimates of its true utilization of a multidimensional niche space. Since this is often very tedious or even impossible, one usually attempts to approximate from separate unidimensional utilization distributions (Fig. 7.3). Provided that niche dimensions are truly independent, with for example any prey item being equally likely to be captured in any place, overall multidimensional utilization is simply the product of the separate unidimensional utilization functions (May, 1975d). Estimates of various niche parameters along component niche dimensions can then simply be multiplied to obtain multidimensional estimates. However, should niche dimensions be partially dependent upon one another (Fig. 7.3), there is no substitute for knowledge of the true multidimensional utilization functions. In the extreme case of complete dependence (with, for example, any particular prey item being found only at a particular height above ground), appearances to the contrary,

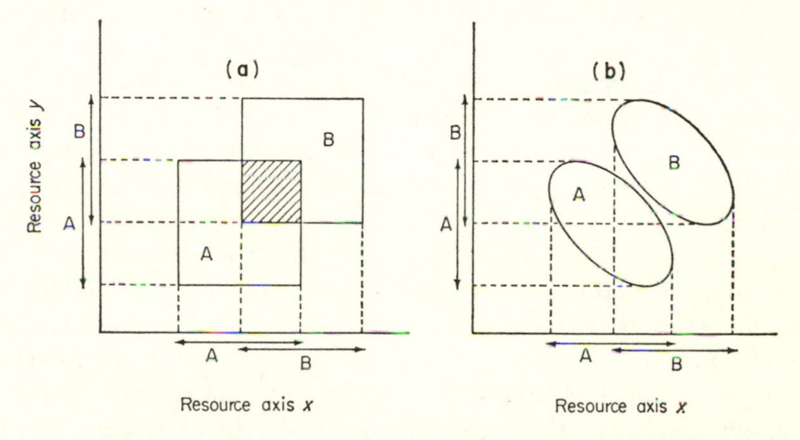

Fig. 7.3. Two possible cases for the use of two resource dimensions (assumed to be constant within boxes or ellipses) for two hypothetical species, A and B. Although unidimensional projections are identical in both case *a* and *b*, true multidimensional overlap is zero in cars *b*. Case *a* illustrates truly independent niche dimensions, with any point along resource axis *x* being equally likely along the entire length of resource axis *y*; under such circumstances, niche dimensions are orthogonal and unidimensional projections accurately reflect multi-dimensional conditions. However, when niche dimensions are partially dependent upon one another (case *b*), unidimensional projections always result in overestimates of true niche overlap.

there is actually only a single niche dimension, and a simple average provides the best estimate of true utilization. Moreover the arithmetic average of estimates of unidimensional niche overlap obtained from two or more separate unidimensional patterns of resource use actually constitutes an *upper bound* on the true multidimensional overlap (May, 1975d). Certain pitfalls in the analysis of niche relationships along more than one dimension have been noted by May (1975d) and Pianka (1974, 1975).

An important aspect of niche dimensionality concerns the notion of the number of neighbours in niche space (MacArthur, 1972). An increased number of niche dimensions results in a greater potential for immediate neighbours in niche space (see Fig. 7.2) and hence may intensify diffuse competition (MacArthur, 1972; Pianka, 1974, 1975).

7.1.3 *Niche overlap and competition*

Because competition coefficients are exceedingly difficult to estimate directly except by population removal experiments (below), measures of niche overlap have often been used as estimates of the alphas in eqs. (7.1) and (7.2) (Pico *et al.*, 1965; Schoener, 1968; Orians and Horn, 1969; Pianka, 1969; Culver, 1970; Brown and Lieberman, 1973; May, 1975d). However, tempting though it may be, equating overlap with competition can be dubious and misleading (Colwell and Futuyma, 1971). Although niche overlap is clearly a prerequisite to exploitative competition, overlap need not necessarily lead to competition unless resources are in short supply. (In a competitive vacuum with a surplus of resources available, niches could presumably overlap completely without detriment to the organisms concerned.) Moreover, interference competition is unlikely to evolve unless there is a potential for overlap in use of limited resources (i.e., exploitation competition must be *potentially* possible). Hence avoidance of competition can lead to entirely non-overlapping patterns of resource utilization (disjunct niches), as for example occurs in interspecific territoriality. In the parlance of the mathematician, niche overlap in itself is neither a necessary nor a sufficient condition for interference competition; moreover, overlap is only a necessary but not a sufficient condition for exploitation competition.

Indeed, the preceeding arguments suggest that there may often be an inverse relationship between competition and niche overlap. If so, extensive overlap might actually be correlated with reduced competition. Such reasoning led me to propose the 'niche overlap hypothesis', which asserts that maximal tolerable overlap should be lower in intensely competitive situations than in environments with lower demand/supply ratios (Pianka, 1972; see also Roughgarden and Feldman, 1975). The impact of predators on competition is little known, but they may often reduce its intensity and hence facilitate coexistence (Paine, 1966; Connell, 1975).

7.1.4 *Temporal variability and fugitive species*

Most existing theory on both interspecific competition and niche relationships assumes that component populations have reached an equilibrium with their resources and differ enough in their use of these resources to coexist. Hutchinson (1951) formulated a somewhat more dynamic and perhaps more realistic concept of a 'fugitive' species, which he envisioned as a predictably inferior competitor, always excluded locally under interspecific competition, but which persists in newly disturbed regions merely by virtue of its high dispersal ability. Hutchinson's mechanism for persistence of a colonizing species by means of high dispersability in a continually changing patchy environment in spite of pressures from competitively superior species has recently been modelled analytically (Skellam, 1951; Levins and Culver, 1971; Horn and MacArthur, 1972; Slatkin, 1974; Levin, 1974). This view of competitive pressures and processes varying in time and space is appealing, although perhaps operationally intractable in many instances. In a later attempt to explain the apparent 'paradox' of the plankton (that is, the coexistence of members of a diverse community in a relatively homogeneous physical environment with few possibilities for niche separation), Hutchinson (1961) suggested that temporally-changing environments may promote diversity by periodically altering the relative competitive abilities of component species, hence allowing their coexistence.

7.2 Empirical studies: general remarks

Competitive effects can be measured in two ways: (1) by perturbation experiments in which rates of population growth of competitors are monitored as they approach equilibrium, and (2) by addition and/or removal experiments, in which the equilibrium population density of a species, N_i^* in eq. (7.2), is monitored at equilibrium both in the presence and in the absence of its competitor, in an otherwise unchanged environment. Either slowed rates of population growth, or an observed reduction in population density, or a niche shift in response to the presence of a competitor constitutes direct evidence of interspecific competition.

Most experimental studies of competition have been performed under relatively homogeneous and constant laboratory conditions. Such studies, briefly summarized below, led to the so-called 'principle' of competitive exclusion of Gause (1934) (see also Cole, 1960; Hardin, 1960; Patten, 1961; DeBach, 1966), which asserts that some ecological difference must exist between coexisting species.

Although competition is the conceptual backbone of much current ecological thought, it has proven exceedingly difficult to study in natural communities, probably partially because reduction or avoidance of competition is always advantageous when possible. Also, the great spatial and temporal variability characteristic of most natural communities demands a dynamic approach to the investigation of competitive interactions. Existing evidence of competition in nature is largely circumstantial (see below) and unequivocal removal experiments hold considerable promise (Connell, 1975). Indeed, because competition lies at the heart of so many ecological processes but has been studied so inadequately, carefully designed and well executed empirical investigations into the precise mechanisms and results of competitive interactions seem virtually certain to be of central importance to the future of ecology.

7.3 Empirical studies: laboratory experiments

Gause (1934) performed some of the earliest competition experiments with several species of *Paramecium*, using laboratory culture media, renewed at regular intervals. Population growth and population densities were monitored both in single-species cultures and in mixtures of two competing species grown together (Fig. 7.4). One experiment clearly demonstrated competitive exclusion of *P. caudatum* by *P. aurelia*. In another experiment with *P. aurelia* and *P. bursaria*, this pair of *Paramecium* species coexisted in a mixed culture, although at lower population densities than when grown in pure cultures of a single species (Fig. 7.4). Competition coefficients, reflecting the intensity of the competitive interaction, are readily calculated from such data.

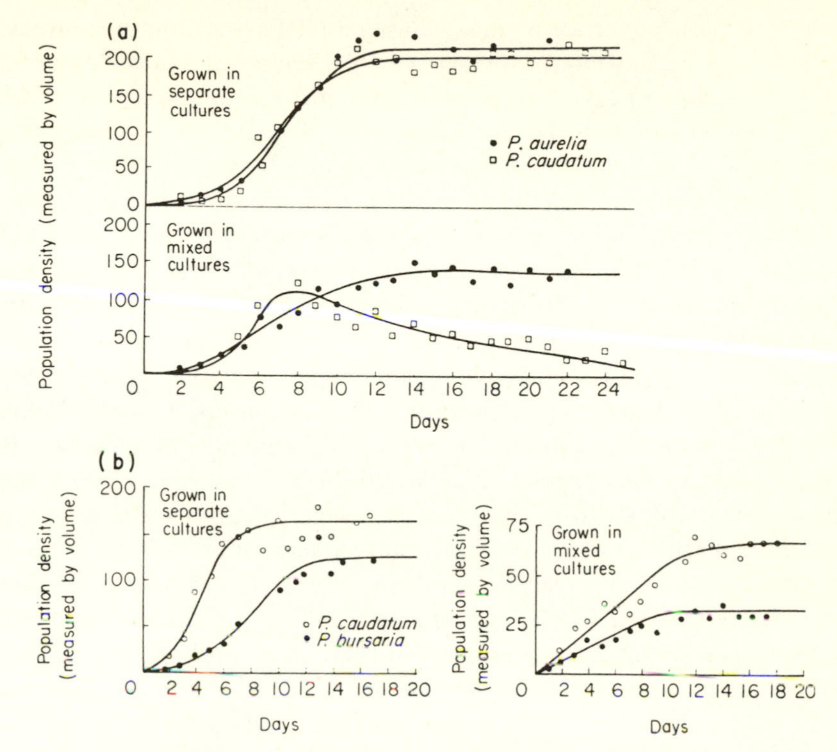

Fig. 7.4. (**a**) When *Paramecium aurelia* and *P. caudatum* are grown together in mixed laboratory cultures, *P. aurelia* excludes *P. caudatum*. Modified from Gause (1934) after Clapham (1973). (**b**) Laboratory competition experiments with two more dissimilar *Paramecium* species, *P. aurelia* and *P. bursaria*, result in coexistence at lower population densities than in pure cultures. From data of Gause, modified from Clapham (1973).

By far the most exhaustive laboratory studies of competition are those of Park (1948, 1954, 1962) and his associates, who worked with flour beetles, especially of the genus *Tribolium*. While Park's studies are much too extensive to review here, they also convincingly demonstrated competitive exclusion. In addition, a series of experiments showed that the outcome of interspecific competition depends upon (a) initial population densities (Neyman *et al.*, 1956), (b) environmental conditions of temperature and humidity (Park, 1954), and (c) on the genetic constitution of the strains of competing species (Park *et al.*, 1964).

Among the numerous other laboratory experiments on competition that have been undertaken since Gause's and Park's pioneering work, perhaps one of the more informative is the elegant recent study of Neill (1972, 1974, 1975). Using aquatic laboratory microcosms containing replicated communities of four species of micro-crustaceans along with associated bacteria and algae under equilibrium conditions, Neill performed a series of replicated removal experiments and allowed the resulting systems to return to equilibrium. Each species of micro-crustacean, as well as each possible pair of species, were removed and estimates of population densities of the various species were made under all possible competitive regimes. Computation of competition coefficients from these equilibrium population densities showed that alphas clearly depend upon community composition. In both Neill's microcosms and in an amphibian community studied by Wilbur (1972), the joint effects of two species upon a third in a 3-species system cannot always be predicted from the separate interactions in the three component 2-species systems. Such results indicate that, if eq. (7.1) is to reflect reality, it must somehow be expanded to include 'interactive' competition coefficients reflecting the joint effects of two species upon a third, as suggested by Hairston *et al.* (1968) and by Wilbur (1972).

7.4 Empirical studies: field observations

Removal experiments under field conditions are usually next to impossible, and have seldom been attempted (but see below). Instead, field studies of competition tend to rely heavily upon 'natural' experiments, in which aspects of the ecology of a species are compared between areas where it occurs alone (allopatry) with other areas where it occurs with another competiting species (sympatry). Provided that the two areas are otherwise basically similar, niche shifts observed in sympatry should reflect the response to interspecific competition. However, as pointed out by Connell (1975), such observations often lack a suitable 'control', since other factors probably differ between allopatry and sympatry.

7.4.1 *Niche shifts and character displacement*

Such a situation occurs among two species of flatworms along temperature gradients in streams (Beauchamp and Ullyett, 1932). Figure 7.5

depicts the distributions of these two species of *Planaria* in streams where each occurs separately and where both exist together. Neither species occupies as broad a range in temperature when the two occur together as it does when it is the only species in the stream (Fig. 7.5). Similar observations of niche shifts in 'incomplete' faunas include studies on salamanders (Hairston, 1951) and birds (Crowell, 1962).

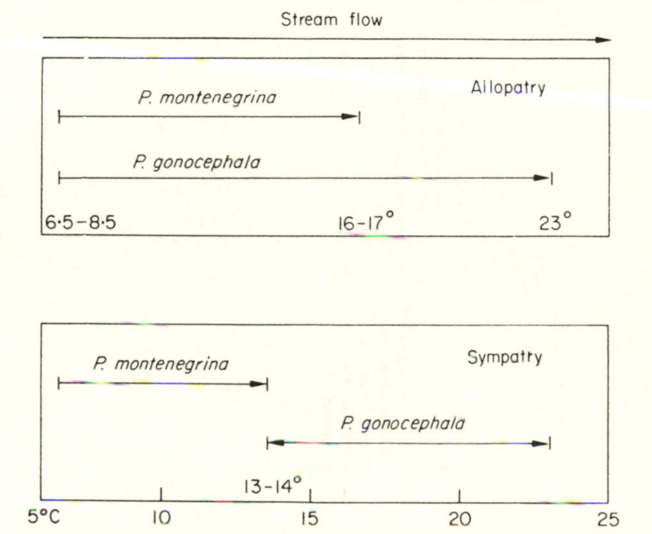

Fig. 7.5. Distributions of *Planaria montenegrina* and *P. gonocephala* along temperature gradients in streams when they occur separately (above) and together (below). Each species is restricted to a smaller range of thermal conditions when in competition with the other. From Beauchamp and Ullyett (1932) after Miller (1967).

A related phenomenon, termed 'character displacement' by Brown and Wilson (1956), sometimes occurs when two wide-ranging species have partially overlapping geographic distributions, with a zone of sympatry and two zones where each species occurs alone in allopatry. Such species pairs are often very similar to one another where they occur in allopatry, but they typically diverge when they occur together (Fig. 7.6).

Hutchinson (1959) first commented on the apparent constancy in the magnitude of morphological character displacement, reporting ratios of mouthpart sizes among coexisting congeneric species of insects, birds and mammals ranging only from about 1.2 to 1.4.

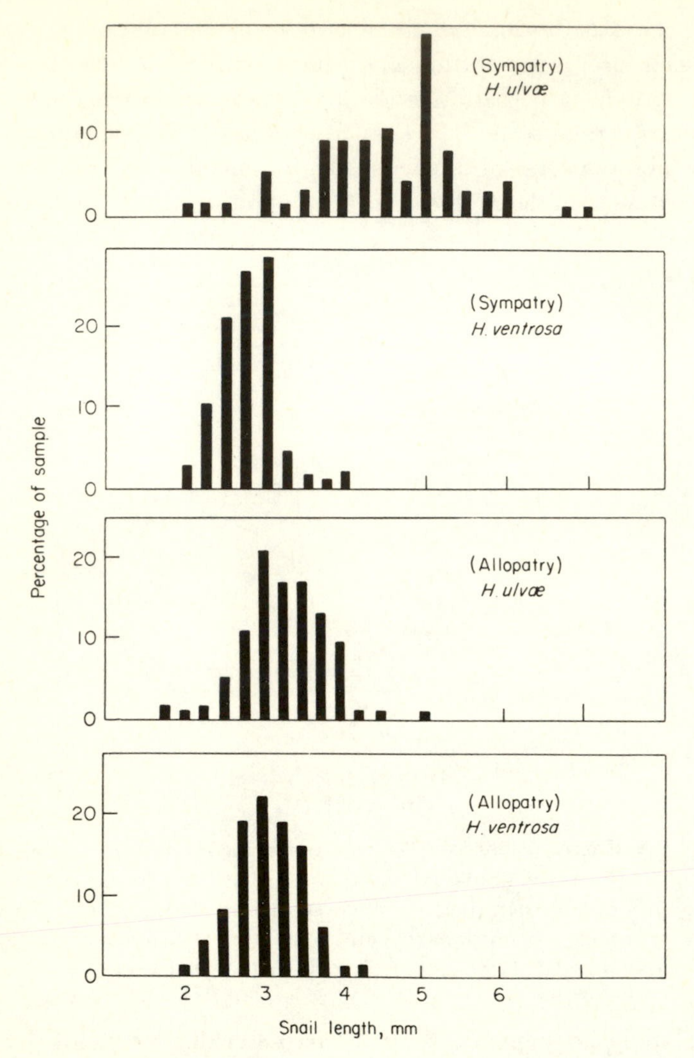

Fig. 7.6. Length-frequency distributions of the shells of two species of snails, *Hydrobia ulvae* and *H. ventrosa*, at a locality where they coexist (upper box) and at two localities where they occur in allopatry (lower two boxes). From Fenchel (1975).

Schoener (1965), Grant (1968) and Diamond (1973) found similar ratios of bill lengths among coexisting pairs of bird species, while Fenchel (1975) obtained comparable but slightly larger character displacement ratios in body sizes of two species of small marine snails. Pulliam (1975) reports very constant, although slightly smaller (about

1.1), bill length ratios among seed-eating sparrows. However, sympatric *Anolis* lizards in the Caribbean often differ by considerably greater ratios, particularly on small islands where the ratio of sizes approaches 2 (Schoener, 1970). Indeed, Schoener (1976a) suggests that niche variance along the food-size dimension and the between-phenotype component of niche breadth may be generally greater among lizards than in birds due both to lack of parental care in lizards and their slower indeterminate growth, which results in greater variation in size among individuals within a population. Cody (1974) gives several interesting possible examples of 'crossovers' in character dis-

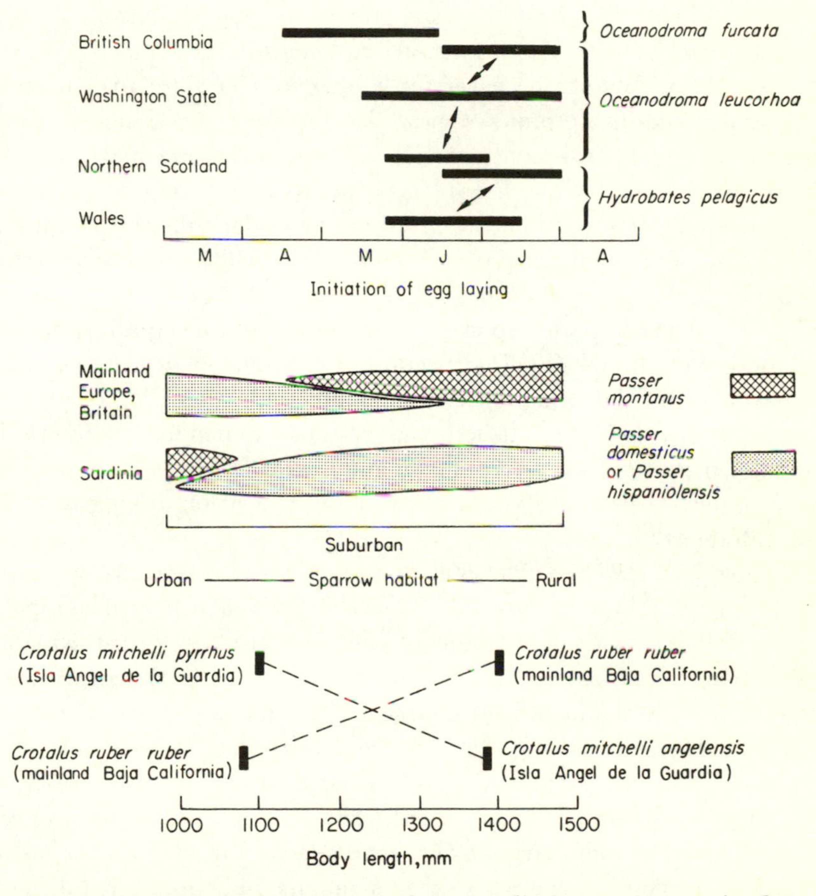

Fig. 7.7. Apparent crossovers in character displacement occur in the breeding seasons of small oceanic petrels (top), sparrow habitats (middle), and body length in rattlesnakes (bottom). Adapted from Cody (1974).

placement (Fig. 7.7). Under circumstances in which there is a potential for intense exploitation competition, niche overlap may be reduced by interference competition leading to interspecific territoriality (Orians and Willson, 1964), which could conceivably select for an actual convergence in phenotypic characteristics involved in recognition and territorial defense (Cody, 1969, 1974). In a critical recent review, Grant (1972a) evaluated existing evidence for and against both character convergence and divergent character displacement; he concludes that not only is the evidence for the ecological bases of character displacement (differential resource use) quite weak, but also that many putative examples of divergent character displacement (including the 'classic' *Sitta* nuthatch case!) could easily represent merely gradual clinal variation associated with various environmental gradients.

Two reasonably strong cases for ecological character displacement have appeared since Grant's review. In one study by Fenchel (1975), mentioned above, two small species of deposit-feeding marine snails, *Hydrobia ulvae* and *H. ventrosa*, have nearly identical size frequency distributions where each occurs in allopatry under different conditions of salinity and hydrography. However, in relatively narrow (and probably quite recent) zones of sympatry along certain salinity gradients, the two species appear to coexist in a stable equilibrium with the more widespread *H. ulvae* becoming conspicuously larger than in allopatry while the other species gets noticeably smaller (Fig. 7.6). Fenchel (1975 and unpublished) has recently expanded these studies and shown particle size selection by snails of different sizes, providing an ecological basis for differential resource utilization arising from the size difference.

The second study of character displacement involves two species of subterranean skinks (genus *Typhlosaurus*), which occur in sympatry throughout the sandridge regions of the southern part of the Kalahari desert (Huey *et al.*, 1974; Huey and Pianka, 1974). Both lizard species eat almost nothing but termites, largely the same few species. Although the smaller of the two species (*T. gariepensis*) is known only from sympatry, the larger species (*T. lineatus*) also occurs in allopatry on adjacent flat sandveld areas. Snout-vent lengths of *lineatus* increase abruptly at the boundary of the sandridges (Fig. 7.8a), suggesting divergent ecological character displacement in sympatry (although the possibility that sandridge habitats differ fundamentally from sandveld ones cannot be entirely discounted). Moreover, heads become *proportionately* larger in sympatry. Correlated with this increase in

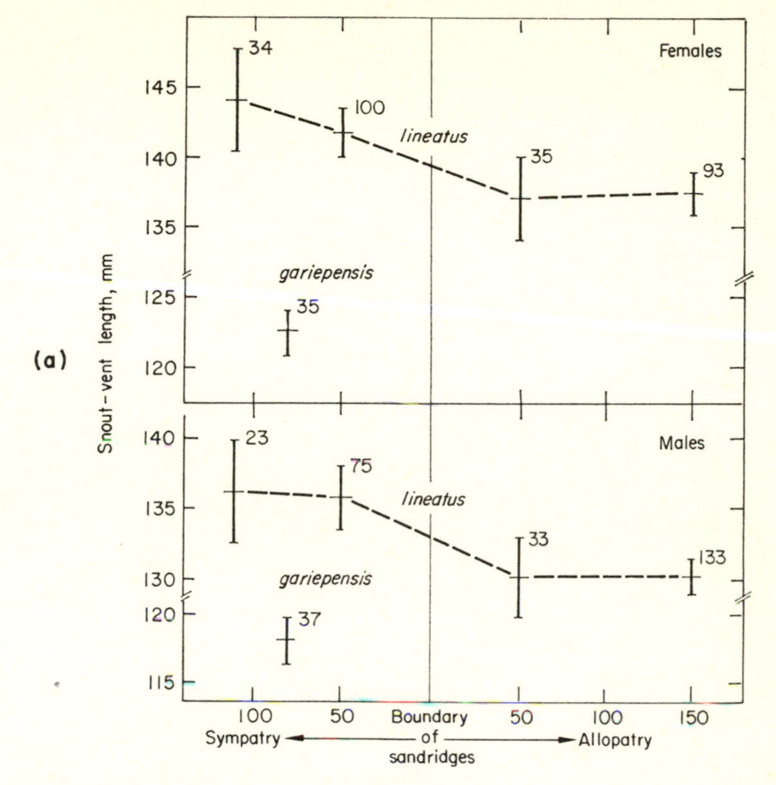

Fig. 7.8. (a) Step clines in mean snout-vent lengths of subterranean legless lizards (Scincidae: *Typhlosaurus lineatus*) associated with the presence of a smaller congeneric species *T. gariepensis*. Head proportions also change with proportionately larger heads occurring in sympatry.

body size and head proportions is a dietary shift to larger castes and species of termites (Fig. 7.8b), which probably reduces overlap and competition with *T. gariepensis*. Ratios of snout-vent lengths of the two species in sympatry are about 1·17, while head length ratios are around 1·5.

Interpreting such niche shifts between allopatry and sympatry is often difficult due to geographical variation in various environmental aspects, such as resource availability (Schoener, 1969; Grant, 1972a; Connell, 1975). Schoener (1975a, 1975b) devised a technique to correct for such changes in resource availability and used it to demonstrate clear niche shifts in response to competition among *Anolis* lizards (Fig. 7.9).

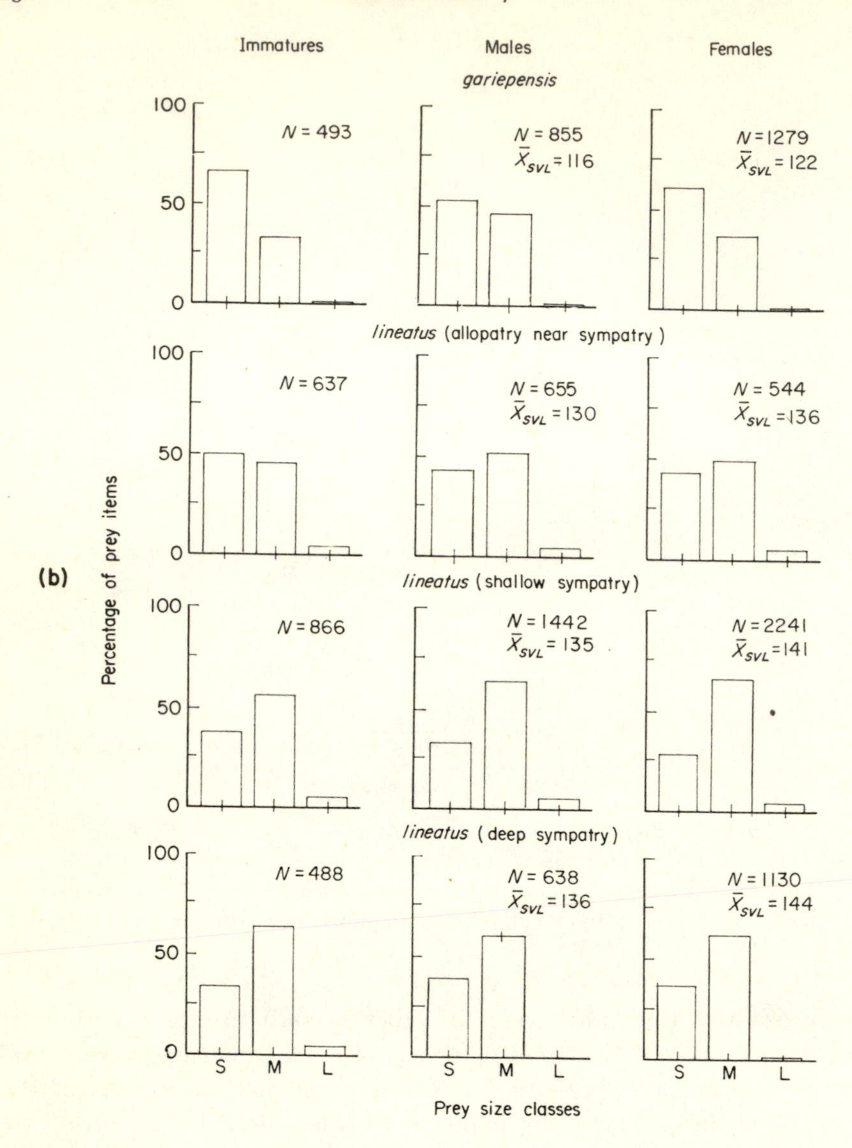

Fig. 7.8. (**b**) Distributions of prey sizes eaten by *Typhlosaurus* immatures, adult males, and adult females under various conditions of allopatry and sympatry. Sympatric *T. lineatus* eat more larger prey items than do allopatric *T. lineatus*. Both figures from Huey *et al.* (1974).

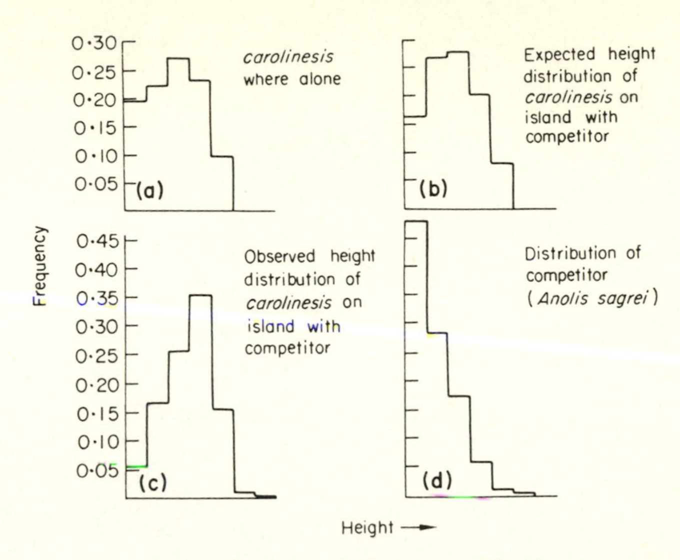

Fig. 7.9. Expected and observed frequency distributions of perch heights of *Anolis* lizards. (**a**) Observed height distribution of *A. carolinensis* where it occurs alone without competitors. (**b**) Expected distribution of perch heights of *A. carolinensis* on another island with different availabilities of various perch heights, assuming no niche shift. (**c**) Observed distribution of *Anolis carolinensis* on the second island with a competitor (compare with *b*). (**d**) Height distribution of the competing species, *A, sagrei*. From Schoener (1975a).

7.4.2 *Overdispersion of niches*

Section 4.3 has given a brief review of several different approaches to the theory of limiting similarity (see also Lawlor and Maynard Smith, 1976; Fenchel and Christiansen, 1976), most of which predict an upper limit on tolerable niche overlap. Schoener (1974) suggests that such limits on niche similarity of coexisting species, coupled with interspecific competition, should result in regular (as opposed to random) spacing of species in niche space. A number of sets of species that differ primarily along a single niche dimension do indeed appear to be separated by rather constant amounts (see, for example, Orians and Horn, 1969; MacArthur, 1972). One celebrated example is Terborgh's observation that various species of antbirds forage at different heights above the ground (Fig. 7.10).

However, as indicated above, ecologies of most potential competitors probably differ along several niche dimensions simultaneously.

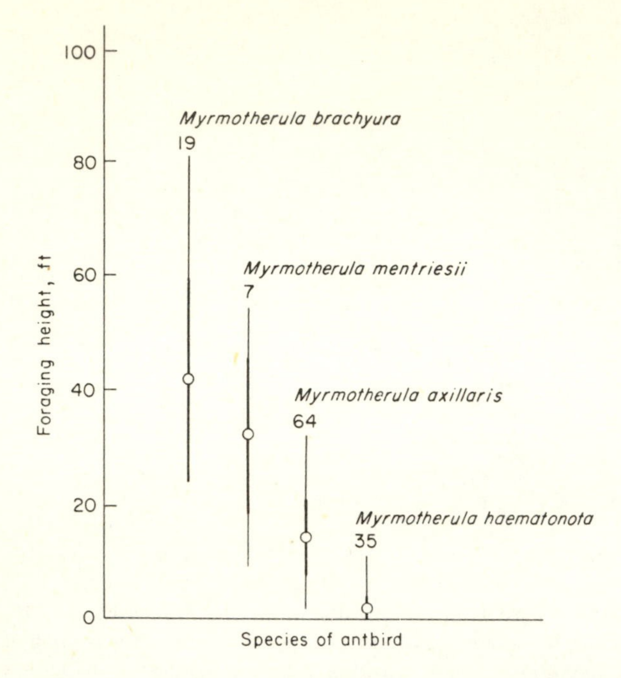

Fig. 7.10. Foraging heights of four species of sympatric antbirds (genus *Myrmotherula*). Means shown with dots and standard deviations by thickened bars. From MacArthur (1972) after data of Terborgh.

Thus, Schoener (1968, 1974) demonstrated that pairs of species of *Anolis* lizards with high dietary overlap tend to separate out in their use of microhabitats, while pairs using similar microhabitats overlap relatively little in prey sizes eaten (Fig. 7.11). Comparable inverse relationships between dietary and microhabitat overlap occur among many pairs of species of nocturnal gekkonid lizards in Australian deserts (Pianka and Pianka, 1976). In a similar vein, Cody (1968) found that grassland birds partition resources along at least three distinct niche dimensions, with the relative importance of various dimensions in separating species differing in various communities (Fig. 7.12). An intriguing degree of constancy in the overall niche separation along all three dimensions suggests that avian niches in these communities are both overdispersed in niche space and that bird species within the communities may have reached some sort of limiting similarity.

My own studies on the niche relationships of desert lizards along three dimensions also suggest overdispersion of niches (Pianka, 1973, 1974, 1975); however, niche separation among these lizard communities

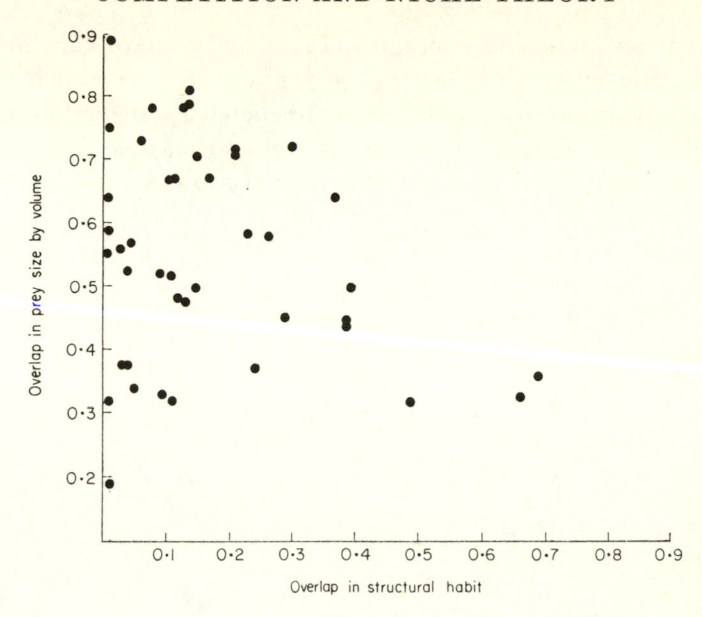

Fig. 7.11. Overlap in prey size plotted against overlap in structural microhabitat among various species of *Anolis* lizards on the island of Bimini. Pairs with high dietary overlap tend to exploit different structural microhabitats; conversely, those with high spatial overlap have relatively little overlap in prey sizes eaten. From Schoener (1968).

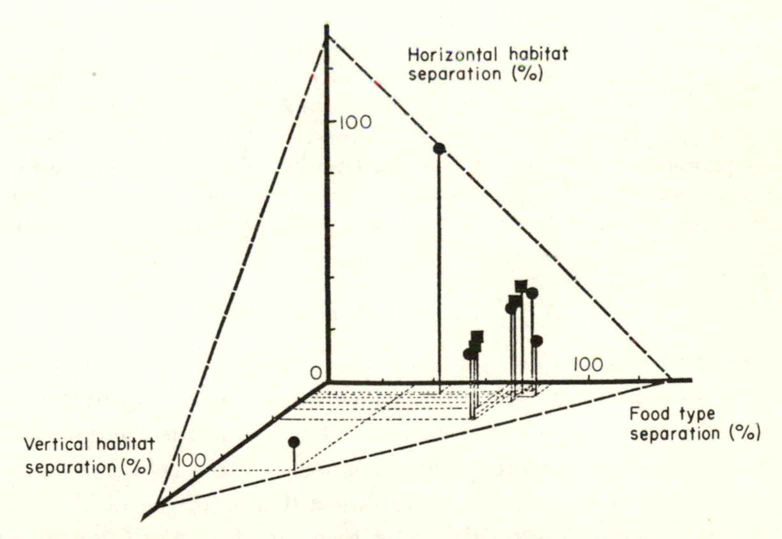

Fig. 7.12. Average niche separation along three dimensions in ten grassland bird communities, suggesting a relatively constant overall amount of separation. From Cody (1968).

is not constant, but both average and maximal overlap vary inversely with the number of lizard species (Fig. 7.13). Such a decrease in overlap with increasing numbers of species is predicted by several theoretical formulations of limiting similarity (above) and may be due to diffuse competition among coexisting species (see also below).

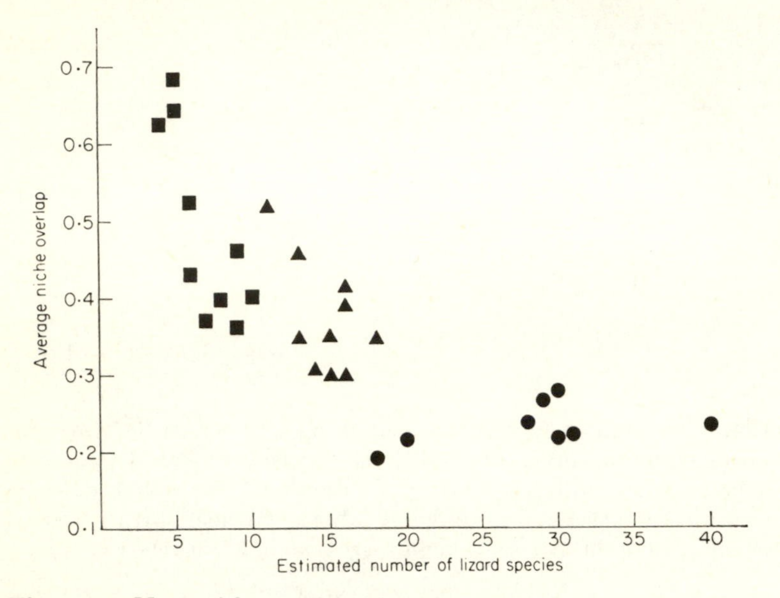

Fig. 7.13. Mean niche overlap plotted against the number of lizard species for 28 study areas on three continents. Squares represent North American sites (Great Basin, Mojave, and Sonoran deserts), triangles are areas in the Kalahari desert of southern Africa, while dots indicate sites in the Great Victoria desert of Western Australia. This inverse correlation is highly significant statistically ($r = -0 \cdot 73, p < 0 \cdot 001$). Several estimates of maximal niche overlap are also inversely correlated with the number of lizard species. From Pianka (1974).

7.4.3 *Experimental manipulations*

Both Connell (1975) and Colwell and Fuentes (1975) have recently pointed out various limitations and shortcomings of natural experiments, especially the lack of a suitable control, and they make a strong case for experimental manipulation of densities in the field. Introduction and/or removal of species with concomitant monitoring of changes in population densities and/or niche shifts before, during, and/or after the experimental manipulation is potentially a very fruitful avenue to studying competition in the field. Not all such experiments can be expected to produce results, however, because niche shifts and/or changes in population densities associated with the presence or absence of potential competitors need not necessarily occur in ecological time unless the populations concerned are periodically released from interspecific competition under natural conditions. A species that experiences strong interspecific competition pressures continually would not be expected to retain the capacity to exploit many of the resources and much of the niche space that is regularly usurped by its competitor; such a species might show only a slight niche shift in ecological time with removal of its competitor. Various such manipulative studies on competition have been undertaken with ants (Brian, 1952; Pontin, 1969), numerous marine invertebrates (Connell, 1961; Paine, 1966; Dayton, 1971; Menge, 1972; Vance, 1972; Stimson, 1970, 1973; Haven, 1973), salamanders (Jaeger, 1970, 1971), lizards (Nevo *et al.*, 1972), birds (Davis, 1973), as well as with many rodents (DeLong, 1966; Koplin and Hoffman, 1968; Grant, 1972b; Joule and Jameson, 1972; Joule and Cameron, 1975). Many of these experiments are reviewed by Connell (1975).

Some removal experiments have documented eompetitive exclusion in pairs of species too similar to coexist. For example, Connell (1961) demonstrated that the barnacle *Chthamalus* persisted in the intertidal rocky zone occupied only by larger *Balanus* barnacles when the larger barnacle species was removed. Noting that the two lizards *Lacerta sicula* and *L. melisellensis* have mutually exclusive geographic distributions on small islands in the Adriatic, Nevo *et al.* (1972) report on the experimental introduction of small populations of each species to islands that supported only the other species; in two of three such experiments, the introduced species went extinct, while on a third island the introduced species appears to be replacing the native form. These authors also performed a reciprocal transplantation introduction experiment with the two species on two small but similar islands for future workers to monitor.

Field experiments on naturally coexisting species have also been informative. Vance (1972) studied competition for empty gastropod shells within and among three sympatric species of intertidal hermit crabs by measuring shell preferences and by manipulating the availability of empty shells. An index of 'shell adequacy' was generated by offering crabs a choice of empty uncontested shells; given such a selection, most crabs moved into larger shells than those in which they were found in nature. Under natural conditions, unoccupied shells tended to be in the smallest shell size categories. Very small hermit crabs tended to have shells close to their preferred size while shell adequacy generally decreased with increasing crab size. Addition of empty shells to the natural environment resulted in increased crab densities, indicating that shells were indeed a limiting resource in short

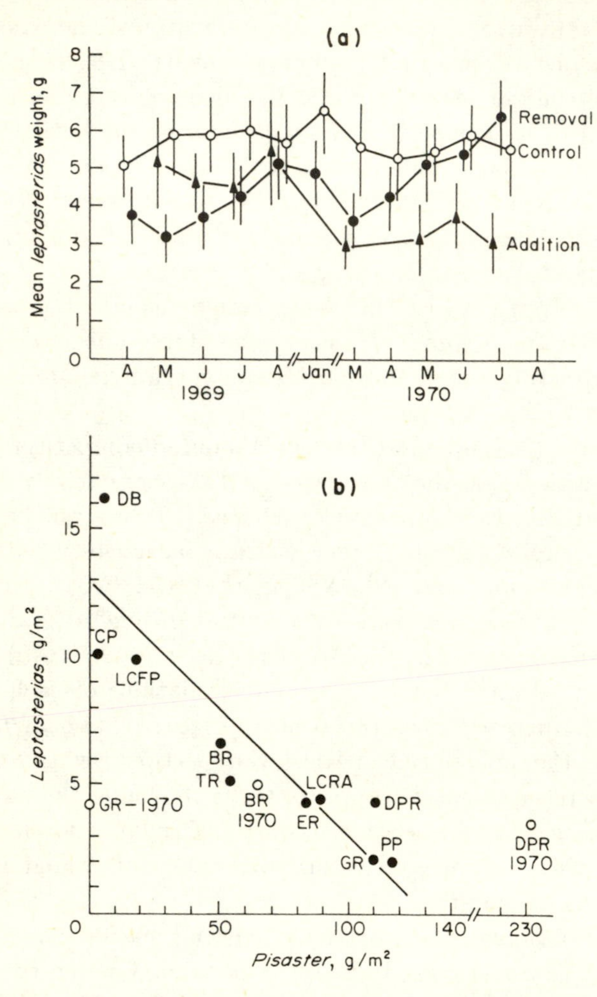

Fig. 7.14. (**a**) Changes in average wet weight of individual *Leptasterias* with removal of *Pisaster* (dots), addition of *Pisaster* (triangles), and in controls with no change in *Pisaster* densities (open circles). (Vertical lines represent 95 per cent confidence intervals of means). (**b**) Plot showing the inverse correlation between the biomass per m^2 of two species of sea stars, *Leptasterias hexactis* and *Pisaster ochraceus* ($r = 0{\cdot}64, p < 0{\cdot}01$). From Menge (1972).

supply. Vance suggested that habitat specificity differences among the three species of hermit crabs could allow the observed coexistence in spite of considerable overlap in their utilization of the limiting resource (shells).

In another field experiment on competition between two broadly sympatric species of sea stars, *Pisaster ochraceus* and *Leptasterias hexactis*, Menge (1972) removed all individuals of the larger *Pisaster* from one small island-reef and added them to another similar reef; a third undisturbed nearby reef supporting both species was monitored as a control (the reciprocal experiment involving removal and addition of *Leptasterias* was not done). Although these two species of sea stars differ greatly in size and reproductive tactics, their diets do overlap broadly (Menge and Menge, 1974). Average weight of individual *Leptasterias* increased significantly with removal of *Pisaster* and decreased with its addition, while the size of control *Leptasterias* did not change (Fig. 7.14a). Moreover, estimated standing crops (biomass/m^2) of the two species vary inversely over the areas sampled (Fig. 7.14b). Although competition coefficients could not be calculated from these data, interspecific competition is clearly implicated. As indicated above, somewhat similar removal experiments have been undertaken with many other marine invertebrates and with small rodents (see Connell, 1974, 1975, for reviews of some of this work).

7.5 Future work: some prospects and problems

Many possibilities still remain for important theoretical work on competitive interactions. Innovative new ways of modelling competition that depart from the traditional Lotka-Volterra equations and the concept of constant competition coefficients will be of great interest. Such approaches should provide fruitful insights into the actual *mechanisms* of competition. Even within the framework of the Lotka-Volterra equations, competition coefficients badly need to be treated as *variables* in both ecological and evolutionary time (see, for example, Leon, 1974 and Lawlor and Maynard Smith, 1976). As explained above, competitive effects between two species may frequently vary with the presence or absence of a third species; theory on interactive competition coefficients is virtually nonexistent. Alphas could also profitably be made density dependent; the actual *shapes* of resource utilization functions might be allowed to change (with an appropriate constraint such as holding the area under the curves constant) either in ecological time due to ethological release or in evolutionary time via directional selection conferring advantages to individual genotypes that deviate from the mean. Roughgarden (1974a, 1974b, 1974c) has made a start on such a theory. Niche overlap theory could be profitably expanded to incorporate effects of diffuse competition (Pianka, 1974, 1975); moreover, the theory of limiting similarity needs to be expanded to many species and to multidimensional resource space.

Fundamental questions emerge from such theoretical considerations. Consider, for example, two competitive communities with similar numbers of species, one of which is composed of several distinct clusters of competitors with competitive interactions among themselves but very weak or nonexistent interactions between members of different clusters. Compare this community with another with the same number of species, but one in which all members interact moderately with all others (i.e., greater diffuse competition). Such a difference between two communities might arise with a difference in niche dimensionality, for instance. What, if any, differences will there be between the two communities in community level properties, such as stability? Will maximal tolerable overlap between pairs of species differ? One might find, for instance, stronger competitive interactions and greater niche overlap between *pairs* of competing species in the first community with fewer immediate neighbours in niche space, but *total* niche overlap summed over all interspecific competitors might well be greater in the second community with greater diffuse competition (see, for example, Pianka, 1974).

Niche breadth theory could profitably be expanded to include age-specific phenotypic changes in resource use. Under such a selection regime, will population structure evolve to make most efficient use of available resources? What selective forces determine the optimal degree of within versus between phenotype components of niche breadth? Roughgarden (personal communication) has suggested that the theory of niche breadth needs to be expanded to combine considerations of optimal foraging tactics with genetic variation among individuals.

Numerous possibilities also exist for important empirical work, of course. For example, the within-phenotype and between-phenotype components of niche breadth have seldom been separated (but see Roughgarden, 1974b). As pointed out repeatedly earlier, field experiments clearly hold great promise, although their precise nature cannot be forseen. Incomplete removal or addition experiments may well allow fairly accurate quantification of competition by measuring the dynamics of population growth. Indeed, gathering useful new data on competition and niche relationships is probably far more challenging (and likely to be more significant) than adding to its existing theoretical foundation.

8
Patterns in Multi-Species Communities

ROBERT M. MAY

8.1 Introduction

For communities with many species, any description of the population dynamics of individual species in terms of the interactions between and within populations usually is very difficult. That is, the methods developed in chapter 2 for single populations and in chapter 4 for pairs of interacting populations are not easily extended to multi-species situations. As pointed out at the end of chapter 4, this is so both because of the proliferation of relevant parameters, and because the dynamical behaviour can become qualitatively more complicated.

One approach is to construct a large computer model, use observations or intuition to infer plausible parameter values, and then simulate the dynamics of the system. Helpful insights are more likely to emerge from such studies if they are subjected to exhaustive 'sensitivity analysis', in which the parameters are varied, one by one, with the aim of discovering how sensitively the dynamics depends upon the values of particular parameters. This is not an easy task. An interesting example of this work is by Harte and Levy (1975), who combine analytic techniques with numerical simulations. There is a growing amount of such work, much of it motivated by the data amassed under the aegis of the International Biological Programme (IBP). I think that these studies will, in time, contribute greatly to our understanding of how ecosystems work; but I equally think that such efforts are in their formative stages, and I shall not discuss them further in this chapter.

An alternative approach is to abandon description of the behaviour of individual species, and focus instead on overall aspects of community structure. As we shall see below, there are many examples where the world appears chaotic and vagarious at the level of individual species, but nonetheless constant and predictable at the level of community organisation. In this spirit, section 8.2 examines some generalities in patterns of energy flow in ecosystems, and section 8.3 considers the sorts of patterns of resource utilisation and food web structure that underlie convergence in the structure and function of geographically separate but ecologically similar communities. Section 8.4 gives a somewhat more technical account of observed patterns in the relative abundance of species in different kinds of communities, and section 8.5 closes the chapter with some diffuse marks on the relation between stability and complexity in ecosystems.

8.2 Patterns of energy flow

In the search for patterns, we may look at the total number of individuals in the different species or trophic levels, or at biomass, or at the way energy flows from one species to another or from one trophic level to another. Stated in this order, these quantities (numbers, biomass, energy flow) are progressively less amenable to casual observation; on the other hand, Table 8.1 suggests they are likely to be progressively more fundamental to the way the system is organised.

In the 1950s and early 1960s, studies of natural lakes (e.g., Cedar Bog Lake in Minnesota and Silver Springs in Florida) and of laboratory aquariums prompted the generalisation that the efficiency at which energy is transferred from one trophic level to the next is around 10 per cent. That is, about 10 per cent of the nett production (measured in calories) of plants ends up as nett production of herbivores, and about 10 per cent of the nett production (in calories) of herbivores makes its way into nett production in the first level of carnivores, and so on (see, e.g., Slobodkin, 1961).

Table 8.1. Population density, biomass, and energy flow for five very different primary consumer populations (after Odum, 1968).

Population	Approximate density (m^{-2})	Biomass (g/m^2)	Energy flow (kcal/m^2/day)
Soil bacteria	10^{12}	0·001	1·0
Marine copepods (*Acartia*)	10^5	2·0	2·5
Intertidal snails (*Littorina*)	200	10·0	1·0
Salt marsh grasshoppers (*Orchelimum*)	10	1·0	0·4
Meadow mice (*Microtus*)	10^{-2}	0·6	0·7
Deer (*Odocoileus*)	10^{-5}	1·1	0·5

Although this intriguing generalisation tends to be true for lakes and aquariums (see, e.g., Kozlovsky, 1968, or the review in Krebs, 1972, ch. 22), it can be much less reliable for terrestrial ecosystems. Table 8.2 shows the percentage of nett productivity of various terrestrial and marine plants that is consumed by their herbivores: since the material consumed is not all assimilated (a fraction being eliminated), and part of that which is assimilated is dissipated as respiration, the transfer efficiency from nett plant production to nett herbivore production will be less than indicated by the percentages in Table 8.2. It is notable that the three highest ratios are for aquatic plants and their herbivores, and for phytoplankton or algae and zooplankton; these aquatic systems will have overall plant-herbivore transfer efficiencies in the neighbourhood of 10 per cent. The terrestrial systems have consumption efficiencies which are lower, and sometimes considerably lower, than the marine ones. In part, this may be because in the sea much of the nett production at one level is indeed consumed by the next level, whereas on land a good deal of it goes directly to decomposers (rather than on up the trophic ladder). This point is made by Heal and MacLean (1975; see also Golley, 1968, and Petrusewicz, 1967) in their review of microbial and faunal production in terrestrial, freshwater and marine ecosystems: they discuss transfer efficiencies of different classes of organisms and at different levels in the trophic chain, and note that in terrestrial systems 90 to 99 per cent of secondary production is by decomposer organisms in the soil or leaf litter. MacLean has underlined the point by remarking that if humans must consume at the secondary rather than exclusively at the primary level (eat meat instead of only plants), they should at least consider eating earthworms rather than beef.

Smith (1976) has elaborated the distinction between marine and terrestrial or intertidal ecosystems, observing that in the latter the struggle to occupy a 2-dimensional surface is a major organising force in the system, which force is not so strong in the open water. He suggests that this tends to produce marine food webs in which nutrient turnover times are short for the primary producers, becoming longer as one moves up the trophic ladder, whereas the constraint of being tied to a stable surface produces the inverse effect in terrestrial and intertidal food webs, with primary producers being relatively long lived, and turnover times characteristically decreasing as one moves up the trophic ladder. These are suggestions, rather than established facts, and deserve further exploration.

Table 8.2. The percentage of nett productivity of various plant hosts that is consumed by feeding-animal species (from Pimentel *et al.*, 1975, in which references are given).

Food-plant Host	Feeding-animal Species	Percentage of Productivity Consumed
Beech trees	Invertebrates	8·0
Oak trees	Invertebrates	10·6
Maple-beech trees	Invertebrates	6·6
Maple-beech trees	Invertebrates	5·9
Tulip poplar trees	Invertebrates	5·6
Grass + forbs	Invertebrates	4–20
Grass + forbs	Invertebrates	<0·5
Alfalfa	Invertebrates	2·5
Sericea lespedeza	Invertebrates	1·0
Grass	Invertebrates	9·6
Aquatic plants	Bivalves	11·0
Aquatic plants	Herbivorous animals	18·9
Algae	Zooplankton	25·0
Phytoplankton	Zooplankton	40·0
Marsh grass	Invertebrates	7·0
Marsh grass	Invertebrates	4·6
Meadow plants	Invertebrates	14·0
Sedge grass	Invertebrates	8·0

The overall transfer efficiency of energy from one level to the next is also significantly dependent on the metabolic strategy of the consumer organism. Warm-blooded animals (homeotherms), such as mammals or birds, typically devote 99 per cent of assimilated energy to respiration: almost all the assimilated calories go to keep the metabolic machinery ticking over, with only about 1 per cent allocated to nett production, in the form of growth and reproduction. Cold-blooded animals (poikilotherms), such as reptiles or arthropods, have less need to be so prodigal with the energy devoted to respiration, and can allocate a higher fraction, typically 10 to 30 per cent, of the assimilated energy to nett production. Such differences in ecological strategies have, for example, the consequence that poikilotherm predators will have a higher ratio of predator/prey standing crop than will homeotherm predators, other things (prey and predator sizes and lifetimes, hunting strategies, etc.) being equal. An interesting application of these general ideas is due to Bakker (1974, 1975a, b), who has used the systematic trends in predator-prey ratios deduced from fossil records as one tine on his multi-pronged argument to the effect that the later dinosaurs were warm-blooded: see Fig. 8.1. This is an example of the contemporary interplay between theoretical ecology and paleontology, which theme is developed further in chapter 12.

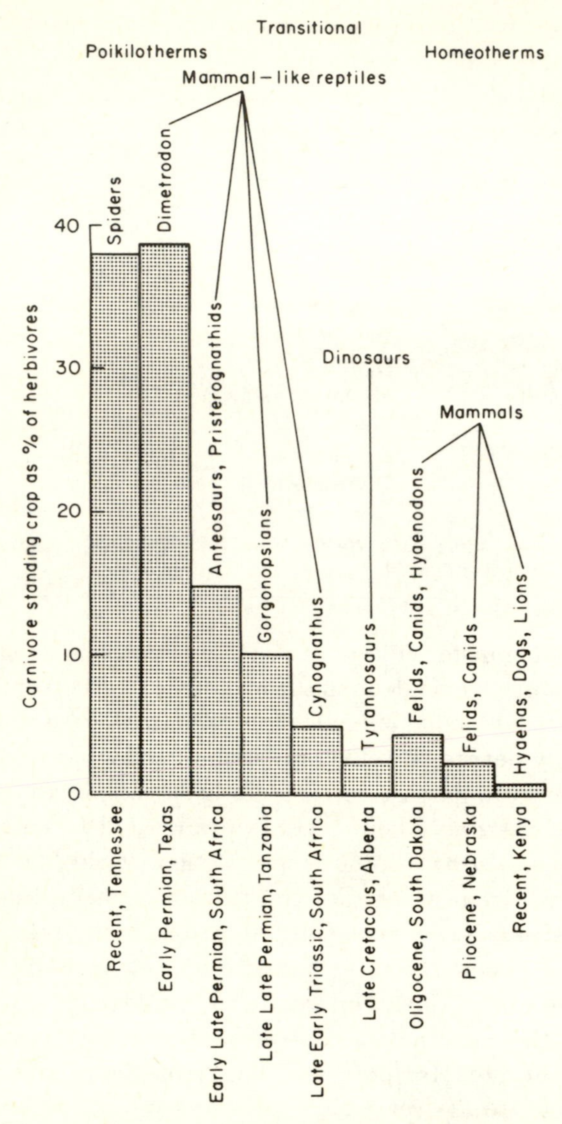

Fig. 8.1. The standing crop of predators expressed as percentage of the standing crop of potential prey, in various groups of living and fossil animals. The figure carries the implication that the later dinosaurs had predator/prey ratios more characteristic of warm-blooded than of cold-blooded animals. From Bakker (1974).

It is thus clear that there is no simple and universal rule setting energy transfer efficiencies between trophic levels at around 10 per cent, but instead there is an array of such figures, depending on the details of the environments and organisms involved. As this growing amount of information is put in order, there are two things to be done with it.

One is to understand it, as it were, from *below*. That is, to understand how the thermodynamics of metabolic processes, in conjunction with the bionomic strategy (in the sense of chapter 3) of the organisms involved, combine to put limits on the efficiency of energy transfer. For accounts of current work in this vein, see Morowitz (1968) or Gates and Schmerl (1975).

The other task is to make *upward* application of this knowledge of transfer efficiencies, to understand why food webs have the structure they are observed to have, and not some other. The essence of such studies is contained in the excessively naive statement that 'if only 10 per cent of the energy is transferred from one level to the next, trophic chains must attenuate rapidly, which is why there are no predators upon lions.' Our knowledge of both transfer efficiencies and of food webs is becoming much more refined than this, and deserves to be put to correspondingly more precise use. The relation between energy flow and trophic structure can be very complicated and non-obvious: mutualistic interactions such as plant and pollinator or plant and seed-disperser may appear inconsequential in terms of overall energy flow, yet play a central role in determining community diversity and structure.

As a postscript to this section, it is revealing to consider current estimates of the world's yearly production of plant material. One of the earliest estimates of terrestrial primary productivity is due to Leibig around 1850: 100×10^9 metric ton per year. Ten years ago the widely accepted figure was $40–50 \times 10^9$ ton per year. The best contemporary estimate (Leith, 1975; Whittaker, 1975) draws together many different strands of investigation, essentially all of IBP origin, to arrive at an estimate back near Leibig's, namely $110–120 \times 10^9$ ton per year. There are still uncertainties, mainly as to productivity in the tropics, and the Russians hold to a more generous figure of 170×10^9 (Rodin *et al.*, 1975). Around 1960, the total marine primary productivity was thought to be about 140×10^9 ton per year; contemporary surveys (Leith, 1975; Rodin *et al.*, 1975) agree on an estimate around $50–60 \times 10^9$. Thus the past decade has seen almost a factor of 10 revision in the terrestrial/marine primary productivity ratio. Estimates of total secondary productivity in various types of ecosystem are less precise, and in general only order-of-magnitude accuracy can be claimed (Heal and MacLean, 1975). These uncertainties in such basic quantities emphasise just how short we are of the sort of detailed knowledge necessary for rational management of our environment.

8.3 Patterns in community structure

Patterns of ecosystem organisation may also be discerned in terms of the actual species composing the community. To do this, we describe the constituent species according to the ecological functions they fulfil, rather than according to taxonomy or phylogenetic ancestry; we describe the villagers by the way they make their living, rather than by their family names or history. As mentioned at the beginning of section 8.2, such patterns have a somewhat more immediate appeal than the patterns in relatively abstract quantities such as energy flow.

A classic study of this kind is due to Simberloff and Wilson (1969), who eliminated the fauna from several very small mangrove islets in the Florida Keys, and then monitored the recolonisation by terrestrial arthropods. In all cases the total number of species on an island returned to around its original value, although the species constituting the total were usually altogether different. Heatwole and Levins (1972) have recently subjected this data to closer analysis, listing for each island the number of species in each of the trophic categories herbivores, scavengers, detritus feeders, wood borers, ants, predators and parasites. The essentials of their findings are contained in Table 8.3. This Table shows vividly that, in terms of trophic structure, the pattern is one of remarkable stability and constancy. On the other hand, in terms of the detailed taxonomic composition of the community of arthropod species on a particular island, there is great variability. The total number of species encountered in the study was 231, whereas a glance at Table 8.3 shows individual islands to have a subset of around 20–30 species; these 20 or so species vary greatly from island to island, or before and after defaunation on the same island.

Similarly, Cody (1968) has analysed how ten grassland bird communities in North and South America are organised with respect to percentage of horizontal habitat selection, vertical habitat selection, and food selection: see Fig. 7.12 (page 135). Eight of the ten communities are clustered around one point in this 3-dimensional figure, which suggests an orderly and repeatable pattern for such bird communities, despite the particular species being very different (even to the extent that the avifauna in North and South America have different ancestry). Fager (1968) has shown that the community of invertebrates in a decaying oak log has a definite structure, although each log will have its own particular collection of species. MacArthur and MacArthur (1961) showed how the diversity of species within bird communities in North America is closely correlated with the amount of diversity of foliage (particularly foliage height), and Recher (1969) showed that this same relation applied directly to foliage height diversity versus bird species diversity in Australia.

Table 8.3. Evidence for stability of trophic structure.

	Trophic classes								
Island	H	S	D	W	A	C	P	?	Total
E1	9 (7)	1 (0)	3 (2)	0 (0)	3 (0)	2 (1)	2 (1)	0 (0)	20 (11)
E2	11 (15)	2 (2)	2 (1)	2 (2)	7 (41)	9 (4)	3 (0)	0 (1)	36 (29)
E3	7 (10)	1 (2)	3 (2)	2 (0)	5 (6)	3 (4)	2 (2)	0 (0)	23 (26)
ST2	7 (6)	1 1()	2 (1)	1 (0)	6 (5)	5 (4)	2 (1)	1 (0)	25 (18)
E7	9 (10)	1 (0)	2 (1)	1 (2)	5 (3)	4 (8)	1 (2)	0 (1)	23 (27)
E9	12 (7)	1 (0)	1 (1)	2 (2)	6 (5)	13 (10)	2 (3)	0 (1)	37 (29)
Total	55 (55)	7 (5)	13 (8)	8 (6)	32 (23)	36 (31)	12 (9)	1 (3)	164 (140)

The table is after Heatwole and Levins (1972). The islands are labelled in Simberloff and Wilson's (1969) original notation, and on each the fauna is classified into the trophic groups: herbivore (H); scavenger (S); detritus feeder (D); wood borer (W); ant (A); predator (C); parasite (P); class undetermined (?). For each trophic class, the first figures are the number of species before defaunation, and the figures in parentheses are the corresponding numbers after recolonisation. The total number of different species encountered in the study was 231 (the simple sum 164 + 140 counts some species more than once).

These are a few among a growing number of analogous studies of the structure of terrestrial, intertidal and aquatic communities.

The examples all point to an underlying community pattern, a trophic skeleton, which is stable and predictable. From the standpoint of the populations of individual species, one may gain an impression of ceaseless change and flux, dominated by environmental vagaries or the accidents of history. At the same time, from the standpoint of trophic structure (as measured for instance by the number of species playing given broad roles, as in Table 8.3) the picture may be one of steadiness and pattern.

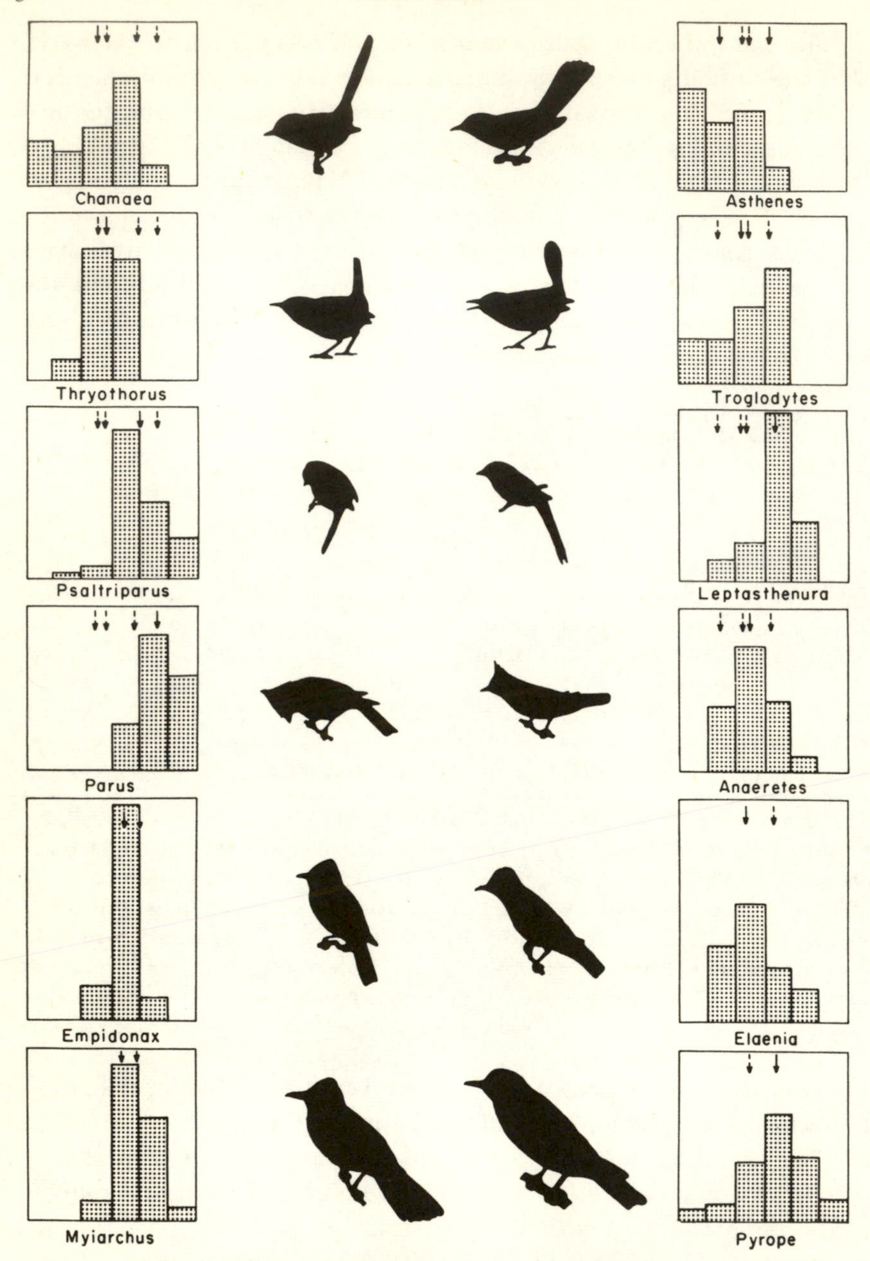

Fig. 8.2. Distributions of foraging heights, and silhouettes of four canopy insectivores (upper) and two sallying flycatchers (lower) in Californian chaparral (left) and Chilean matorral (right). The foraging height intervals are, from the left: ground; ground–6″; 6″–2′; 2′–4′; 4′–10′; 10′–20′. Arrows above each distribution indicate its mean, and those of the frequency distributions of neighbouring species. Silhouettes are all to the same scale. From Cody (1974).

The phenomenon of *convergence* is most familiar at the level of individual species, as displayed in museum showcases: the New World toucan and the Old World hornbill; hummingbirds and sunbirds; anteaters as represented in Neotropical, Oriental, Ethiopian and Australian realms by edentate antbears, pangolins, aardvarks and echidnas, respectively. But the considerations developed above find their expression in the convergence of the structure of entire communities of plants and animals in spatially separate but climatically similar regions.

One illustration is provided in Fig. 8.2, which shows the convergence in the physical appearance of members of guilds of insectivorous birds, one from Californian chaparral and the other from Chilean matorral. The physical similarities derive from underlying similarities in the way resources are divided up within the guild, as also indicated in Fig. 8.2. Cody (1975) has since shown that on a third continent, in South African macchia, equivalents of exactly the same six insectivores are found, including the four foliage species and the two sallying flycatchers. Such examples may be extended to encompass the fauna and flora of different regions not only at the present time, but even across the sweep of geological time. Thus Simpson (1949, 1969) has pointed to the convergence between the mammalian (and marsupial) communities in North and South America prior to the formation of the Panama land bridge in the mid-Pleistocene; up to this time the two faunal assemblies had followed separate evolutionary trajectories for 100 million years or so, each on its own island continent. In a more speculative fashion, Wilson (1975; see also Wilson *et al.*, 1973) has discussed the convergence in community structure between the dinosaur fauna of the Cretaceous, and typical mammalian faunas of today: see also chapter 12.

For a more detailed account of efforts at quantifying such community patterns, and using them to make predictive statements, see MacArthur (1972), Cody (1974, 1975) and the work discussed in chapter 9.

8.4 The relative abundance of species

A more technical aspect of community patterns lies in the relative abundance of the various species.

For any particular group of S species, we may express the number of individuals in the ith species, N_i, as a proportion, p_i, of the total number of individuals, $N = \sum N_i$:

$$p_i = N_i/N. \tag{8.1}$$

This information may then be displayed on a graph of relative abundance versus rank, as exemplified by Fig. 8.4 below; such figures show the patterns in the relative magnitudes of the constituent populations, from most to least abundant. Equivalently, the information may be incorporated in a distribution function, $S(N)$, as in Figs. 8.3 and 8.5 below: here $S(N)dN$ is the number of species comprising between N and $N+dN$ individuals.

Some of the main distributions of relative abundance which occur in natural systems are discussed below. A more complete account, both of the distributions themselves and of the various equivalent ways they are conventionally displayed, is in Pielou (1975) or May (1975f). A separate source of statistical complication lies in estimating the magnitude of the various populations in the first place: the nuts and bolts of pertinent sampling techniques are well reviewed by Southwood (1976).

8.4.1 *Lognormal distributions*

Once the community consists of a relatively large and heterogeneous assembly of species, the observed distribution of species relative abundance, $S(N)$, is almost always lognormal; i.e., there is a bell-shaped gaussian distribution in the logarithms of the species' abundances. This lognormal distribution has been documented for groups of organisms as disparate as diatoms, moths, birds or plants, provided always that the sample is large enough to contain a good number of species (see, e.g., Whittaker, 1972, 1975). Fig. 8.3 provides a typical example.

What is the explanation for these pervasive patterns? Given a largish group of species, it is likely that their relative abundances will be governed by the interplay of many more-or-less independent factors. It is in the nature of the equations of population dynamics that these several factors should compound multiplicatively, and the statistical Central Limit Theorem applied to such a product of factors implies a lognormal distribution. That is, the lognormal distribution arises from products of random variables, and factors that influence large and heterogeneous assemblies of species indeed tend to do so in this fashion. This broad statistical argument similarly suggests lognormal patterns for the distribution of wealth in the USA, or for people or GNP among the nation states of the world. Such is in fact the case.

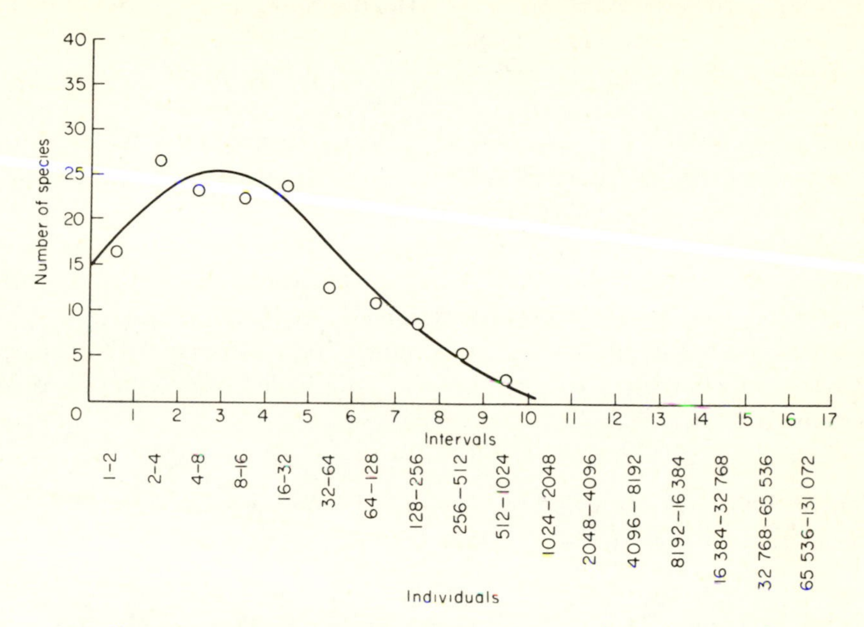

Fig. 8.3. Illustrating the lognormal distribution of relative abundance of diatom species in a sample taken from an undisturbed community in Ridley Creek, Chester County, Pennsylvania. The abundances are, as indicated, plotted as logarithms to the base 2: for further discussion, see the text. From Patrick (1973).

More specifically, ecologists usually write the lognormal distribution as

$$S(R) = S_o \exp(-a^2 R^2). \tag{8.2}$$

Here S_o is the number of species at the mode of the distribution, which species have populations around N_o; a is an inverse measure of the width of the distribution; and, following the convention established by Preston's (1948) early work, R expresses the abundance as a logarithm to base 2, so that successive 'intervals' or 'octaves' along the x-axis correspond to population doublings,

$$R = \log_2(N/N_o). \tag{8.3}$$

The two parameters S_o and a specify the distribution uniquely. In particular, the total number of species S is, to a good approximation, $S \simeq \sqrt{\pi}\, S_o/a$. One further useful parameter may be introduced. Define

R_{max} as the expected R-value, or 'octave', for the most abundant species, and define R_N to be the location of the peak of the distribution in total numbers of individuals [i.e., the distribution of $N \times S(N)$]: the parameter γ is then defined to be

$$\gamma = R_N / R_{max}. \tag{8.4}$$

Any pair of the three parameters S_o, a and γ may now be used to characterize the distribution. The three are approximately related by $a^2 \gamma^2 (\ln S_o) \simeq 0 \cdot 12$ (see May, 1975f).

The purpose of this brief notational frenzy is to discuss two empirical 'laws', which have provoked much speculation in the ecological literature. The first is the observation, made by Hutchinson (1953) and since supported by an increasing amount of data (see, e.g., Whittaker, 1972; Colinvaux, 1973, ch. 37), that the parameter a usually has a value around 0·2,

$$a \simeq 0 \cdot 2. \tag{8.5}$$

The second is Preston's (1962) 'canonical hypothesis', which says the usual value of γ is around unity,

$$\gamma \simeq 1. \tag{8.6}$$

This rule holds true for a large body of data (Preston, 1962), and has been used to discuss species-area relations by MacArthur and Wilson (1967).

Before reading too much ecological significance into eq. (8.5), it is well to note that this rule also applies to the other lognormal distributions mentioned above (wealth, people and GNP of nations). To begin, assume that indeed $\gamma = 1$: the single parameter a now specifies the lognormal distribution, and there is a unique functional relationship between a and the total number of species S. The details of this relation (May, 1975f, Fig. 3) are such that as S varies from 20 to 10,000 species, a varies from 0·29 to 0·13. The ingenious reader can use the mathematical relations sprinkled two paragraphs above to establish the approximate proportionality $a \sim 1/\sqrt{\ln S}$, which makes explicit the very weak dependence of a on S. More generally, for lognormal distributions it can be seen that there is a relation between the total number of species and individuals, S and N, and the values of the parameters a and γ, and that the enormous range of communities with S ranging from 20 to 10,000 and N ranging from $10S$ to $10^7 S$ is characterised by values of a in the range 0·1 to 0·4 and of γ in the range 0·5 to 1·8 (May, 1975f, Fig. 5).

In short, the enigmatic rules (8.5) and (8.6) are mathematical properties of the lognormal distribution for large S. And the lognormal itself reflects statistical generalities.

8.4.2 *Other Distributions*

In simple communities, with attention focussed on a relatively small and homogeneous group of species, distributions of species relative abundance other than lognormal are often seen. In these circumstances, the distribution may reveal biological facts rather than statistical generalities.

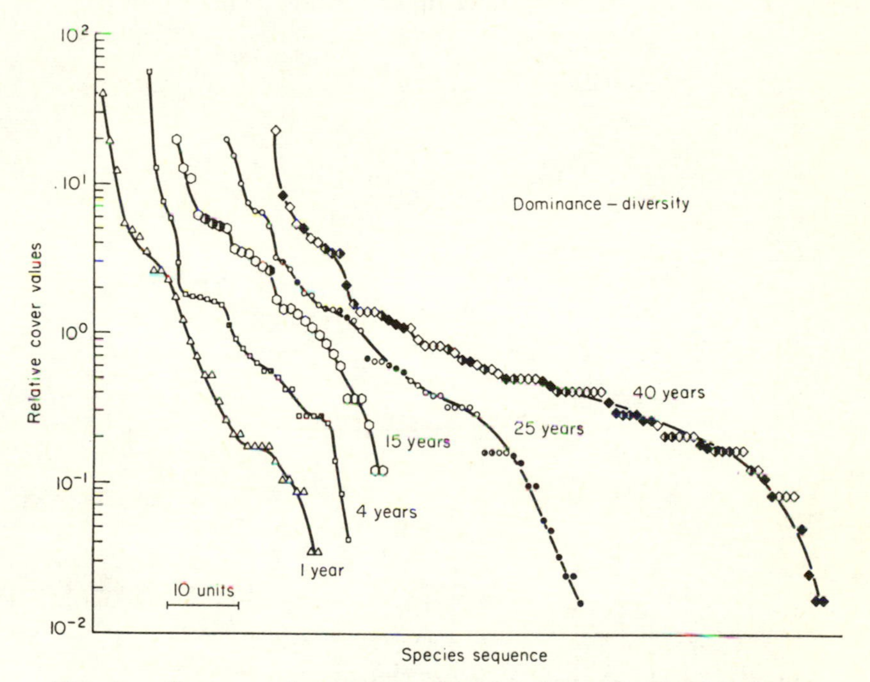

Fig. 8.4. Patterns of species relative abundance in old fields of five different stages of abandonment in southern Illinois. The patterns are expressed as the percentage that a given species contributes to the total area covered by all species in a community, plotted against the species' rank, ordered from most to least abundant. Symbols are open for herbs, half-open for shrubs, and closed for trees. From Bazzaz (1975).

One example is the patterns that various people have described for early successional plant communities (see, e.g., Whittaker, 1975; McNaughton and Wolf, 1973). In the early stages of succession, plant communities will tend to consist of a handful of weedy, pioneer, r-selected species that happened to get there first. The species arriving first will tend to grow rapidly, pre-empting a fraction (k say) of the available space or other governing resource before the arrival of the next species, which in turn will tend to pre-empt a similar fraction of the remaining space before the arrival of the third, and so on. The consequent pattern will show up, very roughly, as a geometric series when the relative abundances are arranged hierarchically, from most to least abundant. A typical illustration is provided by the earlier successional stages in Fig. 8.4: in this figure, relative abundance is plotted logarithmically against rank, so that the geometric series 'niche pre-emption' pattern shows up as a straight line.

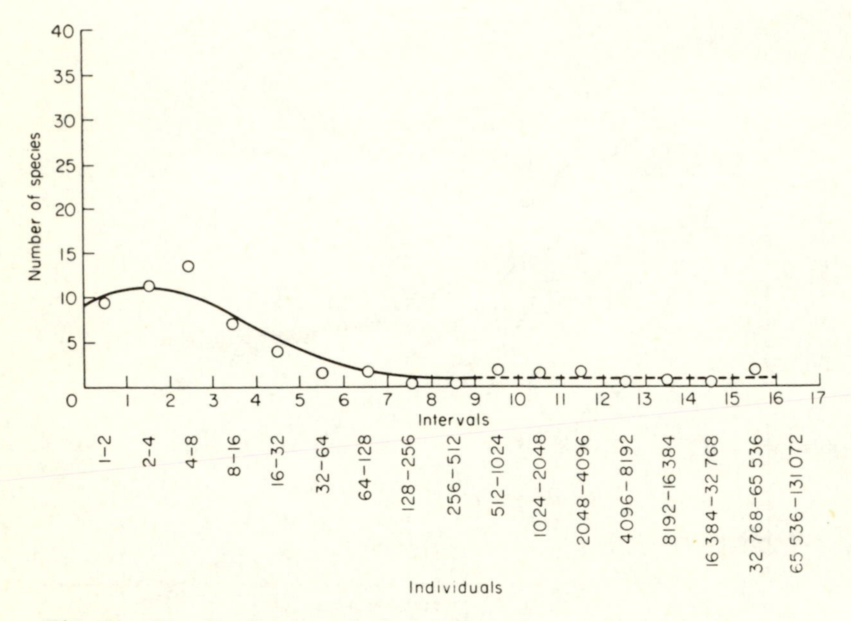

Fig. 8.5. The distribution of relative abundance of diatom species in a community subjected to organic pollution (Back River, Maryland). The features are as discussed in the text. From Patrick (1973).

As succession proceeds, things become more complicated, and a multitude of ecological dimensions are likely to be relevant to the ultimate composition of the community. As discussed above, this leads to a lognormal distribution of species relative abundances: on an abundance-versus-rank plot this produces the sort of S-shaped curve seen for later successional stages in Fig. 8.4, with a preponderance of 'middle class' species. The later, lognormally distributed community tends to be an egalitarian socialist society compared with the feudal hierarchy characteristic of early succession.

Something akin to a reversal of these successional patterns takes place when mature communities become polluted. This has been shown particularly for diatoms in streams and lakes subject either to 'enrichment' by waste heat, sewage, or other organic materials, or to toxic pollution by heavy metals or other poisons (Patrick, 1973, 1975). The initial equilibrium diatom community essentially always shows the classic lognormal distribution. When polluted, the community typically shows a pattern in which a few species become exceptionally common, with their relative abundances tending to exhibit a geometric series on a rank-abundance plot; it is plausible that the cause is an early successional type of r-selection for success in handling the all-important new factor. Closer analysis may, or may not, reveal a lognormal distribution of relatively rare species hanging on down in the tail of the distribution. Fig. 8.5 gives a typical example, which is to be contrasted with the undisturbed community of Fig. 8.3.

8.4.3 *Richness and diversity indices*

One single number that goes a long way toward characterising a biological community is simply the total number of species present, S. This is sometimes called 'richness'.

Often one will wish to go beyond this, yet stop short of describing the full distribution $S(N)$, by constructing some second number to characterise the 'diversity' or 'evenness' of the community; some single number to distinguish between the community with 30 equally abundant species and that with one common species and 29 rare ones.

The statistically-minded will see all such numbers simply as moments of the $S(N)$ distribution, and will be inclined to choose the *variance* of the distribution as a sensible measure of its 'diversity'. Several of the indices which have been proposed in the ecological literature are of this kind, involving $\sum p_i^2$ in one form or another (as $\sum p_i^2$, or as $1-\sum p_i^2$, or as $1/\sum p_i^2$, with p_i defined by eq. (8.1)]. More fashionable is the Shannon diversity index,

$$H = -\sum p_i \ln p_i, \tag{8.7}$$

which is linked by an ectoplasmic thread to information theory. H is a very insensitive measure of the underlying distribution of relative abundance (see, e.g., May, 1975f, Figs. 6, 7). If some such diversity index is to be used, I much prefer one of the 'variance' measures (involving $\sum p_i^2$ in some shape or form), or even the naive 'dominance' index provided by the ratio of the most abundant population to the total population (i.e., the value of p_i for the most common species).

The connections between patterns of relative abundance and degree of pollution, which we noted in section 8.4.2, have sparked interest in the possibility of using some index of diversity to measure how much an ecosystem has been perturbed by human activity. One such study has used the diversity index H of eq. (8.7) as a phenomenological measure of gross toxicity in San Francisco Bay (Armstrong *et al.*, 1971). Clearly, an engineering-style index of this kind would simplify tasks of environmental management.

Taking diatom communities in polluted streams as one among many possible examples, we may note that any such single number will be dominated by the distribution of the handful of common species. But the time scale for recovery of the pristine ecosystem (and, indeed, whether it can ever recover) may depend on the lognormal tail of species which are uncommon in the polluted community, and whose presence will not show up in any overall diversity index. This is a cogent argument for describing the community by its full distribution of species relative abundance, and not trying to condense information into a single diversity index, which may mislead (see also Williamson, 1973).

8.5 Stability and complexity

One aspect of community ecology that has attracted much attention over the years is the relation between the complexity of a food web and its stability. Here increased 'complexity' is usually loosely taken to mean more species, more interactions, and hence more parameters characterising the interactions. Increased 'stability' may be identified with relatively lower levels of population fluctuation, or with ability to recover from perturbations, or simply with persistence of the system.

Elton (1958), among others, has drawn together a collection of empirical and theoretical observations in support of the conclusion that complexity implies stability. This notion has tended to become part of the folk wisdom of ecology, and has sometimes been elevated to the status of a mathematical theorem (e.g., Hutchinson, 1959). But the empirical evidence is at best equivocal: there are many examples of simple natural systems that are stable (in all of the above senses), and of complex ones that are not. Among the growing number of contemporary reviews, I particularly like that by Watt (1968, pp. 39–50).

Here we shall dwell upon only one aspect of this question, namely the answers that emerge from theoretical models.

8.5.1 *Complexity and stability in model ecosystems*

In assessing the stability character of any ecosystem model, various factors enter. *First*, equilibrium configurations will exist only within certain restricted ranges of the interaction and environmental parameters (such as birth rates, assimilation efficiencies, predator attack rates, etc.; the r_i, K_i and α_{ij} in eq. (7.1), for example). Outside this restricted region of parameter space, the equations will describe a collapsing ecosystem in which some, or all, of the populations are fated for extinction. As was illustrated in chapters 2, 4, 5 and 6, mapping out this stable region of parameter space is a standard mathematical exercise. *Second*, even when the parameters are fixed at values which do admit of an equilibrium solution, this equilibrium may be unstable to large perturbations in the population numbers: this involves the ideas of 'resilience', and of global versus local stability, as discussed in section 4.2.3.

These two questions are intertwined, both mathematically and biologically. For parameter values near the centre of the domain of stable values, the dynamical landscape in population space may in general be thought of as a comparatively wide and deep valley; all but the most extreme disturbances to the populations will eventually return to the equilibrium configuration. As the parameter values are chosen nearer the edge of the domain of stable values, the dynamical landscape in population space becomes more like a shallow valley nestled in the top of a volcano; modest disturbances to the populations will see the system spill out from the volcano top. Finally, for parameter values outside the stable domain, there is no valley at all; the volcano tip has given way to a rounded hilltop. Moreover, in the real world, we do not deal with fixed parameter values, but rather the environmental and interaction parameters are themselves fluctuating, in turn driving the population perturbations.

In short, a system which is stable only within a comparatively small domain of parameter space may be called *dynamically fragile*. Such a system will persist only for tightly circumscribed values of the environmental parameters, and will tend to collapse under significant perturbations either to environmental parameters or to population values. Conversely, a system which is stable within a comparatively large domain of parameter space may be called *dynamically robust*.

When these kinds of studies are made, a wide variety of mathematical models suggest that as a system becomes more complex, in the sense of more species and a more rich structure of interdependence, it becomes more dynamically fragile.

This is an important theoretical result, and it comes from models at three different levels of abstraction. At the most concrete level, we may take specific models such as the multi-species generalisation of the simple Lotka-Volterra prey-predator model [eqs. (4.1) and (4.2)], and contrast the typical dynamical behaviour of systems with few and with many species: the 'complex' systems are never more, and are usually less, stable than the 'simple' ones. Other studies, of a generalisation of eq. (7.1) in which the interaction coefficients α_{ij} are assigned randomly, show that as the number of species increases the probability to find a 'feasible' equilibrium (i.e., one where all populations are positive) decreases: see Roberts (1974). More abstract models deal only with general expressions for the way a population's fluctuations about its equilibrium value [$x_i(t) = N_i(t) - N_i^*$ for the ith population] are governed by its interaction with other populations, as measured by interaction coefficients a_{ij}:

$$dx_i/dt = \sum_j a_{ij}\, x_j. \qquad (8.8)$$

Here studies of 'randomly assembled' communities, in which the array of coefficients a_{ij} are assigned according to some random process, show that dynamical stability typically decreases with increase in the number of species, or in the number or strength of interactions between species (Gardner and Ashby, 1970; May, 1972b; McMurtrie, 1975). At a yet more abstract level, similar conclusions emerge from studies which describe only the topology or 'loop structure' of the food web, independent of the actual magnitude of the interactions (Levins, 1975). For a more complete account of all this, see May (1975a, ch. 3).

Thus, as a mathematical generality, increasing complexity makes for dynamical fragility rather than robustness.

This is *not* to say that, in nature, complex ecosystems need appear less stable than simple ones. A complex system in an environment characterised by a low level of random fluctuation and a simple system in an environment characterised by a high level of random fluctuation can well be equally likely to persist, each having the dynamical stability properties appropriate to its environment.

Moreover, if we regard evolution as an existential game, where the prize to the winner is to stay in the game (Solbodkin, 1964), we may conjecture that ecosystems will evolve to be as rich and complex as is compatible with the persistence of most populations. In a predictable environment, the system need only cope with relatively small perturbations, and can therefore achieve this fragile complexity, yet persist. Conversely, in an unpredictable environment, there is need for the stable region of parameter space to be extensive, with the implication that the system must be relatively simple.

In brief, a predictable ('stable') environment may permit a relatively complex and delicately balanced ecosystem to exist; an unpredictable ('unstable') environment is more likely to demand a structurally simple, robust ecosystem.

8.5.2 *Some implications*

These general arguments, which see the paradigms of trophic complexity such as the tropical rainforest or Lake Baikal as being the evolutionary products of stable environments, march with the more down-to-earth discussions of r- and K-selection in chapter 3. r-Selection is associated with a relatively unpredictable environment and a simple ecosystem, K-selection with a relatively predictable environment and a complex, biologically crowded community: a system comprising r-selected species, with their relatively large clutch size or seed set, is well adapted to bounce back after disturbance, whereas a K-selected assembly may be unable to recover after an unusually severe disturbance. In a similar vein, we saw in section 4.4 both that mutualism tends to be a dynamically fragile relationship, and that it is a much more common phenomenon in stable tropical environments. Such biological particularities provide the detailed mechanisms which underlie the dynamical generalities enunciated above. These themes are pursued further in some of the papers in the volumes edited by van Dobben and Lowe-McConnell (1975) and by Farnworth and Golley (1974).

On this view, there is no reason necessarily to expect simple natural monocultures to be unstable. And, indeed, there are many instances of robustly enduring natural monocultures. The marsh grass *Spartina* is one conspicuous example. Another is bracken which in recent years, partly as a result of hilly areas grazed by cows being given over to sheep, has shown itself to be a robust, even aggressively invasive, natural monoculture over increasing areas in Britain (Lawton, 1974). The instability of so many man-made agricultural monocultures is likely to stem not from their simplicity, as such, but rather from their lack of any significant history of coevolution with pests and pathogens.

An important general conclusion is that the large and unprecedented perturbations imposed by man are likely to be more traumatic for complex natural systems than for simple ones. This inverts the naive, if well-intentioned, view that 'complexity begets stability', and its accompanying moral that we should preserve, or even create, complex systems as buffers against man's importunities. I would argue that the complex natural ecosystems currently under siege in the tropics and subtropics are less able to withstand our battering than are the relatively simple temperate and boreal systems.

9
Island Biogeography and the Design of Natural Reserves

JARED M. DIAMOND AND ROBERT M. MAY

9.1 Introduction

The flora and fauna of islands have played a central role in the development of ecological thought, from the early formulations of evolution and biogeography by Darwin and Wallace, through Mayr's demonstration of the role of geographic isolation in speciation, to the analytical theory of island biogeography pioneered by MacArthur and Wilson. Some reasons why islands have been well suited to provoking or testing theoretical ideas are that they have definite boundaries, come in many different sizes and heights and remotenesses, often have relatively simple communities of plants and animals, and serve as ready made evolutionary laboratories offering replicate 'natural experiments' in community assembly.

Islands may be real islands in the ocean, or they may be virtual islands such as hilltops (where for many species the surrounding lowland presents a distributional barrier), lakes, or wooded tracts surrounded by open land. In particular, the natural reserves and wildlife refuges that are set aside from large areas bent to man's purposes may be thought of as islands in a sea of altered habitat. In view of the manifest destiny of much of the world's tropical rain forest, we may ask such questions as: How many species of Amazonian plants and animals will survive if only 1 per cent of the Amazonian rain forest can be preserved? At what rate will species be extinguished? Which species will be likely to survive in the reserve, and which will most likely be lost? These are pressing questions, to which the theory of island biogeography holds at least some of the answers.

This chapter first treats 'static' aspects of the equilibrium biota, discussing empirical and theoretical relationships between the island area, A, and the number of species present, S. Second, 'dynamic' aspects of the equilibrium are examined; equilibrium is seen as a balance between immigration and extinction, and the rates at which the system approaches the equilibrium configuration from below, from the neighbourhood, and from above are discussed. Next we discuss *which* species tend to be present on a given island at a given stage in its history, and conclude with some speculations as to the emergent principles for the design of natural reserves.

9.2 Species-area relations

9.2.1 *Empirical relations*

There have been many studies which compare the number of species, S, on islands of different area, A, but with similar habitat and in the same archipelago or island group. For both plants and animals, and for

Table 9.1. The species-area exponent z.

Organism	Location	z	Source
beetles	West Indies	0·34	Darlington
reptiles and amphibians	West Indies	0·30	Darlington
birds	West Indies	0·24	Hamilton, Barth, Rubinoff
birds	East Indies	0·28	Hamilton, Barth, Rubinoff
birds	East-Central Pacific	0·30	Hamilton, Barth, Rubinoff
ants	Melanesia	0·30	MacArthur and Wilson
land vertebrates	Lake Michigan Islands	0·24	Preston
birds	New Guinea Islands	0·22	Diamond
birds	New Britain Islands	0·18	Diamond
birds	Solomon Islands	0·09	Diamond and Mayr
birds	New Hebrides	0·05	Diamond and Marshall
land plants	Galapagos	0·32	Preston
land plants	Galapagos	0·33	Hamilton, Barth, Rubinoff
land plants	Galapagos	0·31	Johnson and Raven
land plants	World-wide	0·22	Preston
land plants	British Isles	0·21	Johnson and Raven
land plants	Yorkshire nature reserves	0·21	Usher
land plants	California Islands	0·37	Johnson, Mason, Raven

Values of z in eq. (9·1), as deduced from observations on various groups of plants and animals in various archipelagoes. For original references, see May (1975f).

a variety of taxonomic groups from birds to beetles and ants, such studies commonly lead to a relation of the form

$$S = cA^z \qquad (9.1)$$

The dimensionless parameter z (the slope of the regression line on a log S versus log A plot) typically has a value in the range 0·18–0·35; c is a proportionality constant, which depends *inter alia* on the dimensions in which A is measured and on the taxonomic group studied. The data are sometimes better fitted by a relation of the form $S = a + b \log A$: for a fuller discussion see, e.g., Diamond and Mayr (1976).

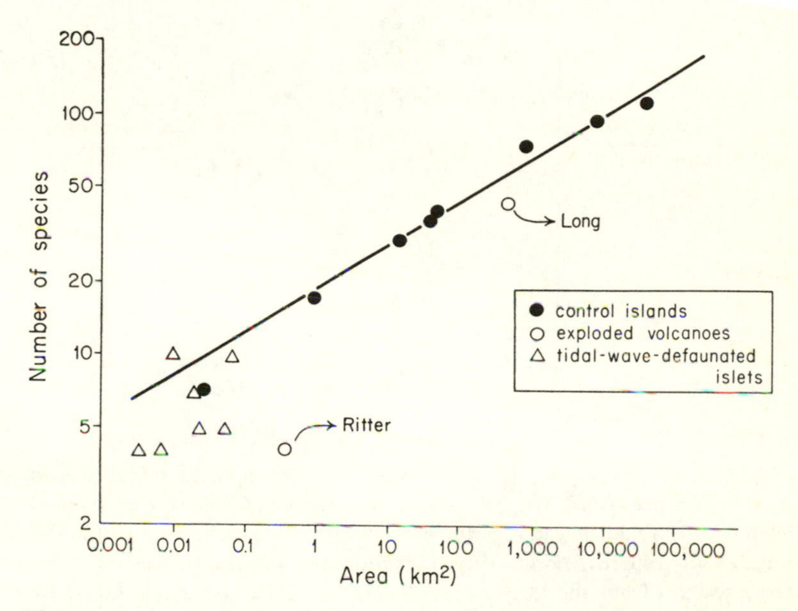

Fig. 9.1. An example of the relation between species number and island area in an archipelago: number of resident, nonmarine, lowland bird species S on islands in the Bismarck Archipelago, plotted as a function of island area on a double logarithmic scale. The solid circles represent relatively undisturbed islands, and the straight line $S = 18{\cdot}9\,A^{0{\cdot}18}$ was fitted by least-mean-squares through the points for the seven largest islands. The open circles refer to the exploded volcanoes, Long and Ritter, where species number is still below equilibrium, especially on Ritter, because of incomplete regeneration of vegetation. The open triangles refer to coral islets inundated by the Ritter tidal wave in 1888. (From Diamond, 1974).

Table 9.1 gives a list of the values of z in the S–A relation (9.1) for groups of plants and animals in various parts of the world. Figure

9.1 illustrates the relation, for number of bird species on islands of the Bismarck Archipelago near New Guinea.

A rough rule, which summarizes the $S-A$ relation (9.1) with values of z in the range typically observed, is that a tenfold decrease in area corresponds to a halving of the equilibrium number of species present.

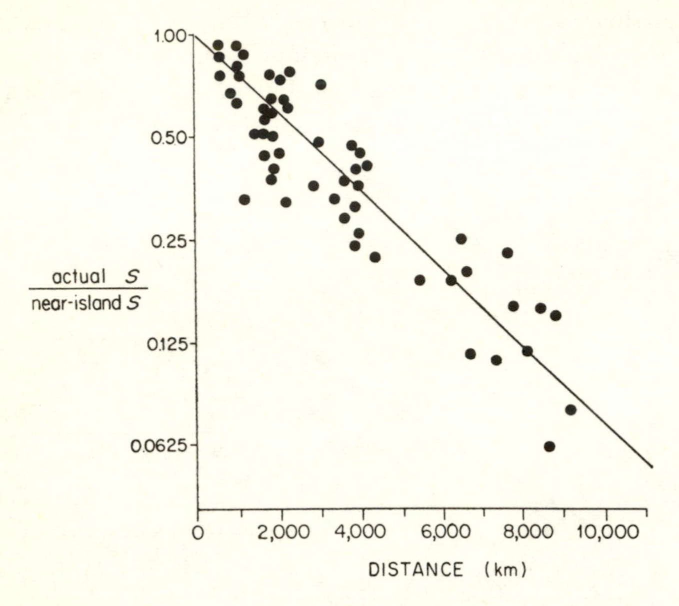

Fig. 9.2. An example of the relation between species number and island distance from the colonization source, for birds on tropical islands of the Southwest Pacific. The ordinate (logarithmic scale) is the number of resident, nonmarine, lowland bird species on islands more than 500 km from the larger source island of New Guinea, divided by the number of species on an island of equivalent area close to New Guinea. The abscissa is island distance from New Guinea. The approximately linear relation means that species number decreases exponentially with distance, by a factor of 2 per 2600 km. (From Diamond, 1974).

Thus, to answer one of the paradigmatic questions in the introduction, saving 1 per cent of the Amazonian rain forest might correspond, very roughly, to saving 25 per cent of the original species. Such relations are admittedly crude and neglectful of detail, but they provide an informed first guess at the relation between the area of a reserve and the number of species which are eventually likely to be preserved in it.

A further empirical rule is that, if one compares islands of similar area, S decreases with increasing distance D from the colonization source. Figure 9.2 illustrates this trend for birds on tropical Pacific islands colonized from New Guinea.

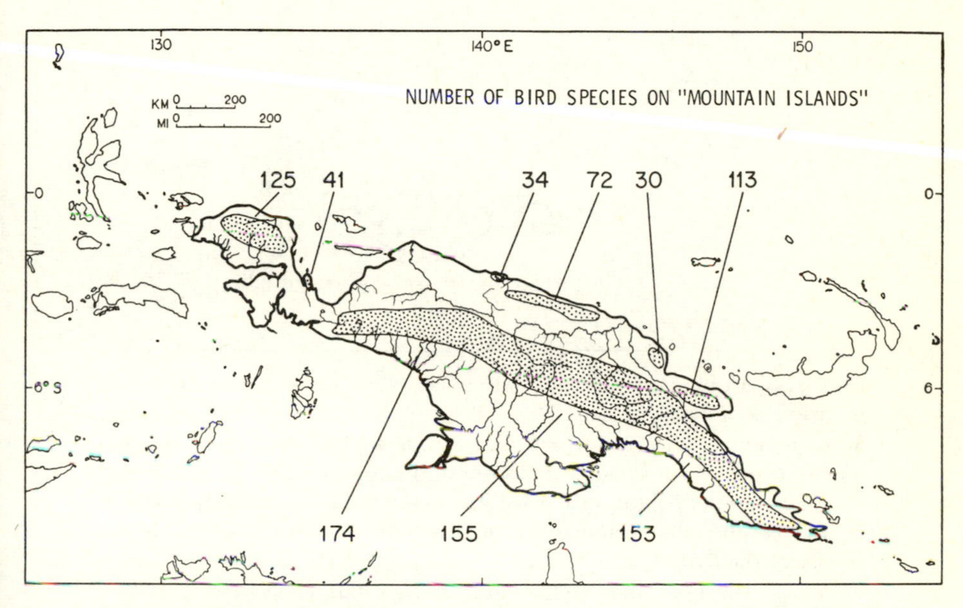

Fig. 9.3. An example of the relation between species number and area of a 'habitat island'. New Guinea consists of a large central mountain range plus about six smaller mountain ranges along the north coast (dotted areas), separated from each other by a 'sea' of intervening lowlands. There are many New Guinea bird species that are confined to higher elevations in the mountains, and for which New Guinea itself therefore behaves as an island archipelago. Numbers on the map give the number of such montane bird species on each small range and at three different locations on the central range. Note that the larger ranges have more montane species. Most of the variation in S not correlated with variation in A is correlated with variation in altitude.

Dependence of S on A is also observed for 'habitat islands' within continents or other islands. Figure 9.3 and 9.4 are two from among the many illustrative examples that could be chosen; see also MacArthur and Wilson (1967), Vuilleumier (1970), Brown (1971), and Diamond (1975a).

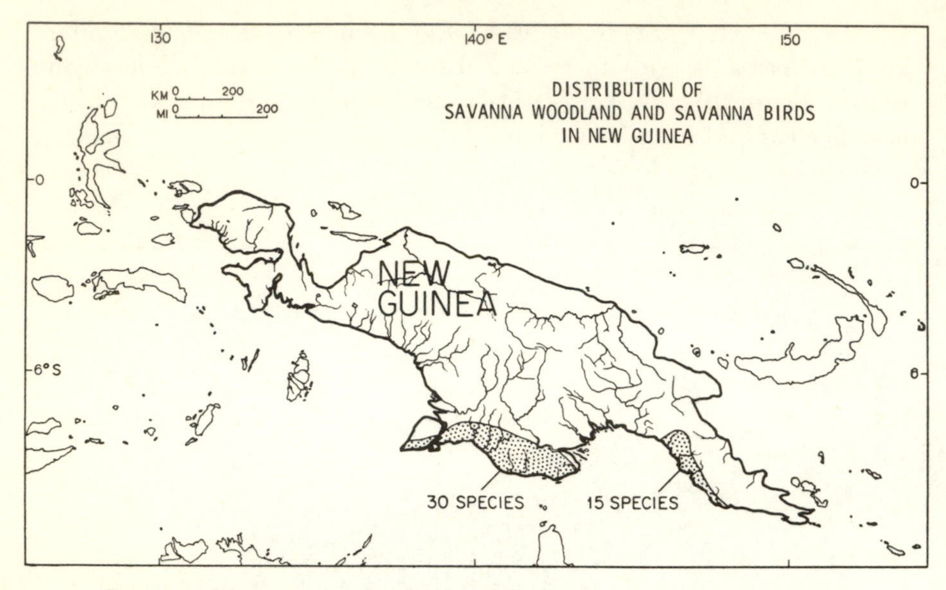

Fig. 9.4. An example of the relation between the area *and* degree of isolation of a habitat island, and the number of species it contains. Most of New Guinea is covered by rain forest, but two separate areas on the south coast (shaded in the figure) support savanna woodland. The characteristic bird species of these savannas are mostly derived from Australia (the northern tips of which are just visible at the lower border of the figure). The Trans-Fly savanna (left) not only has a larger area than the Port Moresby savanna (right), but is also closer to the colonization source of Australia. As a result, the Trans-Fly supports twice as many bird species characteristic of savanna woodland (about 30 compared with 15) as does the Port Moresby savanna. (From Diamond, 1975a).

9.2.2 *Theoretical Explanations*

The phenomenological eq. (9.1) is based on observational data, and is useful as such. This section attempts to provide the equation with a theoretical underpinning.

In the preceeding chapter, we saw that for any large and fairly heterogeneous assembly of species the distribution of relative abundance is likely to be of lognormal form. Going further, Preston (1962) observed that this lognormal is commonly of 'canonical' form, corresponding to the parameter γ (see chapter 8.4) having the special value $\gamma = 1$. In this event, there is a unique relation between the number of species, S, and the total number of individuals in the biota, N say. This was noted by Preston (1962) and by MacArthur and Wilson (1967), who both went on to add a second assumption, namely that there is an approximately linear relationship between the total number of individuals, N, and the island area, A. The underlying implicit biological assumption, that total density of individuals is independent of area (and of S), is not strictly valid, but the deviations do not greatly affect the predicted species-area relation. When put together with the assumption of a canonical lognormal species relative abundance (which relates S to N), the assumption relating N to A leads to a unique relationship between S and A. This mathematical relationship between S and A is complicated (May, 1975f, p. 112), but for large S ($S > 20$ or so) it is increasingly well approximated by eq. (9.1) with $z = 0 \cdot 25$ (May, 1975f; this analytic result is to be compared with the numerical curve-fitting previously employed by Preston, 1962, to get $z = 0 \cdot 262$ and by MacArthur and Wilson, 1967, to get $z = 0 \cdot 263$). At low values of S, the exact S–A relation obtained under the above assumptions exhibits a downturn, of just the kind exhibited by the data (solid circles and open triangles) in Fig. 9.1.

The details of these derivations depend on the rather mystical 'canonical' assumption that $\gamma = 1$. More generally, as we explained in section 8.4, we expect some lognormal distribution of species relative abundance, with the parameter γ in the neighbourhood of unity (say 0·6 to 1·7) for a wide range of values of S and N: see May (1975f, Fig. 4). When coupled with the assumption that N is proportional to A, this leads to a 1-parameter family of $S - A$ curves, depending on the explicit value of the parameter γ. As for the special 'canonical' case, these relations are well approximated by eq. (9.1) once S is relatively large, with z now a function of γ. Specifically, $z = 1/(4\gamma)$ for $\gamma > 1$ and $z = (1 + \gamma)^{-2}$ for $\gamma < 1$, so that the plausible range of variation of γ leads to eq. (9.1) with values of z in the range 0·39 to 0·15 (May, 1975f).

This provides a detail-independent explanation of the empirical eq. (9.1), based on statistical generalities along with the assumption that biomass is roughly proportional to area. However, as discussed in section 8.4, the statistical arguments which lead to the lognormal distribution and a z value around 0·25 are no more than plausible generalities, and there are many circumstances (e.g., early succession, disturbed habitat, etc.) where those arguments do not apply. Any attempt to explain the systematic differences in z-values among the groups of species listed in Table 9.1, or other fine details of S–A relations, will demand that more attention be paid to biological details.

One such more fundamental approach is to derive the S–A relation from an understanding of extinction and immigration rates (see below): this work has been initiated by Schoener (1976) and Gilpin and Diamond (1976). It emerges from such a treatment that z decreases with increasing immigration rates, as exemplified by the progressive decrease in z for birds of New Guinea islands, New Britain islands, Solomon islands, and the New Hebrides (Table 9.1).

9.3 Rates of approach to equilibrium

9.3.1 *Extinction and immigration*

Preston (1962) and MacArthur and Wilson (1963, 1967) have pointed out that the number of species on an island is set by a dynamic balance between immigration and extinction. For any particular island, the nett extinction rate will increase as the total number of species present increases; conversely, the nett rate at which new species are added—the immigration rate—will decrease as S increases. This situation is illustrated schematically in Fig. 9.5 (see also Figs. 9.6 and 9.7). The equilibrium number of species, S^*, is that at which extinction and immigration rates are equal.

This perception gives insight into the relation between an island's size and degree of isolation, and the equilibrium number of species on it. Species immigrate onto an island as a result of dispersal of colonists from continents or other islands: the more remote the island, the lower the immigration rate (i.e., the shallower the dashed curve in Figs. 9·5, 9·6 and 9·7). Species established on an island run the risk of extinction due to fluctuation in population numbers: the smaller the island, the smaller is the population and the higher the extinction rate (i.e., the more steeply rising the solid curve in Figs. 9.5 and 9.7). Area also affects immigration and extinction in several other ways: through its relation to the magnitude of spatial and temporal variability in resources; by being correlated with the variety of available habitats, as stressed by Lack (1973); and by being correlated with the number of 'hot spots', or sites of locally high utilisable resource production for a particular species (Diamond, 1975b). All in all, the larger and less isolated the island, the higher is the species number at which it should equilibrate.

To illustrate how these ideas can be elaborated in more quantitative fashion, we consider the unrealistically simple case where all species have the same, constant immigration rate, μ, and extinction rate, λ.

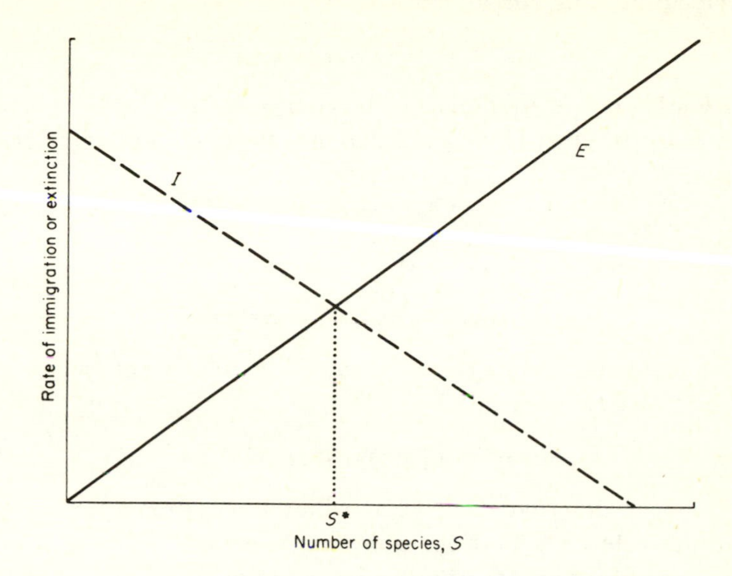

Fig. 9.5. Schematic illustration of an island's extinction curve E (solid line), and immigration curve I (dashed line), as functions of the number of species on the island, S. This figure is for the unlikely case when each species has the same extinction rate and immigration rate, so that the nett rates (expressed as number of species per unit time) are linear functions of S: see eqs. (9.2) and (9.3). The island equilibrium number of species is S^*, where the curves intersect.

Needless to say, this circumstance never prevails in the real world, but it provides a simple model which is useful for exposition; it may also be a sensible approximation to more general nonlinear models (e.g., Fig. 9.7 below) in the neighbourhood of equilibrium. If all species have the same extinction rate, λ, the nett extinction rate, E, expressed as number of species extinguished in unit time, is

$$E = \lambda S. \tag{9.2}$$

Similarly, the nett immigration rate, I, again expressed as number of species newly immigrating in unit time, is

$$I = \mu(S_T - S). \tag{9.3}$$

Here S_T is the total number of potential immigrants in the mainland pool, so that $(S_T - S)$ is the number of candidates for immigration to an island on which S species are already present. Equations (9.2) and (9.3) give the curves illustrated in Fig. 9.5. At equilibrium $E = I$, which gives the equilibrium species number, S^*, as

$$S^* = [\mu/(\lambda+\mu)]\, S_T. \tag{9.4}$$

If the island is not in equilibrium, the change in the number of species in unit time is given by the difference between immigration and extinction:

$$dS(t)/dt = I - E. \tag{9.5}$$

That is, from eqs. (9.2) and (9.3),

$$dS(t)/dt = \mu S_T - (\lambda+\mu)\, S(t). \tag{9.6}$$

Equation (9.6) may be integrated. If the initial number of species on the island is $S(0)$ at $t = 0$, then

$$S(t) = S^* + [S(0) - S^*]\, e^{-(\lambda+\mu)t}. \tag{9.7}$$

This expression describes the rate at which the system approaches the equilibrium value S^*. Setting aside the excessive simplicity of the underlying assumptions, this expression could be fitted to observational data to obtain estimates of λ, μ and S^*.

However, it is grossly unrealistic to assume that all species have the same λ and μ values, hence that extinction and immigration rates vary linearly with S [eqs. (9.2) and (9.3)]. There are two major reasons why nett extinction rates should be expected to increase faster than linearly with S, and likewise why nett immigration rates should fall faster than linearly as S increases: species differ greatly in their values of λ or of μ; and competition causes, for each species, μ to decrease and λ to increase with S. An extreme view, which is implicit in some of the writings of Lack (1973, 1976), is that there are a certain number S^* of species permanently resident on the island, and that any other immigrants will fail to breed; there are a particular S^* species that are ecologically appropriate to the island. This situation is illustrated in Fig. 9.6, which shows the corresponding ultimately steep extinction curve: the extinction rates for the S^* species which, as it were, 'belong' on the island are effectively zero; for any other species, the extinction rate is effectively infinite.

The real situation almost invariably lies between the extremes depicted in Figs. 9.5 and 9.6. In general, as more and more species are packed in, the island approaches ecological saturation, and the

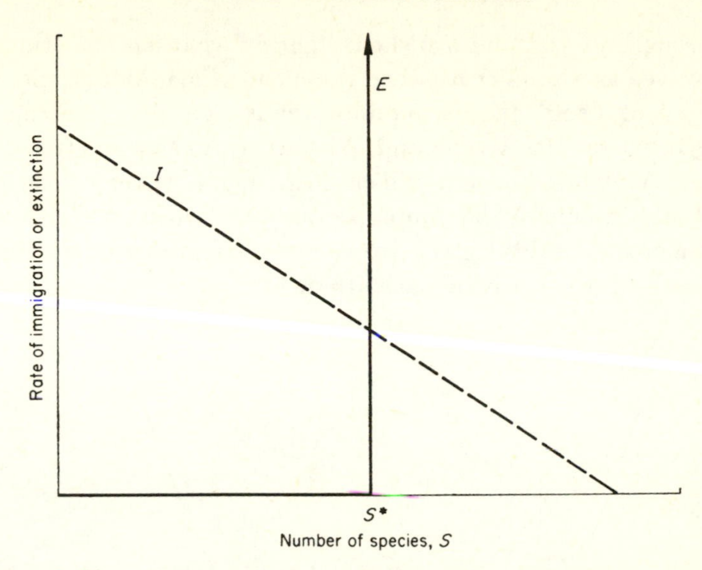

Fig. 9.6. Extinction (solid line) and immigration (dashed line) curves for an island, as functions of S, as in Fig. 9.5. The difference is that here there are S^* species which 'belong' on the island, having effectively zero extinction rate, and *no* other species are capable of breeding on the island. This results in an infinitely steep extinction curve E, as illustrated [i.e., eq. (9.8) in the limit $n \to \infty$].

overall extinction curve, E, steepens rapidly. This may be represented by writing

$$E(S) = \epsilon(S/S_0)^n, \tag{9.8}$$

where n is now a parameter which describes the steepness of the curve. The simple case of eq. (9.2), constant extinction rate for species, corresponds to the limit $n = 1$; the opposite extreme of Fig. 9.6 corresponds to the limit $n \to \infty$ (whence $E \to 0$ for $S < S_0$, $E \to \infty$ for $S > S_0$). Observational data for birds on real islands suggests values of n around 3 or 4 (Gilpin and Diamond, 1976; Schoener, 1976; Diamond and Jones, 1976; see also Terborgh's, 1974, work below). Analogous expressions may be used to parameterize the nonlinear behaviour of most curves for nett immigration, $I(S)$, which for birds prove to be even steeper than the extinction curves ($n \sim 6$: Gilpin and Diamond, 1976). Figure 9.7 illustrates such extinction and immigration curves, for lowland bird species on a typical island in the Solomon Archipelago, in the tropical Southwest Pacific (Gilpin and Diamond, 1976).

The comparison between the realistic Fig. 9.7 and the idealized extremes of Figs. 9.5 and 9.6 sheds light on what has sometimes been misperceived as a conflict between the views of MacArthur and Wilson and those of Lack. Species equilibrium is indeed a dynamic thing (as suggested by MacArthur and Wilson). However, as stressed from the start by MacArthur and Wilson (e.g., Fig. 4 of their 1963 paper), the actual extinction and immigration rates which describe species turnover are, in reality, given by curves that are closer in character to Fig. 9.6 (in the spirit of Lack) than to Fig. 9.5.

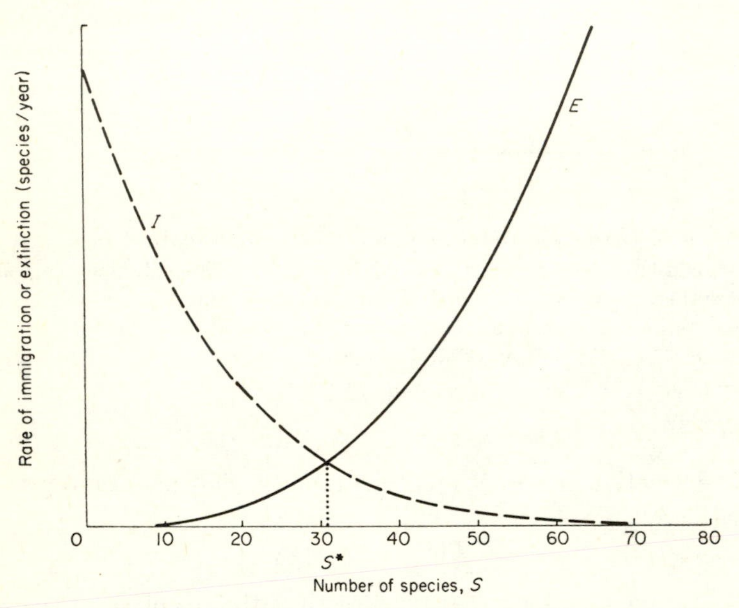

Fig. 9.7. The actual extinction (solid line) and immigration (dashed line) curves for the avifauna on Three Sisters, one of the smaller Solomon islands: the rates are plotted as relative number of species to go extinct, or to immigrate, per year as a function of the number of species on the island. For this island, the extinction curve is of the form of eq. (9.8) with $n = 2 \cdot 75$, and the equilibrium number of species is $S^* = 32$. For further details, see Gilpin and Diamond (1976). These actual curves are to be compared with the idealized extremes depicted in Figs. 9.5 and 9.6.

We now confront these ideas with data on the way the number of species approaches the island equilibrium value from below ($S(0) < S^*$), in the neighbourhood of equilibrium ($S(0) \simeq S^*$), and from above ($S(0) > S^*$).

9.3.2 *Species number increasing toward equilibrium*

One type of study involves observing the increase in species number on islands where the flora and/or fauna have been removed either by natural catastrophe or by experimental manipulation. In these situations, the immigration term I is of predominant importance in eq. (9.5), at least in the initial stages.

The most famous such study was provided by a 'natural experiment', the recolonization of the volcanic island of Krakatoa after its biota had been destroyed by an eruption in 1883 (Docters van Leeuwen, 1936; Dammerman, 1948; see MacArthur and Wilson, 1967, pp. 43–51). Here the number of bird species returned in a relatively short time to the value appropriate to the island's area and isolation, whereas the number of plant species is still rising; the rate of approach to equilibrium obviously depends on the plant or animal group under consideration. Similar 'natural experiments' are provided by the birds of Ritter and Long Islands near New Guinea, whose faunas were destroyed by volcanic eruptions in 1888 and about two centuries ago, respectively (see Fig. 9.1); by the birds of seven coral islets in the same area, where the tidal wave following the Ritter eruption destroyed the fauna in 1888 (Diamond, 1974); and by the initial colonization of a newly created volcanic island such as Surtsey, off Iceland.

As discussed in section 8.3, Simberloff and Wilson (1969) created an analogous 'artificial experiment' by fumigating several tiny mangrove islets off the coast of Florida and observing the recolonization by arthropods.

9.3.3 *Species number around equilibrium*

Another type of test is provided by turnover studies at equilibrium. According to the above interpretation, although the *number* of species on an island may remain near an equilibrium value, the *identities* of the species need not remain constant, because new species are continually immigrating and other species are going extinct. Estimates of immigration and extinction rates at equilibrium for birds have been obtained by comparing surveys of an island in separate years; among such studies are those for the birds of the Channel Islands off California (Diamond, 1969; Hunt and Hunt, 1974; Jones and Diamond, 1976), Karkar Island off New Guinea (Diamond, 1971), islands off the coast of Britain, and Mona Island off Puerto Rico (Terborgh and Faaborg, 1973). Two practical problems in turnover studies are: (a) to estimate man's effect on natural turnover rates (see Jones and Diamond, 1976); and (b) to correct for the precipitous decline in apparent turnover rate with increasing interval between censuses, as a result of extinctions and reimmigrations in the interval remaining unnoticed and cancelling each other's effect (Diamond and May, 1976).

All these turnover studies found that a certain number of species present in the earlier survey had disappeared by the time of the later survey, but that a similar number of other species immigrated in the intervening years, so that the total number of species remained approximately constant unless there was a major habitat disturbance. As expected from considering the risk of extinction in relation to population size, most of the populations that disappeared had initially consisted of few individuals. The turnover rates per year (extinction and immigration rates) observed in these bird studies were in the range of 0·2 to 20 per cent of the island's bird species for islands of area ranging from 400 to 0·4 km^2.

9.3.4 *Species number decreasing towards equilibrium*

The situation of greatest relevance to floral and faunal conservation arises when some fraction of a habitat is set aside as a reserve, and the rest destroyed. Such a reserve will at first be 'supersaturated', containing more species than are appropriate to its area at equilibrium. The situation is the exact converse of an island which has had its biota destroyed (section 9.3.2): equilibrium of species number will be approached from above, and the extinction term E will be of predominant importance in eq. (9.5), at least in the early stages.

A natural experiment of this kind is provided by so-called land-bridge islands. During the most recent ice age, which lasted for an extended period of thousands of years and ended about 10,000 years ago, enough water was locked up in glaciers to make the ocean levels about 100 m. lower than at present. Consequently islands that are now separated from continents or larger islands by water less than 100 m. deep were once attached, and shared the continental biota. Examples of such land-bridge islands are Britain off Europe, Aru and other islands off New Guinea, Trinidad off South America, Fernando Po off Africa, and Borneo and Japan off Asia. Subsequent to these islands being created by rising sea levels about 10,000 years ago, their continental complement of species has slowly decreased towards the equilibrium value appropriate to an island of their modern area.

Terborgh (1974) has made a quantitative such study for birds on five neotropical land-bridge islands. The number of bird species currently on each island is known, and the number present on each island before it was cut off 10,000 years ago may be estimated from the neighbouring mainland species numbers. Terborgh notes that in this relaxation process extinction is predominant [so that I is neglected in eq. (9.5)], and he chooses to describe the nett extinction rate by eq. (9.8) with $n = 2$; thus he fits the data to

$$dS/dt = kS^2. \tag{9.9}$$

The extinction parameter k thus deduced is shown, as a function of island area, in Fig. 9.8. Similar results, and in particular the tendency for extinction rates to decrease as island area increases, have been obtained by Diamond (1972, 1973) for land-bridge islands off New Guinea and elsewhere in the southwest Pacific. Brown (1971) has made an analogous study of the distribution of small mammals in forests which are now isolated on mountaintop 'islands' rising out of the 'sea' of desert in western North America, but which were connected by a continuous forest belt or 'habitat land-bridge' during times of cooler Pleistocene climates (see Table 9.2).

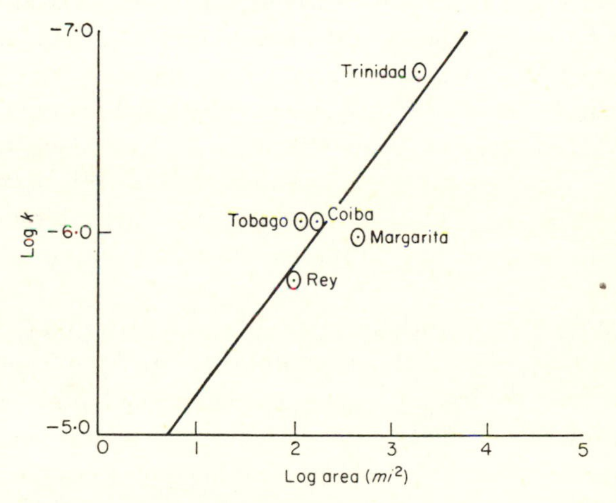

Fig. 9.8. The extinction parameter k of eq. (9.9) as a function of island area, for birds on land-bridge islands in the West Indies. (From Terborgh, 1974).

Terborgh (1974) has made a dramatic application of the calculations summarized in Fig. 9.8, by showing that they correctly predict the extinction rates observed within the present century on Barro Colorado Island. This island was created from a former hilltop by the flooding of adjacent valleys to create Lake Gatun when the Panama Canal was constructed, and since 1923 it has been carefully protected as a wildlife preserve. A meticulous account of the bird species present on the island over the past 50 years has recently been published by Willis (1974). From the k-versus-area patterns for post-Pleistocene relaxation rates on neotropical land-bridge islands, Fig. 9.8, Terborgh makes an estimate of the k value for Barro Colorado Island. Equation (9.9) can then be used to get an estimate of the decline in the island's number of bird species since 1923. This theoretical estimate of 16–17 forest bird species lost is gratifyingly in accord with the actual number of 15.

9.4 Which species survive?

Up to this point the problem of survival has been discussed in statistical terms: what fraction of its initial biota will a reserve eventually save, and how rapidly will the remainder go extinct? We now go on to consider the survival probabilities of individual species.

These considerations bear directly on conservation strategies. If each species had a roughly equal probability of survival, then large numbers of small reserves could be a satisfactory policy: each vest-pocket reserve would lose most of its species before reaching equilibrium, but with enough reserves any given species would be likely to be among the survivors in at least one reserve. The flaw in this strategy is that different species usually have very different area requirements for survival, arising from very different rates of immigration and of extinction.

Consider first the question of immigration. Even if there are many small reserves, a species that is incapable of dispersing from one reserve to another across the intervening sea of unsuitable habitat is doomed to eventual extinction: its lights will wink out, one by one, with no chance of reignition. Conversely, a species capable of dispersing from one reserve to another may persist by virtue of a dynamic balance between local extinction and reimmigration, provided recolonization rates are high enough or extinction rates low enough. Dispersal ability obviously differs enormously among plant and animal species. Flying animals tend to disperse better than non-flying ones; plants with wind-borne seeds tend to disperse better than plants with heavy nuts. The more sedentary the species, the more irrevocable is any local extinction, and the more difficult will it be to devise a successful conservation strategy. These conservation problems will be most acute for the sort of slowly dispersing species found in normally stable habitats, such as tropical rain forest. Even the power of flight cannot be assumed to guarantee high dispersal ability. For example, 134 of the 325 lowland bird species of New Guinea are absent from all oceanic islands more than a few km from New Guinea, and are confined to New Guinea plus associated land-bridge islands. Similarly, many neotropical bird families with dozens of species have not one representative on a single New World island lacking a recent land-bridge to the mainland; and not a single member of many large Asian bird families has been able to cross Wallace's line separating the Sunda Shelf land-bridge islands from the oceanic islands of Indonesia. Such bird species are generally characteristic of stable forest habitats, and have insuperable psychological barriers to crossing water gaps. In short, low recolonization rates may mean either that a species cannot, or that it will not, cross unsuitable habitats.

Given this variation in ability to recolonize, we turn to consider how species vary in extinction rates of local populations. The New Guinea land-bridge islands alluded to in section 9.3.4 offer a convenient test situation (Diamond, 1972, 1975a). As mentioned in the preceding paragraph, there are 134 New Guinea lowland bird species that do not cross water gaps, and consequently post-Pleistocene extinction of these species on land-bridge islands cannot have been reversed by recolonization. Because extinction rates on relatively small islands are high (cf. Fig. 9.8), virtually all these species are now absent from all land-bridge islands smaller than 50 km^2. On the other hand, these 134 species vary greatly in their distribution on the seven larger (450–8000 km^2) land-bridge islands. At one extreme, some species (e.g., the frilled monarch flycatcher, *Monarcha telescophthalmus*) have survived on all seven islands; at the other extreme, 32 species have disappeared from all seven islands. Most of these 32 extinction-prone species fit into one or more of three categories: birds whose initial populations must have numbered few individuals because of very large territory requirements (e.g., the New Guinea Harpy Eagle, *Harpyopsis novaeguineae*); birds with small initial populations because of specialized habitat requirements (e.g., the swamp rail, *Megacrex inepta*); and birds which are dependent on seasonal or patchy food sources, and which normally go through drastic population fluctuations (e.g., fruit-eaters and

Table 9.2. Mammals on mountain islands of the Great Basin.

Number of islands per species	Species	Habitat and diet	Weight (g)	Ruby	Toiyabe	Toquima	White-Inyo	Snake	Schell Creek	Deep Creek	White Pine	Oquirrh	Roberts Creek	Diamond	Stansbury	Grant	Spruce	Spring	Pilot	Panamint
14	*Neotoma cinerea*	GH	317	×	×	×	×	×	×	×	×	—	—	×	×	—	×	×	×	×
14	*Eutamias umbrinus*	GH	57	×	×	×	×	×	×	×	×	×	×	×	—	×	×	×	—	—
13	*Spermophilus lateralis*	GH	147	×	×	×	×	×	×	×	×	—	—	×	—	×	×	×	×	—
12	*Microtus longicaudus*	GH	47	×	×	×	×	×	×	×	×	×	×	—	×	×	—	—	—	—
9	*Marmota flaviventer*	GH	3,000	×	×	×	×	×	×	×	×	—	—	—	×	—	—	—	—	—
8	*Thamomys talpoides*	GH	102	×	×	×	×	—	×	—	×	×	—	×	—	—	—	—	—	—
6	*Sorex vagrans*	C	6·7	×	×	—	—	×	×	×	—	×	—	—	—	—	—	—	—	—
6	*Sorex palustris*	C	14	×	×	×	×	×	—	—	—	—	×	—	—	—	—	—	—	—
4	*Zapus princeps*	SH	33	×	×	—	—	—	—	—	—	×	×	—	—	—	—	—	—	—
4	*Ochotona princeps*	SH	121	×	×	×	×	—	—	—	—	—	—	—	—	—	—	—	—	—
3	*Mustela erminea*	C	58	—	×	—	×	×	—	—	—	—	—	—	—	—	—	—	—	—
3	*Spermophilus beldingi*	SH	382	×	×	×	—	—	—	—	—	—	—	—	—	—	—	—	—	—
1	*Lepus townsendi*	SH	2,500	×	—	—	—	—	—	—	—	—	—	—	—	—	—	—	—	—
	Number of species per island			12	12	9	9	8	7	6	6	5	4	4	3	3	3	3	2	1

flower-feeders). These observations tend to be confirmed by other bird studies, and by the work of Brown (1971), referred to above, on differential extinction rates among mammal species isolated on mountain tops (Table 9.2).

Another natural experiment in differential extinction is provided by New Hanover, a 1200 km² island which in the late Pleistocene was connected by a land-bridge to the larger island of New Ireland in the Bismarck Archipelago. Although today New Hanover has only lost about 22 per cent of New Ireland's species, among these lost species are 19 of the 26 New Ireland species confined to the larger Bismarck islands, including every endemic Bismarck species in this category. That is, those species most in need of protection were differentially lost: as a faunal reserve, New Hanover would rate as a disaster. Yet its area of 1200 km² is not small by the standards of many of the tropical rain forest parks that one can realistically hope for.

The survival prospects for a particular species may be quantified by determining its 'incidence function', $J(S)$ (Diamond, 1975b). We have noted, at least for birds and mammals, that some species occur only on

Footnote to Table on facing page.

Out of the Great Basin desert of the western United States rise 17 mountain ranges to elevations above 10,000 ft. Boreal habitats on the summits of these 'mountain islands' are now isolated from each other by the surrounding sea of desert, but were connected to each other and to the source boreal habitats of the much more extensive Rocky Mountains and Sierra Nevada during cooler Pleistocene periods. On these mountaintops live 13 species or superspecies of small flightless mammals, which cannot cross the intervening desert today but which reached the mountain islands over Pleistocene bridges of boreal habitat. Since the Pleistocene these mammal populations have been subject to risk of extinction without opportunity for recolonization. The table illustrates the regular distributional pattern produced by this differential extinction. The presence (×) or absence (–) of each named species is indicated for each named mountain island, along with the total number of islands inhabited by each species (left-most column) and total number of species inhabiting each island (bottom row; this number correlates well with island area). C = carnivore; GH = generalized herbivore (present in most habitats); SH = specialized herbivore (present in only certain habitats); body weight is given in grams. Herbivores can maintain higher population densities than carnivores, small animals higher densities than large animals, and species of generalized habitat preference higher densities than habitat specialists. Thus, the main patterns of differential extinction emerging from the table are that: rare species survive on fewer islands than abundant species; rare species become confined to the larger islands; and different small islands tend to end up with the same group of abundant species. From Brown (1971).

the largest and most species-rich islands, other species occur also on medium-sized islands, and others also occur on small islands. These patterns may be displayed graphically by grouping the islands into classes containing similar numbers of species (e.g., 1–4, 5–9, 10–20, 21–35, 36–50, etc.), calculating the incidence J or fraction of the islands in a given class on which a particular species occurs, and plotting J against the total species number S on the island. Some such incidence functions for birds in the Bismarck Archipelago are shown in Fig. 9.9; a much more full discussion is in Diamond (1975b). Since S in the Bismarcks is closely correlated with area, these graphs in effect represent the probability that a species will occur on an island of a particular size.

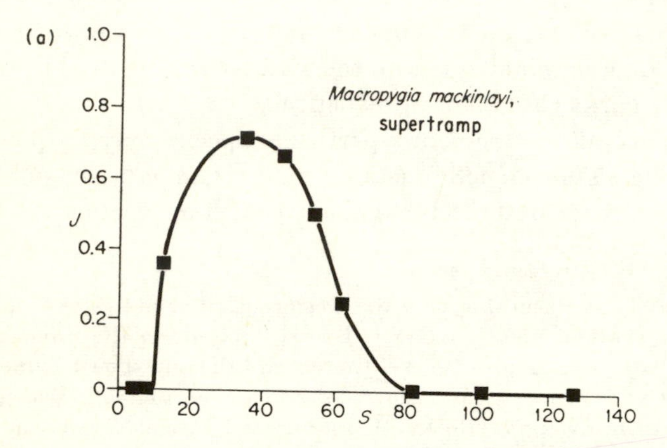

Fig. 9.9a. The incidence function $J(S)$ for the 'supertramp' pigeon *Macropygia mackinlayi*, in the Bismarck Archipelago. The incidence J represents the fraction of those islands, with S values in a given small range, on which the species occurs, as a function of the number of species S on the island. Each point is typically based on 3–13 islands. For a more detailed discussion, see Diamond (1975b).

For most species, J goes to zero for S values below some value characteristic of the particular species, meaning there is no chance of long-term survival on islands below a certain size. These incidence functions can be interpreted in terms of the 'bionomic strategy' of the species, along the lines of chapter 3. Figure 9.9a is typical of species in the so-called 'super-tramp' category (Diamond, 1974, 1975b); as the figure shows, such birds tend to be found on small islands, but *not* on large ones (because of competitive exclusion from species-rich faunas). Figure 9.9b illustrates the incidence function for an intermediate category of species, while Fig. 9.9c is typical of species found only in the presence of many others (i.e., on large islands). In general terms, one may recognize Figs. 9.9a–c as depicting the spectrum from an extreme r-strategy through to an extreme K-strategy (see chapters 2 and 3); the J-versus-S curves, however, go further to give quantitative substance to these notions.

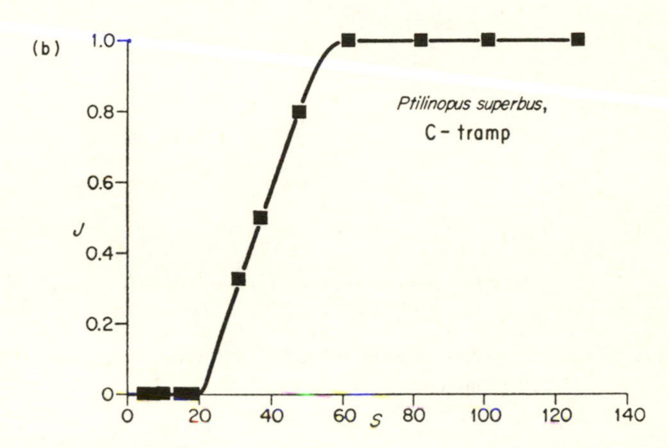

Fig. 9.9b. The incidence function for the so-called 'C-tramp' pigeon *Ptilinopus superbus*, in the Bismarcks. (From Diamond, 1975b).

'Assembly rules' furnish additional information as to which species are likely to be present on a given island. These rules (Diamond, 1975b) codify the observation that, of all possible combinations that could be formed from within a group of related species, only certain combinations exist in nature. The rules are evocative of those of the early days of atomic spectroscopy, with its empirical catalogues of allowed and forbidden spectral lines. Thus there are some pairs or groups of species that never coexist; others that form an unstable combination by themselves, but may form part of a stable larger combination; still others constitute stable combinations that resist invaders that would transform them into a forbidden combination; and so on. For the Bismarck avifauna, a detailed account of the assembly rules for such guilds as the cuckoo-dove, the gleaning flycatcher, the myzomelid-sunbird, and the fruit-pigeon guild has been given by Diamond (1975b, pp. 393–411). Faaborg (1976) has made similar detailed studies of the incidence and assembly of the birds on West Indian islands. In some cases these rules clearly derive from competition for resources, and from harvesting of resources by permitted combinations so as to minimize the unutilized resources available to support potential invaders; other cases are less transparent.

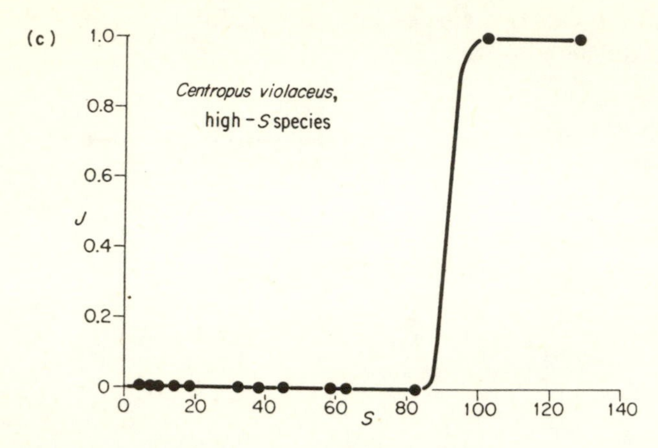

Fig. 9.9c. Incidence function for the 'high-S' cuckoo *Centropus violaceus*, in the Bismarcks. (From Diamond, 1975b).

These detailed observations as to the extinction, immigration and overall incidence $J(S)$ of particular species of course provide the underlying explanation for the patterns of curvature in extinction and immigration functions, as illustrated in Fig. 9.7. Ultimately, the incidence functions and assembly rules are presumably determined by considerations of limits to similarity and niche overlap (see chapters 4, 7) and by the constraints imposed by food web structure (see section 8.3, and in particular the study by Heatwole and Levins, Table 8.3).

The sort of quantitative information that is embodied in incidence functions and assembly rules holds the promise of providing the basis for a predictive science of environmental management.

9.5 Design principles for natural reserves

The work described in this chapter clearly is relevant to questions about the properties of natural reserves: how is the eventual number of species in the reserve related to its area; how do extinction rates vary with area; how do area-dependent survival probabilities vary among species? The answers to these questions prompt the enunciation of certain 'general design principles' for floral and faunal preserves (see, e.g., Willis, 1974; Wilson and Willis, 1975; Diamond, 1975a and 1976; Terborgh, 1974; May, 1975g). As for most generalities, such design principles are useful, but must be applied with caution in any one specific instance. They are no substitute for a painstaking study of extinction and immigration rates, of incidence functions and assembly rules, for each particular conservation project.

In general, a large reserve is better than a small reserve for two reasons: the large reserve can hold more species at equilibrium, and it will have lower extinction rates. By extension of this principle, one large reserve is usually better than several smaller reserves whose areas add up to the same total as the single large reserve. Many species, especially those of tropical forests, are stopped by narrow dispersal barriers: for these species, the effective area of a park may be halved if it is bisected by a major road or a power line, with a consequent reduction in the number of species present once the new equilibrium is reached.

If one must settle for several smaller parks, a way to raise the equilibrium number of species in any one such park is to raise the immigration rate into it. This can be done by careful juxtaposition of the scattered parks, and by providing corridors or stepping-stones of natural habitat between them.

Any given reserve should be as nearly circular in shape as other circumstances allow. Such maximization of the area-to-perimeter ratio minimizes dispersal distances within the reserve, and avoids 'peninsular effects' whereby dispersal rates to outlying parts of the reserve from more central parts may be so low as to perpetuate local extinctions, thus diminishing the reserve's effective area.

These general design principles are subject to many qualifications and equivocations. *First*, several smaller reserves may have the compensating advantage that in an inhomogeneous region each reserve may favour the survival of a different group of species. Even in a homogeneous region, separate reserves may save more species of a set of vicariant similar species, one of which would ultimately exclude the others from a single reserve. This point has been elaborated in detail by Simberloff and Abele (1976). *Second*, the above principles ignore epidemiological aspects of park management: many scattered parks are less susceptible to the ravages of an epidemic disease or analogous disaster. *Third*, there is the obvious point that some 'edge' species, that thrive at the interface between habitats, will prefer several smaller parks, or parks with high perimeter-to-area ratios; conversely, edge-intolerant species will be differentially worse off with several smaller areas, and will be unable to survive once the reserves become too small. (However, many edge species will fare well in non-park habitats transformed or dissected by man, while edge-intolerant species will furnish a disproportionate number of the species most dependent on parks for survival.) *Fourth*, one must consider the dynamical features such as the spatial and temporal patterns of stable oscillation that so frequently characterize populations both in mathematical models and in the real world. The evidence for a roughly 50-year cycle in elephant population eruptions in the area that is today Tsavo National Park in Kenya suggests that such dynamical aspects of natural populations can create management problems when even a very large area is enclosed as reserve.

All these questions need to be illuminated by further research, particularly field work on organisms other than birds. There is clear and present need for the development of techniques to estimate the size and other properties a reserve should possess if it is to fulfil its designated conservation purpose.

10
Succession

HENRY S. HORN

10.1 Introduction

Plant succession has often been idealized as a process whereby a community that has suffered an episodic devastation slowly regenerates a semblance of its former self in the absence of disturbance. These regenerative changes in the community were interpreted by Clements (1916) as emergent properties analogous to the recovery of an organism from injury, and the analysis of properties of the whole community has been continued (Margalef, 1968; Odum, 1969). A recent trend views succession in the context of the adaptations of individual species, independent of any transcendent properties of the whole community (Connell, 1972; Drury and Nisbet, 1971, 1973; Horn, 1974). Tradition attributes the origin of this individualistic attitude to Gleason (1926), but Thoreau (1860) talked about it much earlier as though it were already commonplace.

Over the past 10 years, several workers have converged on representing succession as a Markovian replacement process. A table is set up to show the probability that a given plant will be replaced in a specified time by another of its kind or by another species. With the aid of this table, the current composition of a community can be 'projected' into the future. Markovian models can be applied not only at the level of plant-by-plant replacements (Anderson, 1966; Horn, 1975), but also to transitions among trees of different sizes in a growing stand (Usher, 1966; Leak, 1970; Moser, 1972), local groups of trees (Botkin *et al.*, 1972), woodlots of varying composition (Stephens and Waggoner, 1970), groups of species (Williams *et al.*, 1969), regional forests (Shugart *et al.*, 1973), and even vegetation whose composition is only defined qualitatively (MacArthur, 1961).

An important assumption of most of these models is that the table of replacement probabilities does not change as time passes. Under this assumption, the composition of the modelled community approaches a particular steady state. This state is dependent on the configuration of the table of replacement probabilities, and thus on the biological interactions within and among species, but it is independent of the community's initial composition, and hence independent of the historical vagaries of the devastation that started the succession. Such steady states are notoriously difficult to establish in nature. When very old and 'stable' forests are examined closely, there is always evidence for a history of episodic disturbance. Many trees may be of a particular age, dating from a past devastation by wind, water, landslide, or fire (Maissurow, 1941; Jones, 1945; Loucks, 1970; Heinselman, 1973). The forest may preserve evidence of a sequence of fires (Buell *et al.*, 1954; Houston, 1973) or of hurricanes (Stephens, 1955; Henry and Swan, 1974). Even in the absence of devastation, trees may occur in evenly aged groups dating from episodes that favoured seedling establishment, either climatically (Cooper, 1960) or due to the absence of predators (Peterkin and Tubbs, 1965). Where succession is fast enough to be observed directly, as among herbs or the encrusting beasties of rocky sea shores, the succession may be dominated by chronic predatory devastation (Harper, 1969; Paine, 1974), and alternative quasi-stable communities may develop, whose structure can only be explained by reference to specific historical events (Sutherland, 1974).

The ideal of convergent and undisturbed reestablishment of a stable community is not general in nature, if indeed it exists. Hence there have been periodic controversies over whether succession should imply anything more than the historical description of individual examples (unusually lucid reviews are: Gleason, 1927; Whittaker, 1957; and Drury and Nisbet, 1971, 1973).

In this chapter I analyze the basic linear model of succession to show how chronic disturbances, interspecific competition, and local regeneration affect the rate of convergence of succession. This analysis discloses some of the obstacles to previous theories of succession. The effect of several biologically interesting non-linearities is discussed intuitively, but nonetheless rigorously. In particular, if the presence of vigorous young plants is strongly dependent on nearness to mature plants, or on a highly seasonal rain of seeds, then succession may not be convergent at all. If a community is devastated on a sufficiently regular schedule, its species will be those whose lifespans match the length of the less eventful interludes. These species may then evolve increased sensitivity to devastation late in life, and thereby perpetuate the environmental cycle to which they are adapted.

Although many of the ideas in this chapter arose from a mathematical analysis, I have found that they can be supported by direct appeal to biological assumptions without the intervening mathematical machinery. My examples betray my fondness for trees and forests, and they are illustrative rather than critical or exhaustive. However, the ideas are equally applicable to any communities of sessile plants or sessile animals.

10.2 Forest succession as a tree-by-tree replacement process

By making an accurate map of every tree in a forest and then mapping the same area 50 years later, one can estimate the probability that a tree of each species will live for 50 years, and for each tree that dies, the probability that it will be replaced by another of its own species or by another species. This is a generalization of the familiar process whereby a population's demographic characteristics are summarized in a 'life table' of survivorship and fecundity at each age (see chapters 2, 3). Namkoong and Roberds (1974) have analyzed such a life table to support recommendations for management of Californian coastal redwood forests. Table 10.1 shows probabilities of replacement, calculated in a roundabout way, for Gray Birch, Blackgum, Red Maple, and Beech, four species that can successively dominate abandoned farmland near Princeton, New Jersey (Horn, 1975a).

From Table 10.1 and the current composition of the forest, it is easy to calculate the expected composition of the forest 50 years hence. For example:

$$RM_{50\text{ yrs hence}} = 0\cdot 5GB_{now} + 0\cdot 25BG_{now} + 0\cdot 55RM_{now} + 0\cdot 03BE_{now}. \quad (10.1)$$

Analogous equations can be written for each species, combining the current composition of the forest with the appropriate column of Table 10.1. Then the composition of the forest can be predicted 100 years hence by combining the table according to the same recipe with the calculated composition 50 years hence; and so on for 150 years, 200 years and more. The recipe consists of the rules for multiplication of a row vector (the current composition of the forest) by a matrix (the table of tree-by-tree replacement probabilities). Once the composition is projected far enough into the future, further multiplications by the successional matrix will leave the species' abundances unchanged.

Table 10.1. Fifty year tree-by-tree transition matrix for Gray Birch, Blackgum, Red Maple and Beech.

Now \ 50 years hence	Gray Birch	Blackgum	Red Maple	Beech
Gray Birch	5+0	36	50	9
Blackgum	1	37+20	25	17
Red Maple	0	14	37+18	31
Beech	0	1	3	61+35

Diagonal is % of trees still standing plus % that have been replaced in 50 years by another of their own kind. Off-diagonal terms are % of trees replaced by another species in 50 years. The % of trees still standing was estimated by assuming that trees die at a constant rate such that 5% are left standing after 50 years for Gray Birch, 150 years for Blackgum and Red Maple, and 300 years for Beech. For those trees that die in 50 years, replacement of one species by another is assumed to be proportional to the number of saplings of the latter under a large sample of canopy trees of the former. The forests in which these data were gathered have many more species (Horn 1975b); so this matrix is only an illustrative caricature.

Table 10.2. Predicted composition of a succession.

Age of Forest (years)	0	50	100	150	200 ...	∞	Very Old Forest
Gray Birch	100	5	1	0	0	0	0
Blackgum	0	36	29	23	18	5	3
Red Maple	0	50	39	30	24	9	4
Beech	0	9	31	47	58	86	93

The succession starts with a field full of Gray Birch and then obeys the transitions specified by the matrix of Table 10.1. The stationary composition after an infinite amount of time is gotten by solving the set of equations in which the composition equals the composition multiplied by the matrix. The composition of the 'very old forest' is the proportion of trees of the four species in the canopy of an actual forest that may not have suffered an extensive disturbance for 350 years or more. Several species have been left out of the analysis for simplicity (See Horn 1975b for more details).

Table 10.2 shows an example of these calculations. For a succession starting with Gray Birch several predictions are made. Red Maple should dominate quickly and Beech should slowly increase to predominate later. Gray Birch should be lost rapidly. The forest should become stationary and predominantly Beech, but Blackgum and Red Maple should persist at low abundance, so that the stationary forest is a mosaic of successional patches. All of these predictions are consistent with what apparently happens in a real forest (Horn, 1975a). Similar tests have confirmed the predictive value of similar models for short-term changes in Wisconsin hardwood forests (Moser, 1972), stands of varying composition in New England (Stephens and Waggoner, 1970; Botkin *et al.*, 1972), and regional vegetation patterns in Wisconsin (Shugart *et al.*, 1973).

10.3 Linear models and their consequences

The fundamental property of a model like the preceding one is that it converges on a stationary composition after enough time. The proportional representation of species in this stationary composition is independent of the initial composition of the forest; it depends only on the matrix of replacement probabilities. Peden *et al.* (1973) have shown this for an elaborate computer simulation of forest growth, but it is true for a much wider class of models (section 10.4.2; Horn, 1975b). Hence there is no uniquely biological significance to the convergence itself. The biologically interesting questions involve the structure of the matrix of replacement probabilities. Three idealized structures are shown in Fig. 10.1, along with a schematic representation of the quasi-real case of Table 10.1. The idealized structures, chronic and patchy disturbance, obligatory succession and competitive hierarchy, differ markedly in their patterns and rates of convergence to the steady state.

10.3.1 *Chronic, patchy disturbance*

Where disturbances that result in the deaths of individual organisms and a few neighbours are frequent, localized, and uniformly distributed in time and in space, interactions between species take the form of a race for unchallenged dominance in recent openings, rather than direct competitive interference. An individual of any species may succumb to a given disturbance and its place may be taken by an individual of any other species. Therefore, if an episodic devastation changes the relative abundances to any two species, they are capable of adjusting their abundances to accord with the stationary distribution in as little time as a single generation. Where a community is dominated by chronic and patchy physical disturbance, succession should be rapid, convergent, and relatively independent of the historical peculiarities of the devastation that started it.

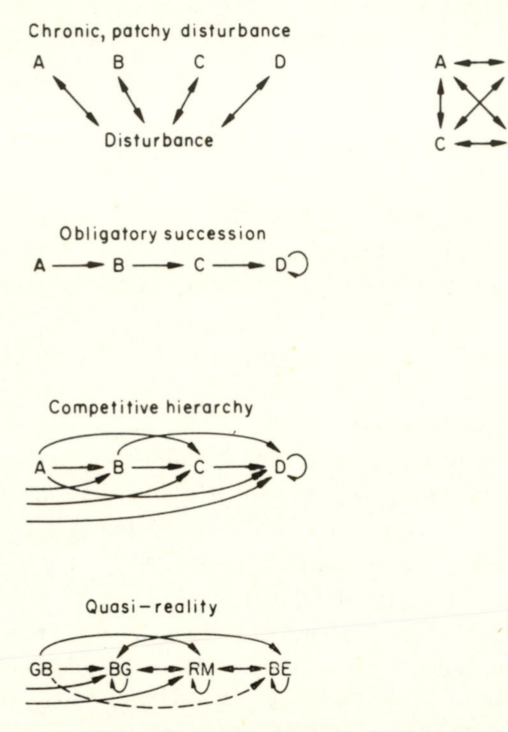

Fig. 10.1. A, B, C and D represent hypothetical species. Where disturbance is chronic and patchy, any species is likely to invade an opening that results from the death of any other species; all conceivable replacements are possible. An obligatory succession results if later species require preparation of their environment by earlier species. In an ideal competitive hierarchy each later species can outcompete earlier species, but can also invade in their absence. The quasi-realistic case illustrates the replacements of Table 10.1, with obvious abbreviations for Gray Birch, Blackgum, Red Maple and Beech.

I suspect that examples of this kind of succession will be found among flood plain forests of the temperate zone, but the most thoroughly studied example that I know is among invertebrates of the rocky intertidal, where some communities are shifting successional mosaics of patches that were battered at various times by gigantic driftwood (Dayton, 1971). Furthermore, Dayton (1975) has found that succession among intertidal algae reaches a stationary state more quickly in areas exposed to waves than in sheltered areas or deep tidal pools.

Patchy disturbance followed by invasion of any of several species makes an important contribution to the celebrated local diversity of tropical and sub-tropical forests (Knight, 1975). Some of this spatial diversity is due to patchy soil (Austin *et al.*, 1972), and some perhaps to deterministic micro-successions that have started at different times, but much can only be attributed to probabilistic accidents of just which species colonizes a recent opening at the appropriate time (Webb *et al.*, 1972). Other mechanisms are involved in the regeneration of these forests. Knight has shown that many of the species that are characteristic of older forests on Barro Colorado Island in Panama are also invaders of recently vacated fields (a 'competitive hierarchy' of section 10.3.3). Webb and friends have found several species whose regeneration is only effective in the immediate vicinity of mature trees of the same species (section 10.4.1). The relative importance of these mechanisms can only be discovered by future studies of the explicit dynamics of tropical forests.

10.3.2 *Obligatory succession*

The obligatory succession is characterized by the necessity of early successional species to pave the way for later species, a prime ingredient in Clements' (1916) theory of the cause of succession. Later species must await the dominance and demise of early species, and the convergence on a stationary state that includes the later species is exceedingly slow, even though it is certain. The classic example of this type of succession is the convergence of communities on sandy or boggy soil. Accumulation of organic matter from appropriately adapted plants increases the water retention of the sandy soil or increases the depth and drainage of peat, in either case modifying the soil in favour of plants with intermediate tolerance of aridity and dampness. Despite the logic of this textbook example, a careful look at the communities that are cited to support it shows that differences between communities are due largely to secular changes in the level of the water table (Olson, 1958; Heinselman, 1963; Raup, 1975). A major role of successional plants remains to be demonstrated critically, beyond a change in composition of only two species. A very clear and well-documented example is the mineral soil that is left by retreating glaciers, which gains fertility from nitrifying bacteria in root nodules of early successional alders (Crocker and Major, 1955). Other early successional plants have this effect and in addition insulate the soil, allowing permafrost to form, which in turn prevents drainage and allows the development of tundra tussocks of sphagnum and sedge (Viereck, 1966).

10.3.3 *Competitive Hierarchy*

In the case of the competitive hierarchy, late successional plants are increasingly capable of dominating eventually in crowded competition with early successional species, but the late successional species are also capable of invading the earliest stages of succession. This succession may or may not reflect the historical accidents of a particular devastation, depending on how early in the succession the later species invade. The succession can begin to converge on dominance by the later species as soon as they do invade. Indeed, careful studies of many north temperate forest successions have shown that late successional species are present almost as soon as early successional species invade a vacated field (Drury and Nisbet, 1973; Oliver, 1975; Heinselman, 1973). The early successional species grow quickly to dominate early, but they cannot survive crowding. Of course crowding also suppresses the later species, but they are at least able to endure to dominate later.

I have examined the mechanism behind a competitive hierarchy relating growth rate of trees to light intensity (Horn, 1971). Most leaves are able to photosynthesize at 90 per cent of their maximal rate with as little as 25 per cent of full sunlight. Therefore a tree with leaves scattered throughout its volume can have a very large total leaf area, with interior leaves that pay for themselves as long as the tree is in the open. When such a multilayered tree is in shade, however, the interior leaves may respire more than they photosynthesize. In shade, the optimal tree has a monolayer of leaves in a shell about its periphery, and all leaves intercept light at its highest intensity. A monolayered tree can grow in the sunlight, but not as quickly as a multilayered tree. If a monolayered and a multilayered tree simultaneously colonize an abandoned field or an opening in the forest, the multilayered tree will grow faster to dominate until it is crowded by its neighbours, when the monolayered tree gains a competitive advantage. If a monolayered tree reaches the opening first, it grows slowly, but its shade prevents invasion by multilayered trees. If a multilayered tree reaches the opening first, it grows quickly, but its understory is open to invasion by a monolayered tree. Although the monolayered tree is suppressed in the understory, it eventually dominates. As a result, the multilayered tree, like Diamond's 'supertramp' birds (chapter 9), must send its seeds flying to attain unchallenged dominance in newly vacant areas.

10.3.4 *Quasi-reality*

The tree-by-tree replacements infered for four species in a New Jersey forest (Quasi-reality of Fig. 10.1) have the overall pattern of a competitive hierarchy. Patchy disturbance plays a role as well, as when a Red Maple or a Blackgum replaces a Beech, or a Blackgum replaces a Red Maple. There is also a hint of obligatory succession, since Beech has not been commonly observed invading open fields in this locale. An additional feature is the tendency of individuals of three species to be replaced by new individuals of their own species. Such self-replacement can be added to the successional species in the idealized obligatory succession and in the competitive hierarchy, where it further slows convergence since early species give way less readily to later species. Even among late successional species, copious self-replacement slows convergence on the stationary state. The recovery of the stationary state from even small disturbances in the relative abundances of late successional species will be sluggish if those species tend to perpetuate themselves locally, rather than sorting out statistically with each other.

10.3.5 *When is succession rapidly convergent?*

There are three paradoxes in the analysis of Fig. 10.1 that help to explain some of the difficulties of previous theoretical approaches to succession. Natural selection favours copious self-replacement whenever it is possible. Therefore evolution tends to slow successional convergence and to destroy the environmental context of selection for adaptations to a particular successional status for a particular species. Drury and Nisbet (1973) go so far as to argue that there are no adaptations to succession itself, but that succession occurs because species that are adapted to geographically local niceties of soil and climate attain short-lived dominance in a temporal gradient of physical conditions that is sharpened by interspecific competition.

Not only is convergence slowed by self-replacement, it may even be destroyed if the amount of self-replacement depends heavily on a species' local abundance. This effect will be defended in section 10.4.1. It has the consequence that succession is certainly slow and dependent on historical accidents, perhaps not even convergent, where the supposed biological cause of succession is most apparent, namely interspecific facilitation in the obligatory succession or interspecific competition in the competitive hierarchy.

On the other hand, succession is rapid and predictably convergent in communities that are dominated by frequent, patchy, and essentially random disturbances. This underscores the fact that successional convergence, where it occurs, is a statistical phenomenon, rather than a uniquely biological one. Only in the competitive hierarchy is succession rapid and convergent enough to be observed and blatant enough to be biologically interpreted. The very reason that this is so is that a prime ingredient of Clements' (1916) successional recipe is missing; early successional species are not needed to pave the way for later species.

Although these results have been presented in the context of tree-by-tree replacements, they are applicable at any other level from size classes within a species to regional blocks of vegetation. Indeed they are applicable to any successional patterns that are generated by linear replacements of one state by another with characteristic probability, whether this probability is measurable in practice or not. Furthermore, the contrasts of Fig. 10.1 can be largely infered from the topology of arrows connecting the species and the number of heads on each arrow. The potential convergence of a succession can be predicted from primitive and qualitative observations of which species are capable of locally replacing which other species.

10.4 Biologically interesting non-linearities

10.4.1 *Recruitment dependent on local density*

The rain of seeds or of other propagules at any stage of succession depends on what species are within the normal distance of dispersal. If in addition the numerical success of propagules is increased in proportion to the local abundance of conspecific plants, succession need not converge at all, even in the absence of disturbance. I have presented an analytical model of this situation (Horn, 1975b), but its major consequence can be argued intuitively. Any species whose abundance has increased as a result of a devastation becomes more efficient at replacing both itself and other species, especially another species whose abundance has decreased below what it was prior to the devastation. Therefore any departures from a stationary distribution tend to be not only self-perpetuating, but even self-augmenting. The composition of the community depends on the historical vagaries of which species were made temporarily more abundant during the last episode of major disturbance.

Since most temperate trees set far more seed than ever reach even the sapling stage (Harper and White, 1974), recruitment of a particular species depends on its presence, but may not be strongly influenced by its local abundance. Nevertheless if a sufficiently large area is clearcut, recruitment to the opening will be less representative of the previous composition of the forest, than will be recruitment to the smaller openings of a patch-cut forest. (Foresters call this 'group selection', but the term has been usurped by other ecologists in an entirely different context). The extensive removal of all merchantable trees of one or more species, so called high-grading, may also alter the pattern of recruitment. Indeed foresters recognize that patch-cutting should produce a faster and more dependable regeneration of the previous forest (Twight and Minckler, 1972), but the process may be immediately uneconomical, or it may invite further devastation by deer or by wind.

If recruitment of young plants is generally proportional to the local abundance of conspecific adults, the consequences for successional theory are profound. Succession would not be necessarily convergent. Alternative stable communities, dependent on accidents of history, could persist side by side in an otherwise uniform environment. The dynamic study of recruitment in a diverse forest is a very promising area for future research. The studies reviewed by Harper and White (1974) are still insufficient to allow a guess of whether forest succession should or should not generally be convergent.

10.4.2 *Direct and persistent effects effects of devastation*

Devastation itself can be represented as a plant-by-plant replacement matrix in the same way that patchy disturbance is shown in Fig. 10.1.

Then succession over a long time interval can be represented by multiplying the initial composition of the forest by a matrix like *SSASSSBSSSSASCSS*, the product of many successional matrices (S) interspersed with various devastational matrices (A, B, and C). Such a succession might be dominated by the pattern of devastation or it might be dominated by a community whose composition is nearly stationary. The balance between these extremes is set by three properties of the matrices and their pattern of occurrence: the frequency of devastation relative to the speed of convergence in the interludes, the number of distinctly different communities resulting from devastation (A, B, C), and whether the effects of a devastation are persistent (represented e.g., by *SSABBBBSS*). The first two of these have already been analyzed intuitively in section 10.3; and the next paragraph is an optional technical note that distils the essence, or at least the fragrance, of that analysis.

The initial composition of a forest can be represented as a linear combination of the eigenvectors of S (also called latent or characteristic vectors of S). Then succession, which multiplies this composition (c_o) by successively higher powers of S, is represented as the same linear combination with eigenvectors (v_i) multiplied by their eigenvalues (λ_i, the latent or characteristic roots of S), each raised to a power (t) that measures the passage of time since the last devastation. Thus:

$$c_t = c_o S^t = (\textstyle\sum_i a_i v_i) S^t = \sum_i a_i v_i \lambda_i^t \qquad (10.2)$$

where c_t is the composition after t of the periods of time over which S is measured, the a_i are appropriate constants, and the i range over the number of rows or columns in S. As time passes, the composition of the forest will be increasingly dominated by the eigenvector corresponding to the largest eigenvalue, which is unity for an exhaustive probability matrix like S, because $\lambda^t = 1$ forever if $\lambda = 1$. The contribution of other eigenvectors to community composition depends on the magnitudes of their eigenvalues, decaying rapidly for those with small eigenvalues, and decaying slowly for those with large eigenvalues, because λ^t tends toward zero as t gets very large for $|\lambda| < 1$. The size of the second largest eigenvalue is therefore an inverse measure of the speed with which succession converges on the stationary distribution. Since the eigenvectors of most matrices like S are orthogonal, the number of eigenvalues that are nearly as large as the largest is a measure of the number of 'degrees of freedom' with which Nature may perturb a community so that subdominant eigenvectors put up a lengthy fight against the dominant one. Hence the number of markedly different and possibly persistent devastations (the A, B, C above) is measured by the dispersion of eigenvalues of S. I have not been able to develop a uniformly interpretable and simple measure of this dispersion. However, as R. Levins has pointed out to me, a whole class of measures might be based on the sum of squares of the eigenvalues, which is the trace of S^2 and thus the sum of products of the elements of S with their transposes. The sum receives contributions only from the self-replacements and two-headed arrows of Fig. 10.1, which, happily, have the biological interpretations already discussed in section 10.3. The only seemingly counter-intuitive notion of this analysis is the demonstration that the number of significantly different kinds of devastation is a property of the successional matrix, rather than a property of either the cause or the effect of devastations.

Of course if the direct effects of devastation are themselves persistent or self-reinforcing, then understanding the structure of a community requires a knowledge of its ancient history. A clear and yet subtle example is Blydenstein's (1967) interpretation of some patchy savannah-like vegetation on the Llanos of Columbia. Apparently fires have swept the area in the distant past, leaving small patches unscathed. The soil drained readily in severely burned areas, which now support only grassy vegegation which dries and burns frequently, intensifying the drainage of the soil, and so on. The less severely burned areas retained their moisture, and they have developed dense patches of forest, which retain enough moisture to remain relatively resistant to the fires that frequently sweep the plains around them.

10.4.3 *Adaptation to regular devastation*

If devastation occurs on a sufficiently regular schedule, this schedule gives a competitive advantage to early successional species whose lifespans match the intervals between catastrophes. Examples span a wide spectrum of intervals. Eastern White Pine is adapted to fill the interval of about 100–300 years between fires (Loucks, 1970; Heinselman, 1973) or severe hurricanes (Henry and Swan, 1974). Chaparral in California (Hanes, 1971) and Pitch Pine in New Jersey (McCormick and Buell, 1968) thrive with a more frequent burning on a cycle of about 10–50 years. Annual weeds are adapted to tillage (Harper and White, 1974), and flood plain herbs to spring flooding. Giant Sequoia has the characteristics of an early successional species, despite the great age of remaining stands (Biswell *et al.*, 1966; Rundel, 1971). Several of the famous groves are on the moraines and outwash of mountain glaciers of the past (Rundel, 1972; Matthes, 1965). I enjoy the fantasy of a venerable Sequoia biding its time in anticipation of the next local ice age. At the opposite end of the spectrum, the colon bacterium is marvelously adapted to devastations of 24 hour regularity.

The occurrence of physical disturbance need not be regular to cause a highly regular pattern of devastation. This is easiest to demonstrate for fires, but comparable models could be developed for other disturbances. Figure. 10.2 shows a hypothetical model in which the intrinsic inflammability of vegetation increases as plant material accumulates after a fire. When the inflammability reaches a threshold, which is high in wet periods and wet climates and low during drought, common accidents of ignition result in fires. In a very wet climate (Fig. 10.2b), only an unusual drought results in a fire, and the temporal distribution of fires is irregular. In a very dry climate (Fig. 10.2c), the timing of a fire may be little changed by wide variations in the timing of weather, and a very regular pattern of fires may result (Fig. 10.2e). The details and predictions of Fig. 10.2 correspond closely to interpretations of cyclic occurrence of fires by Loucks (1970), Rowe and Scotter (1973), and Heinselman (1973), though Heinselman also found that geographically distant fires were synchronized by extensive droughts.

The temporal pattern of fires depends as much upon the intrinsic shape of the curve of inflammability as it does on the temporal pattern of droughts that favour fires. In particular, the regularity of fires is determined by the rate of increase in inflammability near the average threshold for conflagrations in a given environment. If the fires occur on a sufficiently regular schedule, early successional species whose lifespans are matched to the interval between fires may increase their local persistence if they became yet more inflammable late in life. This in turn would steepen the curve of inflammability, and the period between fires would become still more uniform. Mutch (1970) has measured the combustibility of vegetation of several species whose persistence in nature depends upon fires, and found that these species are more combustible than are related species that inhabit communities that are rarely burned. Biswell (1974) notes that Chamise chaparral in California remains vigorous and stable only if it is burned every 15–20 years, and he describes a mechanism whereby 15 year old Chamise increases its own inflammability. The terminal twigs die and then

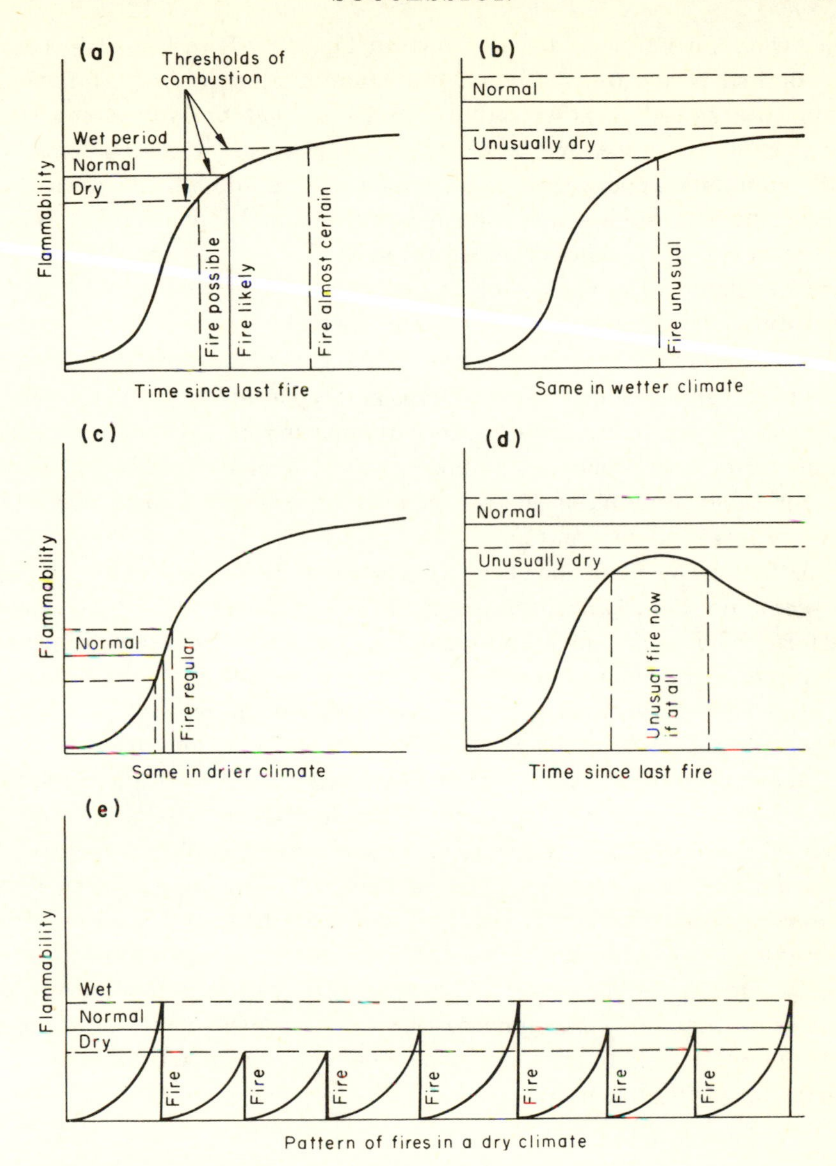

Fig. 10.2. Graphical model for predicting the temporal pattern of natural fires; see text for a discussion.

proliferate by lateral sprouting, and the sprouts repeat this behaviour until the crown of the bush is a well laid bed of tinder.

If accumulated vegetation has a sufficient ability to hold water, inflammability may eventually decrease with an increase in the standing vegetation. Such a situation is shown in Fig. 10.2d. In this case there is a limited period during which the community risks fire. After this period has passed, it is exceedingly unlikely that fire will sweep the community. Historical accidents will then determine which portions of the community appear to be maintained by fire and which appear to escape entirely. Such a mechanism is implied in Blydenstein's (1967) description of the patches of forest in the Llanos of Colombia, and in the early development of some pine forests (Weaver, 1974). Connell (1975) has interpreted alternative and apparently stable associations of intertidal beasties in the context of a model like Fig. 10.2d, with one species of barnacle either being eaten by snails and replaced by another, or else escaping predation by growth, and dominating the stationary community. Here is another class of models in which regular and biologically interpretable patterns of succession may be driven by historical and more or less random events.

Although this model has been developed in the context of fire as the catastrophic disturbance, the same model could be used to discuss patterns of devastation by wind, flood, landslide, predation, disease, or even economic management. The degree of regularity of devastation in each case depends on the intrinsic increase in sensitivity of the vegetation as it grows back following devastation, measured at the threshold of sensitivity at which devastation becomes likely. Such measurements remain to be made (Wright and Heinselman, 1973), even in the case of fire, where appropriate general models would have immediate practical applications in the management of parks and preserves whose major vegetation is naturally dependent on fire (Wright, 1974). Although the techniques for prescribed burning are highly enough developed to allow generalizations about applications to a wide range of environments (Kayll, 1974), there are no general principles that help to predict how a given community will respond to a particular pattern of burning.

10.5 Conclusion and prospect

The only sweeping generalization that can safely be made about succession is that it shows a bewildering variety of patterns. Succession may be rapidly convergent on a stationary vegetational composition, or it may be slow and apparently dependent on accidents of history. Succession may lead to alternative stationary states, depending on the initial composition. The vegetation may continually change in response to more or less random changes in the physical environment, or it may change cyclically by responding to and perhaps reinforcing an environmental cycle.

Analytical models dispel some of the bewilderment by showing that the general pattern of a succession is largely determined by biologically interpretable properties of the individual species that take part in the succession. Where successional convergence is found, it is a statistical phenomenon, rather than a biological one. Whether convergence is rapid or slow depends upon how effectively each species locally regenerates itself and replaces other species. Succession is rapid and convergent where each species can more or less randomly replace any other species directly. Succession is slow and dependent on history if early successional species must modify the environment before later species can invade. Succession will only be directly observable, convergent, and biologically interpretable in those rare cases where the most important species form a competitive hierarchy. Deaths among the competitive dominants may still leave room for the persistence of early successional species, and the 'undisturbed' forest may actually be a mosaic of successional patches.

Whether succession is convergent or not depends critically on how strongly the amount of recruitment of young plants is determined by the local density of mature plants of the same species. This relationship should be explored for a complex community. Until it is known, one cannot predict whether the community will remain stable under human exploitation or will shift radically to a new stationary state in response to a small perturbation.

Some communities are adapted to a more or less regular pattern of natural devastation. More work is needed at both theoretical and empirical levels before generalizations can be made about human management of such communities. In particular the dynamics of a community's developing sensitivity to natural devastation deserve as much attention as the temporal pattern of the physical causes of devastation.

The goals of forest management have recently been extended, from the traditional farming of an economically valuable harvest of wood, to constructing and maintaining forests with a wide variety of specified amenities, to the repair of de facto devastation by exploitation, and even, as Raup (1964) argues persuasively, to managing for flexibility of unanticipated future use. A sound conceptual approach to these problems is a prerequisite to developing experiments in management. It is equally important to discover the conditions under which the preservation or reconstruction of a natural area requires the absence of management.

So far the linear models of forest succession have made gratifying predictions of what was already known. From the general consequences of linear models, I can even predict the range of surprises that remain to be discovered in them. Sections 10.3 and 10.4.2 show that the broad properties of succession depend on the topology of arrows that show the competitive interactions within and between species. Therefore the effect of any conceivable management depends on whether one helps or hinders these competitive interactions. The only simple treatments that will have dramatic effects are just those practices of seeding, weeding and patch-cutting that enlightened foresters have recommended for centuries. If there are real surprises, they must be hiding in biologically interesting non-linearities like the effect of local density on recruitment or the dynamics of sensitivity to disturbance.

11
The Central Problems of Sociobiology

EDWARD O. WILSON

11.1 Introduction

Sociobiology is defined as the systematic study of the biological basis of all forms of social behaviour, in both animals and human beings (see Wilson, 1975). It has only recently been distinguished as a field of study from the remainder of behavioural biology, and is now being reconstructed to a large extent from the first principles of population biology and ecology. Since sociobiology still has a relatively weak theoretical structure, it presents the entrepreneur with unusual opportunities for discovery. The purpose of this essay is to provide interested biologists and mathematicians with a list of the most important questions as one student of the subject conceives them. I trust my presumptuousness will be forgiven, because while the reader knows that the history of no truly vital science can be predicted and its central problems can at best be crudely characterized, empiricists have an obligation to phrase their intuition in a way that encourages the construction of realistic models.

11.2 Group selection

Virtually all of the major advances in ecology and population biology during the past fifty years have resulted from the enclosure of efficient units (the biota, the ecosystem, the deme, the kin network, and others), accompanied by the invention of an accounting system through which the properties of the units can be evaluated. The same procedure appears to be the key to theoretical sociobiology. The most pervasive basic problem of sociobiology is the evaluation of group selection. In brief, we need to know to what extent and by what means groups of organisms—the extended family groups, societies, whole populations, or whatever—are selected, as opposed to individuals and their immediate progeny. The very interpretation of altruism hinges on this distinction. Does a given so-called altruistic act really constitute altruism, in the sense that the donor sacrifices personal genetic fitness to benefit another? Or is it merely a shrewd manipulation by which the 'donor' places itself in an ultimately more advantageous position within the group? Unless evolution by natural selection is in some unknown way circumvented, the existence of true altruism implies the counteraction of individual-level selection by group selection. Of equal importance is the alternative possibility that group selection reinforces individual-level selection. If this occurs, social evolution can proceed more rapidly than ordinary evolution. Let us consider several concrete problems the solutions of which depend on these distinctions.

11.2.1 *The origin of insect sociality*

The most favourable subjects for the testing of group selection are the social insects. Certain peculiarities in the behaviour and sexual ratios of social Hymenoptera (wasps, ants, bees) can apparently be explained only by a form of selection operating on a complex of kin, as opposed to selection operating on individuals.

As first pointed out by Hamilton (1964; see Trivers and Hare 1976, and Wilson, 1975), the haplodiploid mode of sex inheritance results in females being more closely related to their sisters than to their brothers; in the case of a singly inseminated mother, the coefficient of relationship (ρ) of daughter to sister is 3/4 and of daughter to brother 1/4. The result should be a female-oriented society with strong altruism toward siblings, which is conspicuously the case in nature. Trivers and Hare (1976), by extending Hamilton's argument, showed that a female ant or other hymenopteran can ordinarily improve her inclusive fitness through rearing her sisters and brothers only if she distorts the ratio of investment so as to produce more sisters among the generation of new queens and males. The distortion should come to equal a 3:1 preference of sisters over brothers in the sexual brood (measured, say, in dry weight), because beyond this point the reproductive advantage would tip back toward the brothers. Hence when the mother queen lays all the eggs, and the worker caste is in control of rearing the larvae, the ratio of investment should approximate 3:1. When one or more workers lays eggs, producing sons, the situation is more complex. The laying worker will prefer a larger investment in females, while her coworkers will prefer this bias only to the extent that the mother queen also produces some of the males. The basic scheme is given in Table 11.1. The conclusion drawn by Trivers and Hare is that under a wide range of conditions, the ratio of investment should be distorted away from 1:1 and in favour of females, with 3:1 the rule for species with non-laying workers. Data gathered by Trivers and others fit these predictions remarkably well.

Table 11.1. The coefficient of relationships, ρ, of a worker to close kin under two conditions of reproduction and two ratios of investment in males and females.

Ratio of investment in sexual brood (female:male)	Case 1: Worker lays eggs to produce sons.			Case 2: Queen lays all eggs.	
	Sexual brood			*Sexual brood*	
	son	sister		sister	brother
	$\rho = \frac{1}{2}$	$\rho = \frac{3}{4} \leftarrow$	worker	$\rightarrow \rho = \frac{3}{4}$	$\rho = \frac{1}{4}$
1:1	$\bar{\rho} = \frac{10}{16}$			$\bar{\rho} = \frac{8}{16}$	
3:1	$\bar{\rho} = \frac{11}{16}$			$\bar{\rho} = \frac{10}{16}$	

Kin selection thus appears to be operating in the social Hymenoptera. The key residual problem, which I will arbitrarily label as the first in the series of this article, is the following:

(1) *Given that kin selection now occurs in insect societies, to what extent did it contribute to their origin?*

Lin and Michener (1972), Michener and Brothers (1974), Alexander (1974), and Eberhard (1975) have evaluated various additional advantages of early eusociality: the superiority of mutual nest construction and defense, the opportunities workers enjoy of ovipositing surreptitiously in the presence of the queen or even of taking over when she dies or leaves, and the advantage the queen receives from simply controlling and exploiting the workers at the expense of their own genetic fitness. Michener, Eberhard and their coworkers viewed these factors as being potentially auxiliary to kin selection, while Alexander formulated the extreme hypothesis, now disproved by the investment ratio data, that exploitation is the primary force and kin selection negligible. What we must have next is an accounting system. We would like to be able to write a multiple regression equation incorporating each of the factors, in addition to kin selection, and yielding the inclusive fitnesses of the would-be queens and workers. When the averaged fitnesses reaches a certain *eusociality threshold*, the species is likely to evolve all three of the three basic traits of a 'truly' social (eusocial) insect: cooperative brood care, division of the group into reproductive and sterile castes, and overlap of at least two generations.

Meanwhile, it appears that the enhancement of kin selection by haplodiploidy does add a critical amount to the fitnesses of eusocial species. Eusociality is almost limited to the Hymenoptera, having been known to originate elsewhere only in the single case of the termites, which descended from cockroach-like forms. Other kinds of arthropods, including many beetles, spiders, and orthopterans, build nests, manipulate objects skillfully, and care for their young to varying degrees, but none have attained the eusociality threshold. It is equally true that haplodiploid enhancement alone is not enough. Other haplodiploid insects exist, notably some mites, some thrips, the aleurodid whiteflies, the iceryine scale insects, and the beetle genus *Micromalthus*, which are entirely solitary.

To summarize, plausibility arguments have been made for the existence of several selective forces leading to eusociality in insects. Convincing evidence has also been presented that kin selection in particular plays an important and in most cases crucial role. What is needed now is the means of evaluating the relative weights of all the forces.

11.2.2 *Group selection in human beings*

Evolution by group selection is a stronger possibility in human beings than in other kinds of mammals. The boundaries of groups are defined by a far more precise and sophisticated kin recognition than that found in other mammals, and they are further sharpened by cultural divergence. Human groups are also more conscious of each other. They ponder the outcomes of past contacts and plan aggressive adventures by means of skilled, cooperative manoeuvres. Genocide, enslavement, and genosorption have been commonplace events during at least the last several millenia of man's history. These processes can accelerate the evolution of certain forms of aggressive behaviour. They probably increase representation of the underlying genes through both competition of members within the groups and competition between groups. Simultaneously, they can promote the evolution of altruistic behaviour. Self-sacrifice by individuals during war or economic competition can be more than counterbalanced by the advantage provided the group as a whole. Thus while genes favouring self-sacrifice are declining *within* groups vis-à-vis homologous selfish genes, they can be increasing in the population as a whole due to the advantage they give the benefitted group in competition with other groups that lack altruists.

The resulting theoretical problem of sociobiology has profound ethical implications:

(2) *To what extent is patriotism, and particularly the willingness to fight and risk death for the ingroup, based on group selection that counteracts individual-level selection?*

We need a calculus to separate group versus individual selection with reference to the special human forms of altruism. What is the minimum reward a surviving warrior and his family must receive in order to compensate for the genetic risk he takes? Can such rewards explain his reckless commitment, if we express them solely as a gain in inclusive fitness in competition with other family groups belonging to the same tribe? Or must his behaviour be explained at least in part as the outcome of the growth of the entire tribe (or nation) at the expense of other, commensurate groups? There is a need not only for a complete theory but also for ways to estimate the parameters in human societies.

The solution of this sociobiological problem is a prerequisite for the creation of a fully rationalistic morality. To the extent that the capacity for reckless aggression on behalf of the ingroup has been evolved by individual selection, it should prove easier to curb ethnocentrism, bigotry, and blind patriotism. If, on the other hand, the latter qualities were evolved by group selection, we should expect them to be more tightly programmed and less subject to moderation by improvement in individual status. Some of the descriptive laws of cultural anthropology and sociology suggest a form of conservation in the operation of group conflict. In particular, loyalty and aggression are manifested as though they were constant quantities to be divided among whatever reference groups are important to the individual. LeVine and Campbell (1972) have expressed three of the most relevant laws as follows:

(i) The higher the level of political complexity the less frequent is internal feuding.

(ii) Societies that frequently engage in war with their neighbours are less likely to have feuding than societies that have peaceful external relations.

(iii) Frequency of interethnic warfare and frequency of intraethnic feuding are positively related in stateless societies and negatively related in politically centralized societies.

The full explanation of these phenomena is likely to prove impossible until the genetic models suggested by the second problem have been developed and tested. Relatively quick but limited progress can be expected for the special case of an ultrasimple population structure (see, for example, Boorman and Levitt, 1973; D. S. Wilson, 1975; Gadgil, 1975). If Bronislaw Malinowski's original conception of well-defined, homogeneous tribal units were true, an ultrasimple structure might be a reasonable first approximation. However, more recent studies have shown that human population structure tends to be extremely complex and richly diversified, with a strong dependence on culture. Ethnic boundaries are seldom sharp. Territorial boundaries interpenetrate in complex ways, and they are often overlain by linguistic, cultural, and historical-political areas in complex patterns. Furthermore, cultural and linguistic traits are sometimes found to vary in a continuous manner as one proceeds from one ethnic 'centre' across 'boundaries' to others. Finally, groups shift rapidly in their loyalties, forming alliances in one year and dividing into quarreling factions the next. The most clearly bounded tribal boundaries are actually those imposed on diverse peoples by political states, the rulers of which find it more convenient to deal with such subordinate units instead of the chaotic patterns that originally existed (LeVine and Campbell, 1972; Fried, 1975). These considerations lead to our next key problem in theoretical sociobiology:

(3) *Given that ethnographic qualities tend to be discordant in pattern, continuously varying in space, and constantly changing in time, what are their effects on population structure and how have they affected group selection?*

In summary, human population biology is unique in many respects when compared with that of animal species, and it varies in detail among different cultures. The form and intensity of group selection therefore requires a modeling effort different from that directed at the simpler plant and animal systems. The premises of this new theory must conform closely to ethnographic data.

11.2.3 *Intrinsic constraints on kin selection*

Kin selection has a potentially implosive evolutionary effect. In the absence of constraints there should be a strong selective advantage for inbreeding combined with recognition and cooperation among kin. The ideal circumstance would be incestuous mating between parents and offspring and between brothers and sisters within tightly organized family groups that behave selfishly toward equivalent units. Such a breeding system would facilitate the evolution of extreme altruism and cooperation among the family members, since a large fraction of genes are expected to be shared through common descent by the altruists and the recipients of the altruism.

But of course a severe constraint does exist; it is the well known decline in performance due to loss of heterozygosity. The decline has been well documented and the theory germane to it explored at length by population geneticists (see Crow and Kimura, 1970; Cavalli-Sforza and Bodmer, 1971). The elements of kin selection theory have also been developed more or less independently of this inbreeding theory (Hamilton, 1964, 1972, 1974; Eshel, 1972; Wallace, 1973; Trivers and Hare, 1976). The task before us is clearly the refinement and synthesis of the two elements:

(4) *How are the gains through inclusive fitness and losses through inbreeding depression balanced; and in which life history strategies is this balance struck so as to increase the likelihood of more intense inbreeding and social life?*

11.2.4 *The evolution of sex*

A problem closely related to the preceding is created by the evolutionary origin of sex. The ultimate degree of relatedness is found in the absence of any sex, namely where individuals reproduce themselves by asexual means to create clones of genetically identical individuals. This is the state prevailing in the colonies of most kinds of lower invertebrate animals, and its concomitants are extreme altruism, cooperation, and division of labour among the colony members. In a sense, asexual reproduction is the limiting case of inbreeding. It has certain potential advantages, including rapid population growth, minimization of adult investment due to the reduced need of the parent to mate, and, not least, the enhancement of the genetic prerequisites for advanced social behaviour. Yet asexual reproduction is the exception rather than the rule in the great majority of animal phyla. This paradox can be expressed in the following way:

(5) *What are the actual selective forces that have favoured the repeated origination of sexual reproduction, in spite of the negative effect of such reproduction on social evolution?*

This problem has been carefully considered in recent years but still remains moot. It can be effectively resolved into two competing hypotheses, called by Maynard Smith (1971b) the long-term and short-term explanations, respectively. In essence, the long-term explanation holds that evolution proceeds faster in sexually reproducing populations because favourable genes can be assembled more quickly in individual organisms by the process of segregation and recombination. And because the more rapidly evolving species tend to replace their slower counterparts, sexual reproduction is the prevailing mode in most kinds of organisms. The short-term explanation simply makes the same argument with reference to the progeny of individual organisms in competition with other sets of progeny. Williams (1975) has constructed a persuasive case for the widespread occurrence of the individual-level selection subsumed by the second explanation, and he discounts any importance for the population-level selection implicit in the first, long-term explanation. The matter is complicated by the fact that episodes of natural selection favouring or disfavouring sexual reproduction at the two levels reinforce one another. What is needed is a formulation of theory that will permit measurement of the relative weights of the two processes. Apart from the main controversy, Williams has also shown how development of the theory of sexual evolution can lead to a deeper understanding of diversity in life cycles.

11.2.5 *Adaptive demography and the optimization of caste*

In populations and societies of vertebrates, the statistics of demography represent epiphenomena. Age and size frequency distributions in particular emerge as the total result of many events that affect individual animals in the course of their struggle to survive and to reproduce. These distributions, and the survivorship and natality schedules that yield them, are based on natural selection at the individual level. Virtually all individual vertebrates attempt to reproduce at maturity, and their survivourship and natality schedules can justifiably be interpreted as the most productive or nearly the most productive that individual members of the population expect to achieve

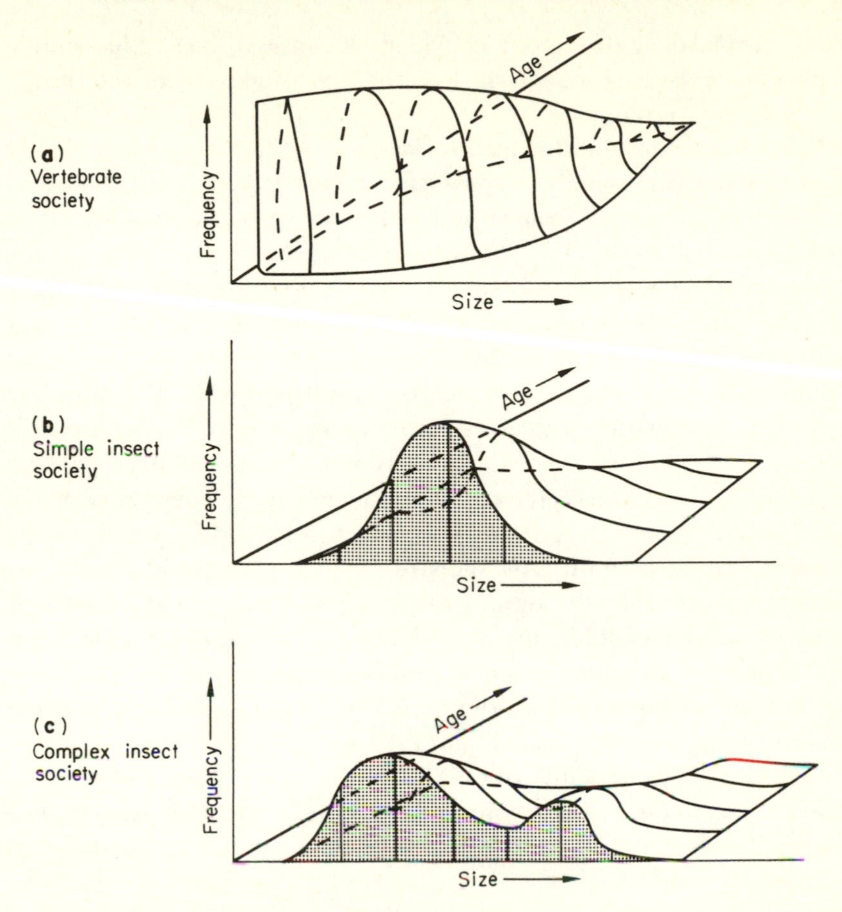

Fig. 11.1. The age-size frequency distributions of three kinds of animal societies. These examples are based on the known general properties of real species but their details are imaginary. (**a**): The distribution of the 'vertebrate society' is nonadaptive at the group level and therefore is essentially the same as that found in local populations of otherwise similar, nonsocial species. In this particular case the individuals are shown to be growing continuously throughout their lives, and mortality rates change only slightly with age. (**b**): The 'simple insect society' may be subject to selection at the group level, but its age-size distribution does not yet show the effect and is therefore still close to the distribution of an otherwise similar but nonsocial population. The age shown is that of the imago, or adult instar, during which most or all of the labour is performed for the colony; and no further increase in size occurs. (**c**): The 'complex insect society' has a strongly adaptive demography reflected in its complex age-size curve: there are two distinct size classes, and the larger is longer lived. (From Wilson, 1975).

in the particular environment in which the species lives. This simple assumption is the basis of classical demography and most of the theory of evolutionary ecology.

In colonial invertebrates and social insects there exists a radically different situation. Genetic fitness is not based on the survival and reproductive capacity of the individual members of the colony. It is instead determined by the number of individuals belonging to the reproductive caste which are emitted by the colony. In termites, for example, fitness is determined by the number of new queens and males that survive to found new colonies. Natural selection is at the level of the colony; only a few members multiply, while the majority are sterile and devote themselves to increasing the welfare and reproduction of the sexual castes. As a consequence, the demographic properties of the sterile castes are not epiphenomena. They are directly adaptive with reference to the colony as a whole, and they are shaped by colony-level selection. An age-size frequency distribution can be changed through the altering of growth thresholds, so that a lower or higher proportion of sterile members reaches a certain weight. Or it can be changed by altering the longevity of particular castes; if soldier ants die sooner, for example, their caste will be less well represented numerically in each moment of time (see Fig. 11.1).

In the setting of truly colonial life the statistical 'moments' of demographic frequency distributions take on a new significance. The means still reflect the rough adjustment of the size and age of individual organisms to the demands of the environment. The variance and higher moments acquire a directly adaptive significance because of their caste structure.

(6) *There is a need for a new form of demographic theory, which can be called adaptive demography as opposed to conventional demography.*

11.3 Accelerating forces in social evolution

Several processes associated with social existence can increase the rate of social evolution. The first is reinforcing selection at two or more levels. When selection favours the genotype at not only the level of the individual but also at the level of the kinship group, the society, or the population, changes in gene frequency will occur much more rapidly than if the selection were applied at the same intensity to individuals alone.

(7) *Formal investigations are needed to evaluate the effects of reinforcing selection on various combinations of selection units.*

It is probable that reinforcing selection has acted at two or more levels in human beings, and that such reinforcement is to some degree responsible for the rapid mental evolution which occurred over approximately the past five million years.

The second process is the one I have referred to elsewhere as the multiplier effect (Wilson, 1975). Social organization is a phenotype in exactly the same sense that the configuration of an enzyme or the length of a bone are phenotypes, yet it is the farthest removed from the genes. Social organization is derived from the behaviour of individuals and the demographic properties. The smallest possible change in an enzyme, namely the substitution of a single amino acid residue, might translate into altered neurons and a substantial change in behaviour and demographic schedules. These alterations could then be translated into a major social effect through the cascading of secondary effects into the multiple facets of social life. An excellent case is found in the differences between man and the chimpanzee (King and A. C. Wilson, 1975). The genetic distance between these two primates, as measured by electrophoretic comparisons of proteins encoded by 44 loci, is very small. It is equivalent to the distance separating sibling species of *Drosophila*, and is only 25–60 times greater than that between Caucasian, Black African, and Japanese populations. Yet the mental and social differences between the chimpanzee and man are immensely greater.

Evolutionary change can result in either divergence or convergence of traits in different populations. Numerous striking cases of both have been reported in social species. Geographically semi-isolated populations of anubis baboons, for example, differ strongly in the details of their social organization, yet individual populations are closely similar to other species and even other genera of primates.

(8) *There is a need to measure the process of amplification of phenotypes, from enzymes to the details of social organization; the general properties of the multiplier effect also merit theoretical examination.*

Finally, the interaction of genetic and cultural evolution is a little explored but important subject with particular significance for the study of human social evolution. Culture can be viewed as a finely tuned device which tracks environmental change faster than genetic evolution. As such, it is capable of greatly narrowing or broadening phenotypic variance and thereby ultimately influencing genetic

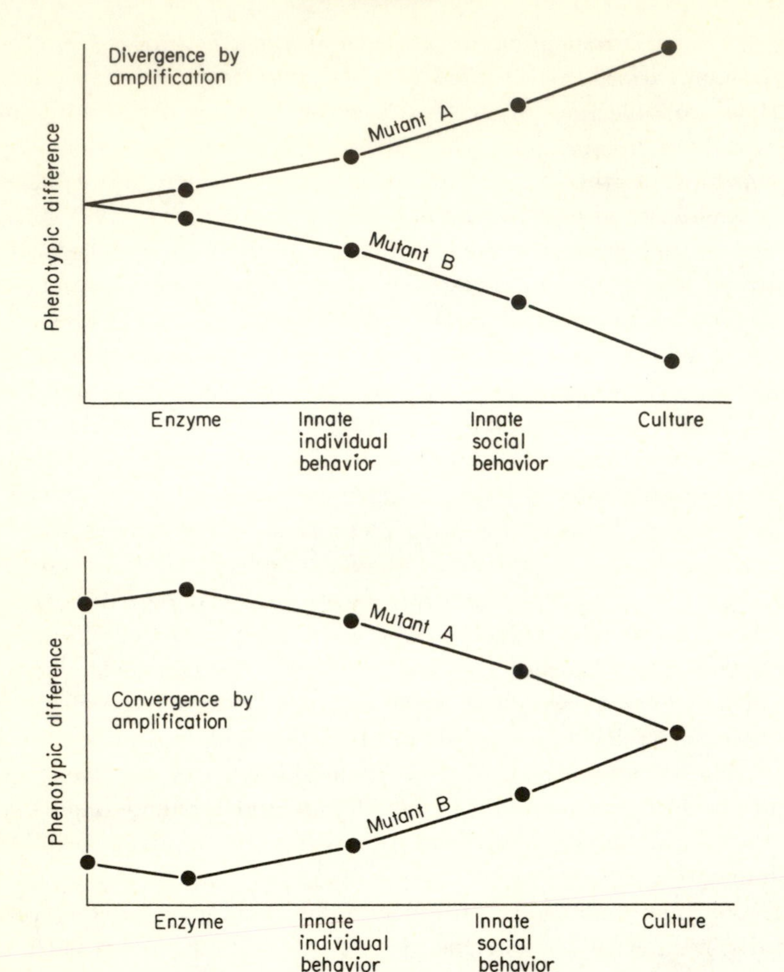

Fig. 11.2. The multiplier effect in phenotypic change. In the upper diagram a small amount of change in a single enzyme, perhaps resulting from one gene substitution, causes two phenotypes to diverge slightly in enzyme structure; this divergence is amplified through successive steps until it becomes a major difference in culture. In the lower diagram the same amount of change at the level of the enzyme results in close convergence in culture. Although such amplification has been documented qualitatively, these two cases are imaginary.

evolution. Consistently high conformity by members of the society will ultimately reduce genetic variance and slow genetic evolution. Low conformity, or high conformity accompanied by rapid cultural change, will increase genetic variance and probably speed genetic evolution by uncovering successful new phenotypes at a higher frequency. Genetic assimilation can occur, in which a modified cultural environment enables unusual phenotypes to appear and to flourish, gradually increasing the genes that most bias individuals toward their manifestation, and eventually making appearance of the phenotypes more likely even if the culture shifts to a type previously unable to produce the phenotypes. Conversely, a small amount of genetic evolution can result in an amplified cultural change, as already suggested. The properties of cultural change and dissemination have begun to be expressed in a form that permits analogies with genetic evolutionary change (see Cavalli-Sforza and Feldman, 1973). Such research might now be extended to address the following question:

(9) *What are the modes of interaction of genetic and cultural evolution, and under what conditions do they accelerate or slow one another?*

11.4 Conclusion

The nine 'central problems' suggested here do not form a complete catalogue; others can no doubt make longer and more sophisticated lists. The problems have been put forward in order to illustrate the fertility of the emerging discipline of sociobiology and the immediate challenges it offers both theoreticians and imaginative empiricists. Furthermore, I am confident that any such formulation of basic issues will arrive at the following and even more important conclusion: the division between biology, particularly population biology, and the social sciences no longer exists.

12
Palaeontology Plus Ecology as Palaeobiology

STEPHEN JAY GOULD

12.1 Introduction

Palaeontology ought to be one of the world's most exciting subjects: dinosaur mania among children indicates a primal fascination. Yet palaeontology, as traditionally pursued by professionals, has been the poor sister, and often the laughing stock, of the sciences. It has been portrayed, not unfairly, as the dullest variety of empirical cataloguing practiced by the narrowest of specialists. In 1969, the editors of *Nature* wrote (anonymous, 1969, p. 903): 'Scientists in general might be excused for assuming that most geologists are palaeontologists and most palaeontologists have staked out a square mile as their life's work. A revamping of the geologist's image is badly needed.'

Traditional palaeontology was the handmaiden of stratigraphic geology. In 1947, a famous geologist defined the profession this way (Kay, 1947, p. 162): 'Paleontology is a means by which rocks are better classified in time and in environment of origin.' As an auxiliary to geologists, it became a profession of atheoretical, taxonomic specialists: a budding palaeontologist indentured himself to a biological group and a small section of geologic time; he then learned all the names, coined some new ones, and refined the stratigraphy of his chosen section. But a remarkable transformation has taken place during the last decade. In short, palaeontology has allied itself with evolutionary biology, and has experienced all the challenge and excitement of its most rapidly developing subdiscipline—theoretical ecology.

The change in emphasis has been dramatic. I began my subscription to the *Journal of Paleontology* in 1964. Of the 18 articles in my first issue, 11 had titles in the form: 'New species of .... from the .... (age) of .... (place).' Of the seven remaining, 4 were purely taxonomic. In 1975, the Paleontological Society founded a new journal, *Paleobiology*, to receive more theoretical papers, thus preserving the documentary role of the *Journal of Paleontology*. Nonetheless, of 12 articles in the latest *Journal of Paleontology* (September, 1975), only 3 list taxon, time and place. The others focus upon biological questions and cast descriptive sections in their light.

Traditional palaeontology was preeminently a geological discipline, but it did not ignore biology completely. It extolled the importance of 'evolution' but used the term to mean little more than the tracing of phylogeny and the identification of macroevolutionary trends in morphology. Moreover, it did not, on the face of it, deny the importance of ecology—'palaeoecology' was the fad subject of the 1940's and 1950's. But 'palaeoecology' bore little relationship to anything an ecologist would recognize; as a creation of the petroleum industry, it remained largely geological. The earth's strata are not arranged in a layer cake; rocks of the same age record different environments and some fossils, restricted to environments rather than time planes, do not serve traditional stratigraphy well. Thus, a limited autecology nearly exhausted the content of traditional palaeoecology: (1) environments (rock types) preferred by given taxa were recorded; (2) for fossils entombed in more than one rock type, correlations were sought between environment and morphology, largely in the interest of taxonomic refinement.

A glance at the most celebrated document of traditional 'palaeoecology', the 1000 page multi-authored Treatise of 1957 (Ladd, 1957), illustrates its largely atheoretical concern. It begins with an indifferent section of 100 pages on generalities (mostly nonbiological) and then proceeds via the conventional catalogue of times and taxa. Its body of 900 pages contains: (1) a set of case studies in environmental reconstruction ordered not by problem or concept, but by the old layer cake from Precambrian to Recent; (2) a set of essays on taxonomic groups, ordered by the conventional ladder of amoeba to man. Yet, we should not ascribe this atheoretical approach only to the biological naiveté of most traditional palaeontologists. Ecology, at the time, also emphasized highly specialized, inductive and largely atheoretical work. Palaeontologists had looked in (the 1957 Treatise has a companion volume on modern ecology—Hedgpeth, 1957), and they found little to inspire a change in direction.

The transformation of palaeontology during the last decade into a 'creative, chancy young man's game' has many roots (to borrow E. O. Wilson's description of modern evolutionary biology, enlightened by the approach of theoretical population ecology—Wilson, 1969). But I believe that the most important cause has been the direct impact of theoretical population ecology. Palaeontologists have looked in again, and this time they have found insight into their traditional problems of diversity and its history. We make no apology for this borrowing. Scientific change usually involves a fertilization from other disciplines (Kuhn, 1962); new directions rarely arise *sui generis* from the established professionals of a field. Moreover, we are beginning to repay the debt by providing our own theoretical insights for ecologists to consider.

I write this chapter to document the modern interaction of ecology and palaeontology. It is not a review article (for it leaves out far more than it includes); I am only seeking to illustrate in a highly selective way how palaeontology is incorporating ecological theory to produce a conceptual science worthy of the name palaeobiology.

12.2 Specific puzzles

12.2.1 *What are the primary controls on the diversity of fossil communities?*

Geological methods must be used to answer one of the most crucial general questions for any biological study of fossil communities: how well does a fossil assemblage reflect the living community. The biases of differential preservability, burial and subsequent alteration must be recognized and subtracted. In many cases they cannot be measured accurately, and palaeobiologists have sought the ideal situation of rapidly buried communities, entombed *in situ* within sediments that suffered little subsequent alteration.

We find (not surprisingly) enormous variation in numbers of species within fossil samples that represent their living communities adequately. The explanation of these differences occupied much of classical palaeoecology. Favoured modes of explanation went little beyond: (1) historicism, e.g., low diversity in areas only recently available for habitation; and (2) 'physicalism', the notion that species are under the direct control of physical variables, and that communities of low diversity reflect extreme values of a controlling variable—hypersalinity or low temperature, for example. The interaction of species and their evolutionary strategies (beyond morphological adaptation to a substrate) did not figure in classical explanations.

The framework of stability-diversity theory in modern ecology has inspired a new approach to the study of fossil communities. Organisms adapt to the relative stability of environments by altering their life-history strategies; intensity and frequency of disturbance may serve as the main environmental control upon diversity. Since the inevitability of environmental fluctuation (at least in the long run) is a cardinal principle of geology, this new theme has intrinsic appeal to palaeontologists. (It matters little that classical stability-diversity theory may be suffering a spectacular collapse within ecology (e.g., May, 1975a, and section 8.5). The surviving aspects have been emphasized by palaeobiologists; moreover, data from the fossil record have helped to spur the collapse.)

Using the old approach in his early papers, for example, Hallam (1969) invoked 'physicalism' to explain the low diversity of the Boreal in contrast to the Tethyan Realm among European Jurassic faunas. Seas of the Boreal Realm, he argued, maintained abnormal salinities. In later works (Hallam, 1975), he has tied low Boreal diversity to the frequent and severe disturbance of this realm; high diversity in Tethys reflects the stability of its environment. One may fairly ask what has been gained by this shift; does it represent anything beyond a more fashionable explanation for the same old data. I prefer it because it suggests, in the context of ecological theory, much more to test and a richer domain of biological insight (it may, of course, be wrong). With the physicalism of unusual salinity, one can scarcely go beyond two lines of biological evidence: are the fossils related to modern taxa that tolerate unusual salinities; and, do the fossils bear morphological signs (e.g., stunting) that often accompany life in these rigid environments? Modern ecological theory opens the whole field of adaptive strategies, and its criteria of quantitative population dynamics. New properties for study and test include survivorship curves, recruitment, measures of evenness among coexisting taxa and the attributes of r and K strategists preserved in fossils (and utilized, for example, by Snyder and Bretsky (1971) in arguing that paedomorphosis for rapid generation time in unpredictably fluctuating environments furnishes a better explanation than stunting by abnormal salinity for the depauperate 'dwarf' fauna of the Ordovician Maquoketa Formation).

In a pioneering paper, Bretsky and Lorenz (1969) noted that Paleozoic nearshore communities are generally less diverse, but geologically more persistent than offshore communities (data challenged by Thayer, 1973). They argued that nearshore environments suffer severe and unpredictable fluctuations. These environments support a limited number of physically controlled *r*-strategists. Stable (subtidal) offshore environments maintain diverse, biologically accommodated communities (*sensu* Sanders, 1968). But classical stability-diversity theory maintained that biologically accommodated communities should be stable in time. Why, then, do the nearshore, physically controlled communities persist for longer periods? Bretsky and Lorenz argue for an effect of scale. Offshore communities are more stable in the short run of ecological time; a nearshore community cannot survive in a desiccated estuary (though its elements might move to another area). But in the long run of geological time, ultimate control is exerted by the very rare major fluctuation. These rare events can wipe out a complex, biologically accommodated community, but the few species of nearshore communities are *r*-selected for surviving environmental fluctuation and have a better chance of 'weathering the storm'.

The correlation of low diversity with frequent, unpredictable and severe disturbance has not been challenged in the current attack on stability-diversity theory in ecology. This is the correlation that palaeobiologists have exploited most often. Levinton (1970) may have resolved the old dilemma of why some fossil beds are so jammed full with remains of a single species while beds below and above do not record it at all. These accumulations do not just represent another example of current winnowing or differential transport—another depressing example of the biological inadequacy of our fossil record. They are the opportunistic species of physically controlled communities caught in the acme of their transient success. Rollins and Donahue (1975) explain patterns in diversity within Pennsylvanian cyclothems by showing that initial faunas of low diversity are composed of opportunistic species. As sea level rises, these are replaced by a more diverse community of biologically accommodated species. Low diversity is not a function of history (early stages of transgression), but of adaptive strategy in harsh, fluctuating environments. Kauffman (1972) relates the low diversity of North Temperate Cretaceous molluscs to the irregular fluctuations of their environment.

On the other hand, the necessity for a correlation between high diversity and environmental stability is under attack (e.g., by May, 1975a). Highly stable environments may eventually beget low diversity, since a few excellent competitors may eventually assert their dominance (as many natural monocultures testify). Maximal diversity may arise under conditions of intermediate disturbance (too much disturbance eliminates all but a few physically accommodated species; too little permits the complete usurpation of space by best competitors). This reechoes one of Horn's themes from chapter 10 (Horn, 1975a, b). Good evidence for contol by disturbance comes from work on modern communities by palaeontologist R. Osman (1975). At Woods Hole, Massachusetts, he found that maximal diversities characterize rocks of intermediate size (large rocks are too big to be disturbed by turning; little rocks roll over too often). But didn't Bretsky and Lorenz anticipate this revision of ecological theory (admittedly at a different scale)? They challenged the dogma that diverse communities are necessarily stable, but only for the long perspective of geological time with its assurance of an eventual, major disturbance. They did not realize that this lack of stability might also extend into the micro-world of ecological time as well. Since the publication of Simpson's seminal treatises (1944, 1953), palaeontologists have been trying to extrapolate the world of microevolution to their time scales. Perhaps it is time for ecologists to examine the insights of palaeontologists for potential compression into their own worlds.

12.2.2 *How does genetic variability relate to the diversity of communities and the probability of extinction?*

Bretsky and Lorenz (1969) linked their pattern of nearshore-low diversity-long term persistence vs. offshore-high diversity-earlier extinction to a genetic speculation. They proposed that nearshore species are 'heteroselected' for genetic variation needed for flexibility of populations in frequent times of severe fluctuation. Offshore species are 'homoselected' for restricted variation since they adapt so narrowly to their stable habitats. They then claimed that the celebrated mass extinctions of the geological record differentially removed species from homoselected, offshore communities. These species did not have enough genetic variability to meet the adaptive requirements of rapidly changing environments. At times of crisis, in other words, the offshore species became victims of their own previous success.

This palaeobiological speculation has directly inspired some of the most fruitful work in recent years at the interface of ecology and genetics. Palaeontologists cannot study genes directly, but many of them have learned the techniques of electrophoresis to probe modern organisms for insight into the palaeobiological dilemma of extinctions. Is there a relationship between stability of environment, adaptive strategy and the amount of genetic variability present in populations?

Schopf (a palaeontologist) and Gooch (1972) studied genetic variability in 8 deep-sea species (1000–2000 m) and reported their results with the title: 'A natural experiment to test the hypothesis that loss of genetic variability was responsible for mass extinctions of the fossil record.' In this highly stable environment, they found 20–50 percent of loci polymorphic, thus casting great doubt on Bretsky and Lorenz' contention that reduced variability in stable environments favoured the extinction of offshore forms. Palaeontologist J. W. Valentine, with geneticist F. J. Ayala and several colleagues have extended this approach to reach empirical conclusions directly opposite to the speculation of Bretsky and Lorenz (Valentine *et al.*, 1973; Ayala *et al.*, 1973b; Valentine and Ayala, 1974). They studied species on a geographic gradient in trophic resource regimes from stable-tropical to unstable-high latitude; they found consistently that tropical species maintained more genetic variability in their populations than high latitude forms. Their first studies compared the incommensurable (tropical clams, temperate horseshoe crabs and antarctic brachiopods) and permitted no firm conclusions. But Valentine (personal communication) has just shown me data for closely related species of krill along geographic gradients. Again, variability increases steadily towards the tropics.

Bretsky and Lorenz, in retrospect, may have made the fatal error of confusing the physiological plasticity of individuals with the genetic variability of populations. Individuals in physically-controlled communities must be able to tolerate a wide range of environments, but they may achieve this by rigorous selection for genes conferring plasticity; variation among individuals is a different matter. Valentine believes that populations of tropical environments are more variable because the richness of stable habitats provides a wide range of microenvironments to which different subpopulations adapt. In any case, if this correlation between environmental stability and genetic variability holds, it will provide a powerful selectionist argument against the hypothesis of neutralism as a source of most genetic variation. The selectionist-neutralism debate may now be the cardinal controversy in evolutionary biology. A large piece of its resolution may have been inspired directly by a palaeobiological speculation. Valentine and Ayala (1974, p. 70) are justified in praising Bretsky and Lorenz 'for producing an elegant and testable hypothesis'—even though it was probably wrong. It is a good measure of the excitement of ecology that scrupulously correct factual documentation is no longer the only activity that wins praise among palaeontologists.

12.2.3 *Succession as a model for microtemporal change*

Palaeontologists may love to speculate about the general course of life, but most of their work involves historical sequences in local sections encompassing a relatively small amount of time. Changes in taxonomic composition up a section have traditionally been interpreted as sets of evolutionary events or evidences of migration mediated by changing environments. Several recent studies have suggested that such common patterns often represent autogenic successions. Since periodic wipe-out is the geological fate of any community at a given spot, the opportunity for rediversification through succession occurs with great frequency in the fossil record. Bretsky and Bretsky (1975) have interpreted some late Ordovician sequences as successions initiated by the colonization of offshore barren muds by two or three opportunistic deposit-feeding species. Walker and Alberstadt (1975) detect successional sequences in several fossil reefs (Ordovician to Cretaceous in age).

A problem in the Walker and Alberstadt (1975) paper illustrates the potential of ecological theory. They are puzzled by the sharp decrease in species diversity at the final stages of apparently autogenic successions. Since they accept the dogma of classical stability-diversity theory, they are stumped by this 'uppermost domination zone' of drastically reduced diversity (far greater than the orthodox slight reduction in climax communities). They are led to the *ad hoc* suggestion of allogenic control by changed water turbulence to produce 'a physically accommodated community from what was a near climax stage biologically accommodated one' (1975, p. 243). But they might avoid this *deus ex machina* by recognizing May's demonstration (1975a, c) that highly stable environments might eventually yield communities of low diversity. In any case, palaeontology can provide the only record of complete *in situ* successions. The framework of classical succession theory (probably the most well known and widely discussed notion of ecology) rests largely upon inferences from separated areas in different stages of a single, hypothetical process (much like inferring phylogeny from the comparative anatomy of modern forms). We can provide direct evidence to supplement the revisionist discussion in chapter 10.

12.3 Basic questions in the history of life

The history of life has been anything but smooth, and its major episodes have provided the greatest problem of palaeontology since Cuvier's day. A list of theories for the Cambrian 'explosion' and Permian extinction alone would probably fill the London telephone directory. Most palaeontologists (not including myself) also believe that directional patterns can be discerned through all this tumult of pulsating extinctions and radiations, and the causes of generally increasing diversity or morphological complexity have been much debated. During the past five years, ecological theories have inspired new explanations for all these ancient dilemmas; some seem so satisfactory that I might even be led to proclaim certain issues as settled if I did not so well appreciate the short half life of previous 'definite' solutions.

12.3.1 *Pulsations in the history of life*

About 600 million years ago, almost all the phyla of marine invertebrates enter the fossil record for the first time during the short space of a very few million years. Yet, for the previous 2500 million years of earth history, little more than prokaryotic bacteria and algae populated the world, forming extensive stromatolites (algal mats) in their favoured habitats. Most explanations for this Cambrian 'explosion' have been crassly physicalist. The highly touted Berkner-Marshall hypothesis (1964), for example, holds that it marks an attainment of atmospheric oxygen levels sufficient to screen ultraviolet light (by an ozone shield) from shallow waters and provide for respiration—as if organisms were inert billiard balls, responding immediately and automatically to any external stimulus.

Stanley (1973a) has recently proposed an ingenious explanation by importing a key notion of contemporary ecological theory. Contrary to intuition, the introduction of a higher trophic level (a herbivore in a plant community, a carnivore among herbivores) tends to increase rather than decrease diversity at levels below it. 'Cropping' frees space otherwise completely usurped by the one (or very few) dominant competitors of the uncropped system. Precambrian stromatolites look, to Stanley, like a classical uncropped ecosystem of very low diversity. The 'hero' of the Cambrian explosion may have been the first single-celled herbivorous protist. Stanley's thesis does not tell us how this herbivore evolved (in part, the knotty problem of the origin of the eukaryotic cell); but he does provide a reasonable explanation for the rapidity of subsequent change once it began to crop the Precambrian monoculture. The cropping would be geologically instantaneous on a worldwide scale. The combination of available ecospace and the unexploited potential of a recently evolved eukaryotic cell may have determined the greatest burst of evolutionary activity this planet has ever known.

Two hundred seventy five million years later, the debacle came as fully half the families of invertebrate animals became extinct in an equally short period of time at the boundary between Paleozoic and Mesozoic eras (Newell, 1973). This 'Permian crisis', unlike the Cambrian explosion, lies within the part of earth history well documented in the fossil record. Hence, it has received more attention from the profession, and may rightly be called palaeontology's outstanding dilemma (Gould, 1974). Favoured explanations again have tended to be physicalist—extraction of salt from the oceans and explosion of supernovae being among the most popular recent suggestions.

I believe that a solution—or at least a first order control—is now emerging thanks to the union of two theoretical positions that scarcely existed a decade ago: the revolutionary paradigm of plate tectonics in geology, and the equilibrium thinking of the MacArthur-Wilson (1967) approach to theoretical ecology which was discussed in chapter 9. The late Permian represents the singular time (at least since the Cambrian explosion) during which the earth's continents coalesced to the single super-continent of Pangaea. The coincidence of this coalescence with the Permian extinction is not likely to be accidental, but we still must ask why the union of land masses should provoke such a catastrophe for shallow-water marine life.

Schopf (1974) has recently argued that the ecological literature on species-area curves may provide a key insight: the relationship between species diversity and habitat area often conforms to a simple power function with a slope considerably less than unity, as discussed in section 9.2 [see eq. (9.1)]. The fact of correlation is enough for Schopf's argument; I will bypass the issue of whether area *per se* is the controlling variable, or whether it merely acts as a surrogate for such determining factors as habitat diversity. The joining of all major continents produced a marked reduction in the area of shallow seas for two reasons: (1) basic geometry: shallow seas are peripheral to continents, and there is far less periphery around a single large continent than around several small continents of the same total area; (2) the mechanics of plate motion. When continents join, the margins of plates lock together since continents are too light to be subducted. Plate tectonics requires a balance between creation of new sea floor at mid-oceanic ridges and subduction of old sea floor at plate margins. If subduction stops, creation ceases as well. The oceanic ridges collapse as the supply of molten material (previously designated for new sea floor) dwindles. The ridges make up a substantial amount of undersea topography, and their collapse would cause a marked reduction of sea level. This reduction would drain the continental shelves and drastically reduce the area available to shallow water marine animals.

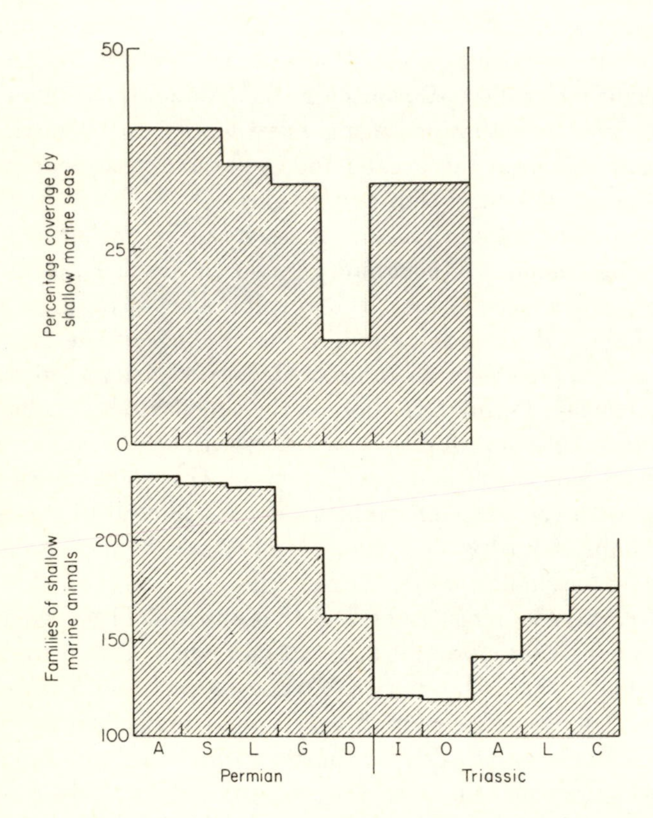

Fig. 12.1. Correlation of faunal extinction with the reduction in area of shallow water seas during the Permian crises. (From Schopf, 1974.)

Schopf (1974) tested his hypothesis of areal control by using the best available evidence of geological maps to measure the reduction in shallow seas during the last three Permian stages (they were reduced, he concludes, from 40 per cent of possible early Permian distribution to 15 per cent for the latest Permian). He also calculated the reduction of taxa by extinction during these stages (Fig. 12.1). In a companion paper, ecologist D. Simberloff (1974) then developed a way of estimating species-area curves from the faunal data of families (for a variety of reasons—including the sins of past taxonomists and the imperfections of the geological record—we can form no reliable estimates for numbers of fossil species). Using Schopf's data for faunal and areal reductions, he then obtained a good fit to standard species-area curves. Schopf (1974) concludes that reduction of area itself served as the first-order control of the Permian extinction.

12.3.2 *Basic patterns in the history of diversity*

If we count the number of fossil taxa recorded in the palaeontological literature, we note a pattern of steady increase modified by the pulsations discussed in the last section. A rapid increase marks the Cambrian explosion. A slow decline sets in towards the late Paleozoic, culminating in the precipitous drop of the late Permian. Since Triassic times, the trend has been steadily upward. (The Cretaceous extinction did not markedly affect marine invertebrates.) Valentine (1970, 1973) has argued that this empirical pattern is a fair representation of true history and that, following the Cambro-Ordovician filling of the ecological barrel, diversity pretty well maps the configuration of continents—moderately high with moderately separated mid Paleozoic continents, declining as the continents join to form Pangaea, and steadily increasing with post Permian continental separation (Fig. 12.2). Opportunity for endemism among shallow water faunas becomes the chief control of global diversity (augmented by such secondary factors as increased climatic stability in a world of small, widely separated continents).

Raup (1972) has challenged Valentine's interpretation by arguing that the empirical curve might represent little more than an artifact of sampling. After all, preservability (more rocks and less alteration) increases markedly as we approach the present day. Raup makes a rough estimate of the biases and demonstrates that the empirical curve can be obtained by sampling a real distribution with the following pattern: rapid rise in the Cambrian, increase to a mid Paleozoic high, slow decline to a late Paleozoic intermediate level, and constancy thereafter.

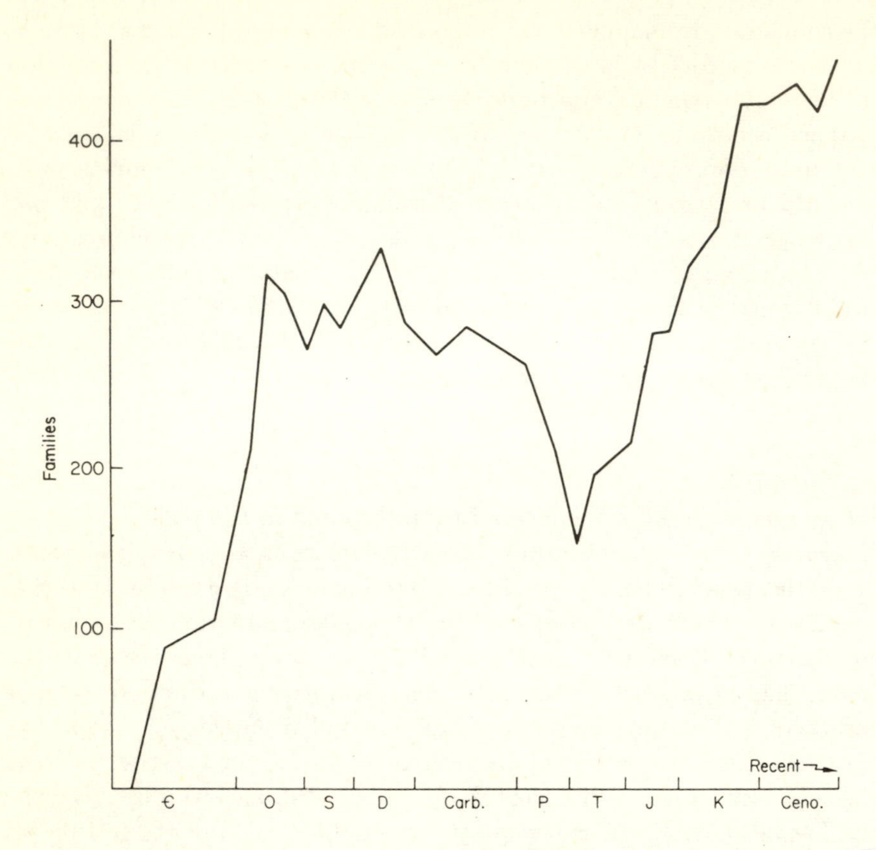

Fig. 12.2. Empirical curve for diversity through time for families of shallow water invertebrates. (From Valentine, 1970.)

Hidden in this debate is one of the most interesting ecological questions we can ask within a palaeontologist's time scale. If diversity increases generally through time, why? One answer, emphasized by Valentine, holds little interest for an evolutionary theorist. If diversity simply maps continental positions, then its vector of increase is an 'accidental' result of historical trends in continental separation; different patterns of plate motion would have yielded a different curve of diversity through time. But Valentine (personal communication) also accepts an 'intrinsic' influence upon increasing diversity. He argues that, all things being equal, average niche breadth should tend to decrease with time—i.e., we should find increased species packing *within* a constant amount of geographical and ecological space. An alternate view (towards which I incline) holds that the ecological barrel filled up quickly following the Cambrian explosion. Limiting similarity is rapidly approached and there are no intrinsic trends in species packing. The same general equilibrium number of species applies to similar environments through time. Actual changes in diversity reflect the influence of external events in altering equilibria by changing environments—e.g., the decrease in equilibrium number caused by draining of the continental shelves in late Permian times.

This debate is merely the latest incarnation of the most fundamental question in palaeontology: does the history of life have an intrinsic direction (towards greater morphological complexity, increased diversity, etc.)? In more general terms, is steady-statism or directionalism a more appropriate metaphor for the history of life? This is an ancient argument, long predating any belief in organic evolution. It was, contrary to conventional historiography, the major issue separating Lyell from the catastrophists (Hooykaas, 1963; Rudwick, 1972; Gould, 1975). Lyell argued for a Newtonian timelessness in earth history—constant overturn of species through extinction and new creation, but no trends in complexity or diversity. The catastrophists, as progressionists, claimed that each new creation approached more clearly the perfected forms of our current earth.

The obvious test for an intrinsic increase in diversity requires a comparison of perfectly preserved faunas in similar environments covering the same area through time. R. Bambach (personal communication) has attempted to do this, although the vagaries of an imperfect fossil record do not permit great confidence in the measure of control (especially over areal extent of a community). Bambach's results are interestingly ambiguous. He divides communities into nearshore, offshore shallow and offshore deep, and times into early Paleozoic, mid Paleozoic, late Paleozoic, Mesozoic, and Cenozoic. For the opportunistic, low-diversity nearshore communities, Bambach finds no trends in species diversity through time. Both categories of offshore communities display the same pattern: low diversity in the early Paleozoic, no statistically significant differences for nearly 400 million years from mid Paleozoic through Mesozoic, but a rise in diversity for Cenozoic communities. The low diversity of early communities surprises neither side: it represents the filling of the ecological barrel following the Cambrian explosion. The long period of subsequent stability seems to favour the steady-statist metaphor; but the Cenozoic increase, if real, demands a special explanation. Bambach believes that it may reflect increasing replacement by narrow niched molluscs of the broader niched dominants of earlier communities (brachiopods and bryozoans, for example).

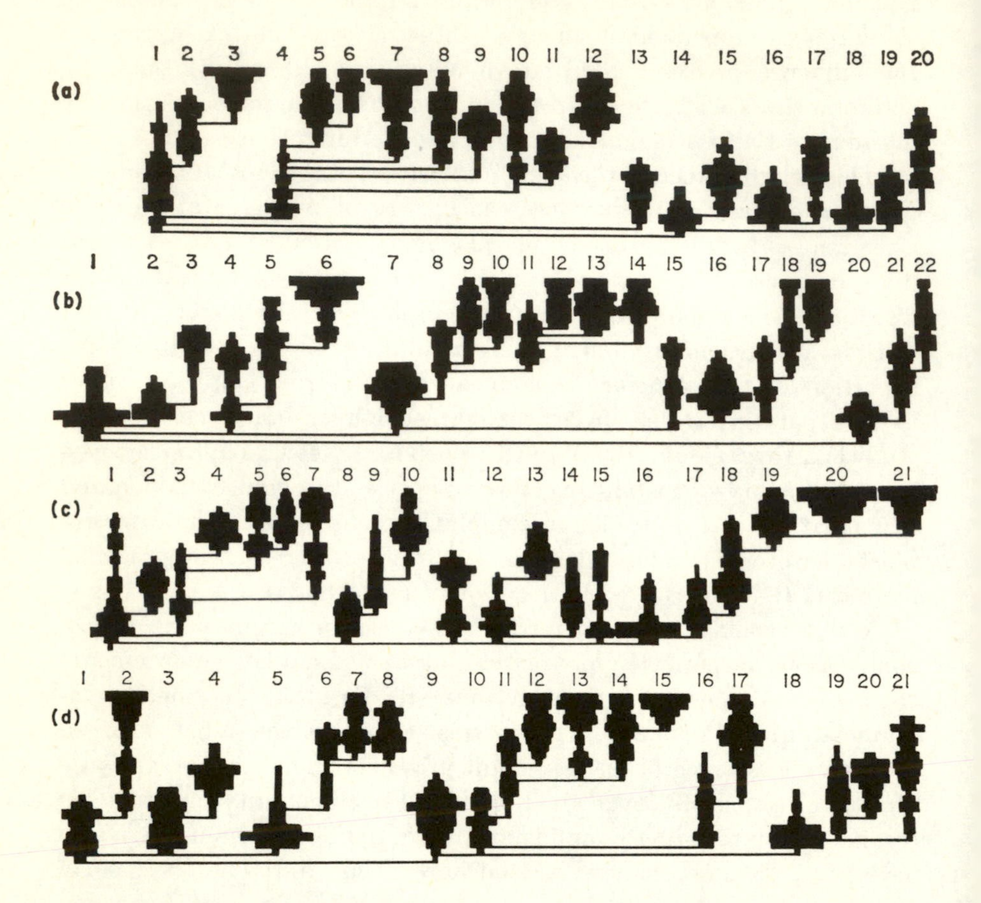

Fig. 12.3. The diversity and form of stochastic clades simulated within an equilibrium system by Raup *et al.* (1973). Only the seed of the random number generator (for decisions on extinction and branching) varies among the runs.

Raup *et al.* (1973) have followed a deductive approach in supporting the steady statist position (Raup and Gould, 1974, have also tried to model the history of morphology with similar techniques). The success of equilibrium thinking in theoretical ecology (MacArthur and Wilson, 1967; Levins, 1968; May, 1975a) has supplied the model for this enterprise. Raup *et al.* simulate evolutionary trees under conditions of equal probability for branching and extinction of lineages, and a constant equilibrium number of coexisting lineages for all time periods. They classify the tree into a smaller number of higher categories (clades) and draw their diversity diagrams through time (Fig. 12.3). The simulations reproduce many features of real clades, universally attributed to directed causes in the past (e.g., strong waning of one clade while another waxes as evidence of competition). Gould *et al.* (in preparation) are now undertaking an extensive statistical comparison of real and stochastic clades.

12.4 Conclusion

12.4.1 *The ecological way of thought*

I have reviewed some specific contributions of ecological theory to the solution of long standing palaeontological dilemmas. Ultimately, however, its greatest impact may lie in a more subtle and pervasive direction: the reorientation of basic attitudes towards evolutionary questions. In discussing adaptation within phyletic sequences, it is legitimate to consider both 'immediate' significance in local situations and 'retrospective' significance for future evolutionary success (since we read phyletic stories after they have occurred). Darwin argued (largely in vain among palaeontologists) that evolutionary change was nothing more than adaptation to local circumstances. A 'trend' can only be a concatenation of these local adaptations; there are no intrinsic directions in evolution. Ecological time is rooted in this Darwinian present; ecological themes encourage palaeontologists to study adaptation for its immediate significance. Traditional palaeontology rarely worked at this level; it focused instead on the meaning of adaptation as a contribution to long-term evolutionary trends. And when it considered the immediate significance of adaptation at all, it did not venture beyond morphology in a 'physicalist' perspective—i.e., to what aspect of the physical environment is this structure fitted? The ecological themes of population dynamics, life history strategies and species interactions were simply not categories for consideration.

If we are now treating immediate significance seriously with these new themes, then almost every empirical study in palaeontology will be conducted differently in the future. I can only offer personal testimony. Two very similar species of land snails inhabit all environments and times of the Bermudian Pleistocene. They are usually found together in the same formation, but one or the other is sometimes absent. Ten years ago, I noted that one species in several localities from western Bermuda seemed to converge slightly upon the other. For years, I could not drop the blinders of traditional thought. I racked my brains for either phyletic meaning in the long run, or for the physical parameter that conditioned it—i.e., I searched for a way in which the sedimentary environment resembled that of samples dominated by the other species. I never considered interaction. Only last year did I realize that these western samples included the only allopatric occurrences of my converging species and that the greater separation in sympatry probably represents character displacement. In an excellent example of shifting perspectives, Stanley (1973) has invoked differences in intensity of competition to explain the more rapid rates of evolution in mammals vs. clams. Simpson (1953) set this issue as a classic dilemma in macroevolution. Previous explanations had been confined to the traditional categories of morphology (e.g., mammals more complex).

To illustrate the blinders of macroevolutionary phyleticism, I submit the following: In a type of paedomorphosis, termed progenesis or paedogenesis (Gould, 1976), descendants retain the juvenile morphologies of ancestors because ontogeny is truncated by precocious sexual maturation. Progenetic organisms are usually very small and rapidly developing. Traditional wisdom (e.g. de Beer, 1958) has denied progenesis any importance in evolution because it is essentially 'degenerate', leading to simplification of structure and shortening of ontogeny—fine for some dead end parasites, but not for anything of phyletic significance. Note first of all, how this denial assumes that retrospective significance is the only thing that matters in evolution—a typical palaeontological chauvinism. Traditional thought could find no general immediate significance, but the ecological theme of life history strategies almost surely provides it (Gould, 1976, chapters 8–9): progenetic organisms are *r*-selected for rapid maturation in unpredictably fluctuating environments. Secondly, palaeontologists have had to face a dilemma of their own making. The origin of many higher taxa has been attributed to paedomorphosis by the sexual maturation of marine larvae. Clearly, the easiest mechanism for such paedomorphosis is progenesis (precocious maturation). But progenesis is supposed to have no evolutionary significance, and speculators have been driven to fanciful arguments for why paedomorphosis in these marine larvae might have been neotenic (delayed somatic development with no shortening of ontogeny). Slowly developing, giant larvae become the fanciful models for new higher taxa. But: (1) we now have a good explanation from ecological theory for the immediate significance of progenesis; (2) immediate and retrospective significance are different categories. I see no reason why a small, rapidly maturing organism living in a precarious environment cannot be the stem species of a successful group.

12.4.2 *Cross fertilization*

Palaeontologists should be grateful to theoretical ecology for the reinvigoration of their field. But we surrender no sovereignty and look for reciprocal influence. I have already suggested that palaeontological data for truly temporal sequences might resolve some of the issues surrounding succession theory and the relationship of diversity to disturbance. I also believe that a bit of palaeontological traditionalism might serve as a good antidote for some excesses of simplification often encountered in ecological models. Too many models treat species as billiard-balls of no recognizable shape (though perhaps of different sizes). Yet adaptive morphology must be consulted for the solution of many problems in diversity. Take, for example, the remarkable species swarm of cichlid fishes in deep African lakes. A theoretical ecologist might look to stability of environments, breeding structures and life histories. But he would miss an important part of the story if he neglected the details of morphology. In an elegant paper, Liem (1973) demonstrates that explosive speciation was permitted by a shift in the insertion of two fourth levator externi muscles. This shift permits the pharyngial jaws to prepare food (rather than just transporting it as before). This frees the premaxillary and mandibular jaws to evolve exuberant specializations for the collection of food—scale scraping, eye biting, etc. As Deevey wrote in reviewing a founding work in modern palaeoecology (1965, p. 593): 'Older ideas of functional morphology are still valid; ecologists can learn much by simply looking at animals, before mentally decomposing them into fluxes of organic carbon.'

Palaeontology will retain its autonomy because the world allows no complete translation across its hierarchical levels of complexity. We should not surrender our uniqueness to the prestige and success of ecology; extrapolation from ecological to evolutionary time remains a dangerous game. When Tappen (1971) refers to patterns of diversity over many millions of years as a 'succession' (if only analogously), I begin to feel very uncomfortable. The concept of succession does not include replacement by evolution. The laws of scaling require different explanations at contrasting levels. An elephant cannot be built to look like a mouse, and millions of years demand their own set of concepts.

Consider evolutionary 'trends' as one phenomenon that probably requires explanation at its own level. In the traditional view of phyletic gradualism, trends occur within lineages and are merely extensions of selection acting for the moment within populations. But Eldredge and Gould (1972) have argued that rapid events of speciation provide the primary input for macroevolutionary change; large populations rarely transform slowly and steadily. Each event of speciation represents a local adaptation, not a stage of a developing trend. Events of speciation provide an essentially stochastic input to evolutionary trends. The trends themselves represent differential preservation and success of a subset of speciations. Trends cannot be explained in the ecological time of speciation itself, but only in the evolutionary time of a higher-order 'selection' of speciation events (Eldredge and Gould, 1972, pp. 111–113; see also Stanley, 1975, who refers to this mechanism as 'species selection'). The laws of ecology will not encompass trends, but they do set the speciations that serve as their building blocks. Ecological time will remain the fundamental level of palaeontological analysis. We give our thanks to modern ecology for teaching us the rudiments of perception in Darwin's own sphere of operation.

13
Schistosomiasis: A Human Host-Parasite System

JOEL E. COHEN*

13.1 Introduction

Human schistosomiasis (pronounced SHIS-to-so-MY-uh-sis, or sometimes, synonymously, bilharzia) is a family of diseases caused primarily by three species of the genus *Schistosoma* of flatworms. The adult worms inhabit the blood vessels lining either the bladder or intestine, depending on the species of worm. Hence the worms are also known as blood flukes.

Jordan and Webbe (1969) provide a monographic review of human schistosomiasis. Malek's (1961) shorter review emphasizes the ecological point of view. The insatiable reader is invited to consult the 10,286 references cited by Warren and Newill (1967), which do not include most of those on which this chapter is based.

Aside from its human impact, schistosomiasis is a marvellous source of puzzles for the biologist. How does this metazoan parasite escape annihilation by human immune responses? How do genetic factors in snails condition susceptibility to infection by the asexual stages? How do worms of opposite sex find each other in the dark?

How does the schistosome cope with successive environments of human faeces, water, snail tissue, water again, and human serum and other tissues? Why is the fraction of snails observed to be infected in natural populations so remarkably low (Warren, 1973)? Are there immunological cross reactions between human schistosomes, as has been suggested for malaria (Cohen, 1973a)?

How do human behaviour and agricultural practices combine to propagate an insidious pest, and how can people balance conflicting demands of economics and health in controlling it?

The worldwide prevalence of schistosomal infections has not been

* This work was supported in part by the U.S. National Science Foundation, under grant number BMS 74–13276.

measured credibly. A figure conventionally cited is 200 million people, or approximately one out of every twenty people on the planet. The disease is virtually unknown in the rich countries of the world. Thus it is concentrated among those people least able financially to defend themselves against it.

According to a foundation which devotes a substantial fraction of its money to research on schistosomiasis, 'preliminary calculations indicate that global control would cost at least $2 billion per year at today's prices with the best current methods!' (Hoffman, 1975, p. 26). The exclamation point is intended, in context, to emphasize how expensive global control would be. But the resulting estimate of $10 per case for worldwide control of schistosomiasis should be compared with the $30 per case spent in the United States for research alone on muscular dystrophy, and with the alternate uses $2 billion have been put to recently.

'There is little doubt that all three schistosomes can cause considerable pathological change, sometimes in a comparatively large proportion of the population, but the evidence suggests that only a proportion of those so affected die of the disease' (Jordan and Webbe, 1969, p. 168). The absence of quantitative information from this assessment of the impact of the infection on health fairly reflects the information available.

After a schematic description of the life cycle and ecology of *Schistosoma mansoni*, this chapter will review several mathematical models of aspects of schistosomiasis. To date no mathematical model of the disease has come close to representing faithfully most of what is known about the natural history of the disease; none, in fact, has even tried. So the field is still open. The juxtaposition of these models will make clear the selective emphasis, and some of the omissions, of each.

13.2 A life cycle of schistosomiasis

The common features of all three schistosome species which are the major parasites of man are summarized in Fig. 13.1 (from Jordan and Webbe, 1969, p. 7, from whom the following brief description is also drawn). The life cycle consists of an obligatory alternation of sexual and asexual generations. The sexual generation occurs in man (or sometimes other mammals). The asexual generation must pass through specific snails.

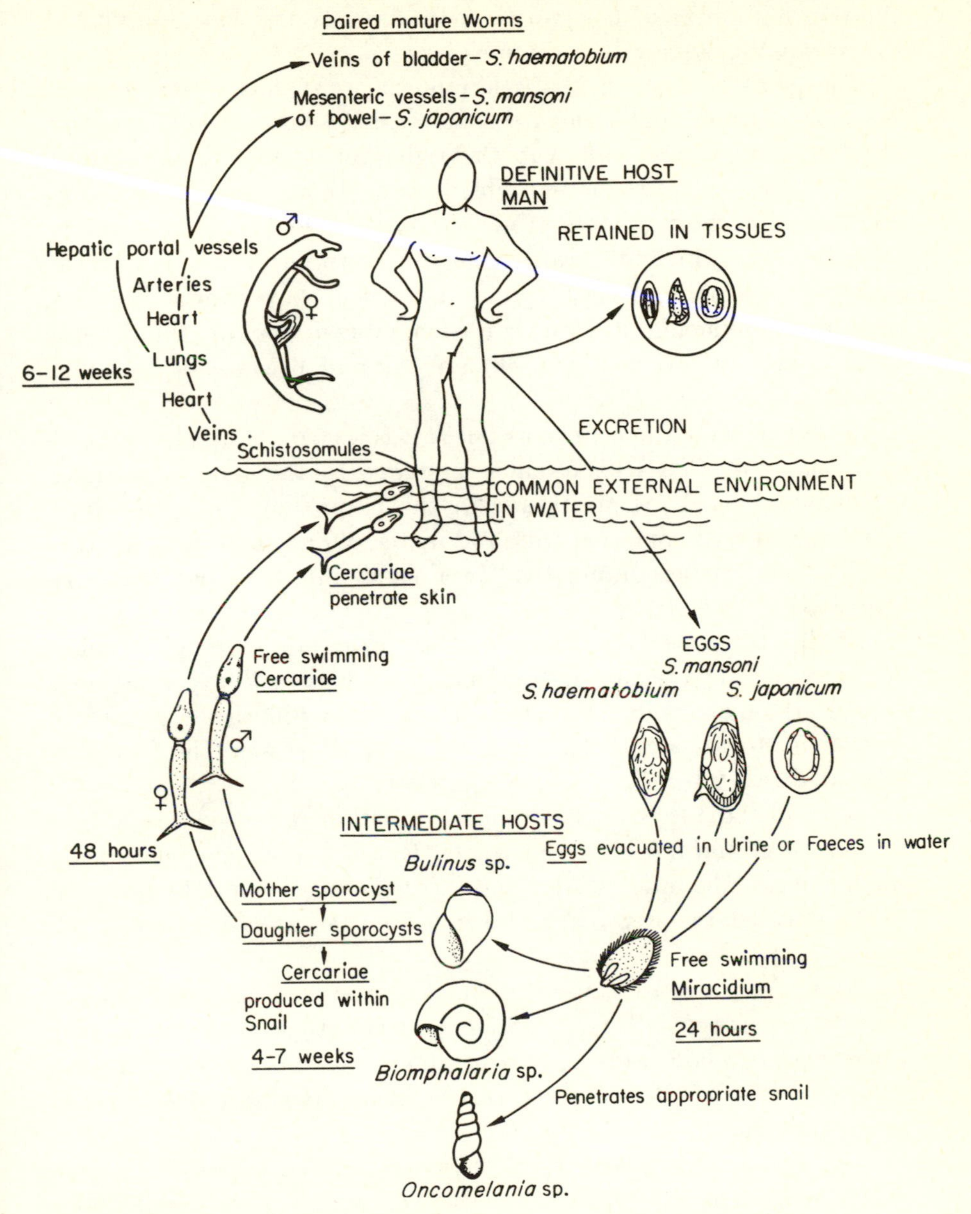

Fig. 13.1. The life cycle. (From Jordan and Webbe, 1969, p. 7.)

Eggs produced by the sexual stage leave people in urine (in the case of *S. haematobium*) or faeces (*S. mansoni* and *S. japonicum*). Eggs that reach water (a stream, an irrigation canal, a pond, puddle, or ritual ablution bath near a mosque) shed their shells and hatch a ciliated free-swimming stage called a miracidium.

The quantitative estimates in the following refer chiefly to *S. mansoni*.

A miracidium that locates a snail within approximately one day penetrates it. If the snail has the right genotype, the miracidium develops through two larval stages into thousands of cercariae. Each cercaria that escapes from the snail, starting 4 to 5 weeks after the initial infection, can swim up to two days. When cercariae penetrate the skin of host species in which they cannot develop, they die, producing a skin irritation. When one of the human schistosomal cercariae enters human skin, it becomes a wormlike 'schistosomule'.

Here we follow the course, as far as it is known, of a schistosomule of *S. mansoni*. The schistosomule migrates to the lung, sometimes producing a cough, then appears in the portal system of the liver, where it reaches sexual maturity and mates. Worm pairs then migrate to the blood vessels lining the lower small intestine and the large intestine.

At this point the happy couple of worms resemble (roughly) a hot dog in a roll. The female, 7 to 17 mm long, lies in the gynecophoric canal of the male, who is 6 to 13 mm long and cylindrically shaped to correspond to the walls of their home, a blood vessel. The female is estimated to lay from 100 to 300 eggs a day.

Some of these eggs work through the wall of the blood vessel, into the lumen of the intestine. Carried by faeces, these eggs again begin the life cycle. The interval from the entry of cercariae into human skin to the first detectable passage of eggs in the faeces varies from 5 to 6 weeks.

Most of the disease caused by the infection results from the eggs that do not escape with faeces. Some eggs get stuck in the tissue near where they are laid, causing fibrosis and granuloma as the host tries to protect itself. Other eggs get washed to the liver and spleen where they cause similar damage.

The medicines available to kill the schistosomes in people have so many dangerous side effects that they must be administered under medical supervision; and they are costly. So they may help rich tourists returned home; but by themselves they do nothing to protect a person in an endemic area against reinfection once he or she leaves the clinic. Even if enough medical personnel were available to treat all the infected population in a single month, the snail (and sometimes other mammalian) reservoirs of infection would persist. The control or eradication of schistosomiasis is a truly ecological problem.

13.3 Proportion ever infected

If present conditions influencing infection with a parasite existed since well before the oldest people alive were born, then every age group should have the same proportion infected as did the community's oldest members when they were at each corresponding age. Assuming constant conditions, a snapshot of the distribution of infection according to age in a community today provides a picture of the life experience of a cohort (a group of people all born at the same time).

The mathematical models described in this section and the next take advantage of that possibility by using data about the prevalence of previous or present infection in a population at one point in time to make inferences about the dynamics of infection over time in a cohort.

Suppose a cohort is entirely susceptible to infection at some initial time, usually taken to be birth. Suppose that this cohort is exposed to a constant force of infection per unit time. 'This force is to be measured in effective contacts per unit time, no matter how complex may be the events leading up to these contacts' (Muench, 1959, p. 16). Here the force of infection a summarizes everything that has to do with the contact between cercariae and people and with the subsequent establishment of a detectable infection. Suppose also that it is possible to classify each individual as ever having been infected or not.

Let N be the total number of individuals initially in the cohort. Let $x(t)$ be the fraction of the cohort that has not been infected, and $y(t)$ the fraction that has ever been infected, by time t. By definition $x(t)+y(t) = 1$. Assume $x(0) = 1$ and $y(0) = 0$. Then $Nx(t)$ is the number of individuals never infected at time t. These individuals are constantly exposed to a force a of infection. So the change per unit time in the number $Nx(t)$ of people never infected is

$$d[Nx(t)]/dt = -aNx(t),$$

or, cancelling N, assumed constant,

$$dx/dt = -ax, \qquad x(0) = 1. \qquad (13.1)$$

Similarly, for the number ever infected,

$$dy/dt = ax = a(1-y), \qquad y(0) = 0. \qquad (13.2)$$

The solution of eq. (13.2) is simply

$$y(t) = 1 - e^{-at}. \tag{13.3}$$

This model assumes no change in the size of the cohort due to death or emigration. Death or emigration will have no effect on the fraction $y(t)$, so long as the death or emigration rates have identical values, $\mu(t)$, for both previously infected and never infected individuals (Cohen, 1972). To see this, we write the appropriate equations, where now the total population $N(t)$ is no longer constant:

$$d[N(t)\,x(t)]/dt = -aN(t)\,x(t) - \mu(t)\,N(t)\,x(t); \tag{13.4}$$

$$d[N(t)\,y(t)]/dt = aN(t)\,x(t) - \mu(t)\,N(t)\,y(t). \tag{13.5}$$

By writing $N(t) = N(0)\cdot\exp\left[-\int_0^t \mu(s)\,ds\right]$ here, we recover precisely eqs. (13.1) and (13.2) for the *fractional* population functions $x(t)$ and $y(t)$.

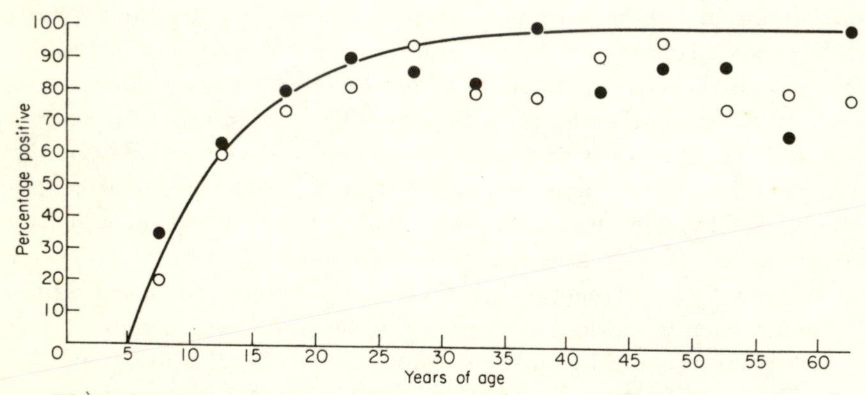

Fig. 13.2. The proportions observed positive in response to *S. japonicum* antigen skin tests as a function of age in the coastal division, Palo, Philippines. Open circles, 1954; solid circles, 1962. Solid line, prediction from eq. (13.3). (Adapted from Hairston, 1965b, p. 172.)

If conditions were constant over the past, and the means of detecting previous infection include individuals infected long ago, then (Hairston, 1965b) a survey of all age groups in a population at a given time should give a graph of the fraction of people ever infected as a function of age that looks like eq. (13.3).

Figure 13.2 takes $t = 0$ as 5 years of age. Infections before that age are neglected. The data are the fractions of people in each age group judged ever to have been infected with schistosomes on the basis of skin tests in 1954 and 1962. Simple in theory, in practice these tests suffer from many difficulties (Kagan, 1968).

Particularly for the younger age groups, the fit of eq. (13.3) to the data is reasonable. The systematic discrepancy between data and observation at the upper ages is plausibly explained by Hairston (1965b), on the basis of other calculations, as due to an insensitivity of the skin test to previous infection if the person has not recently been exposed to female cercariae or has no living female worms.

The numerical value of the parameter $a = 0 \cdot 12$ used in Fig. 13.2 was not obtained by fitting that curve to those data. The parameter was estimated by fitting another eq. (13.9) below to different data, from stool examinations, on the same population. This finding, which I consider one of the most remarkable in the epidemiology of schistosomiasis, suggests that an incredibly simple, and simple-minded, mathematical model can usefully interpret the age distribution of previous infection and yield information about the dynamics of infection which would otherwise be unavailable.

13.4 Proportion currently infected

Female *S. mansoni* worms in human beings live an average 3 to 4 years; other species of human schistosomes are comparable (Hairston, 1965a, p. 52; Jordan and Webbe, 1969, p. 152).

A person in whom all female worms have died no longer excretes eggs. (A person may also no longer excrete eggs because tissue traps the eggs or because living females are unmated. We ignore these complications.) Hence some individuals previously infected may pass from currently excreting eggs to no longer excreting.

13.4.1 *A model allowing loss of infection*

Let $y(t)$ (not the same as in section 13.3) be the proportion of a cohort which is currently giving evidence of infection by excreting eggs and $z(t)$ be the proportion which has been previously infected but is no longer passing eggs. As before, let $x(t)$ be the proportion of the cohort at time t which has never been infected. Assume no death or emigration, so $x(t)+y(t)+z(t) = 1$.

If the cohort is subject to a constant force of infection a, and those individuals currently giving evidence of infection are now further subject to a constant risk of loss of infection b, then under constant conditions the proportions x, y and z behave (Muench, 1959) according to

$$dx/dt = -ax, \qquad x(0) = 1, \tag{13.6}$$

$$dy/dt = +ax - by, \qquad y(0) = 0, \tag{13.7}$$

$$dz/dt = +by, \qquad z(0) = 0. \tag{13.8}$$

All individuals are uninfected initially. The sum of the derivatives is zero, as it must be since the cohort does not change size. Then if $a \neq b$,

$$y(t) = a(e^{-bt} - e^{-at})/(a-b). \tag{13.9}$$

Repeating the argument in section 13.3 shows that if death and emigration occur at equal rates in all three fractions of the cohort, the same equations (13.6) to (13.8) still hold.

In the limit as t gets large, $y(t)$ approaches 0, because eqs. (13.6) to (13.8) really assume that once having lost infection, an individual does not regain it. There is no re-entry from the category measured by $z(t)$ to the category measured by $y(t)$. Hence $z(t)$ must be interpreted as the proportion having lost infection and not exposed to risk of further infection.

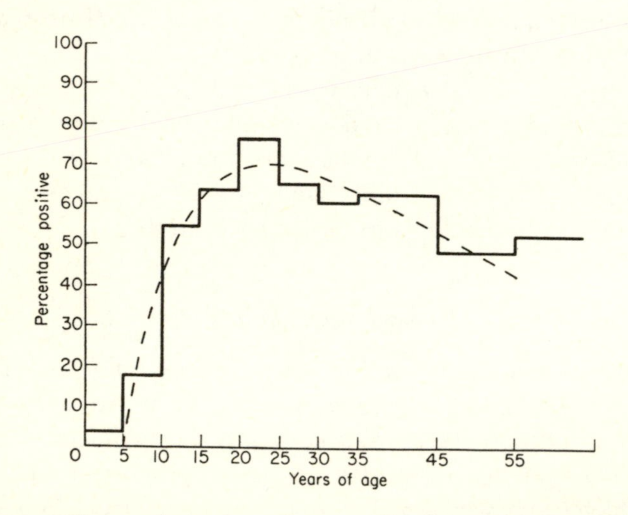

Fig. 13.3. Observed age prevalence (solid line) and theoretical age prevalence (dashed line) from eq. (13.9) of human infection with *S. japonicum* in the coastal division, Palo, Philippines, neglecting transmission before 5 years of age. (From Hairston, 1965b, p. 171.)

As Warren (1973) emphasizes, it is not more plausible to assume that such individuals are immune to further reinfection than it is to suppose that, for cultural and behavioural reasons, they simply have less contact with cercaria-laden water.

Figure 13.3 plots $y(t)$ and the observed proportions with *S. japonicum* eggs in their faeces by age in the same Philippine population pictured in Fig. 13.2 (Hairston, 1965b). The annual rate $b = 0 \cdot 02$ of becoming negative is not the annual death rate of individual female worms because (assuming the eggs are not blocked in the person's tissues) all the females in the person have to die, without replacement, for the person to stop excreting eggs.

Snails too pass through the stages of being never infected, infected and shedding (cercariae, instead of eggs), and (possibly) no longer infected (Fig. 13.4). Sturrock and Webbe (1971) plot proportions of snails shedding as a function of age, which they estimate by size.

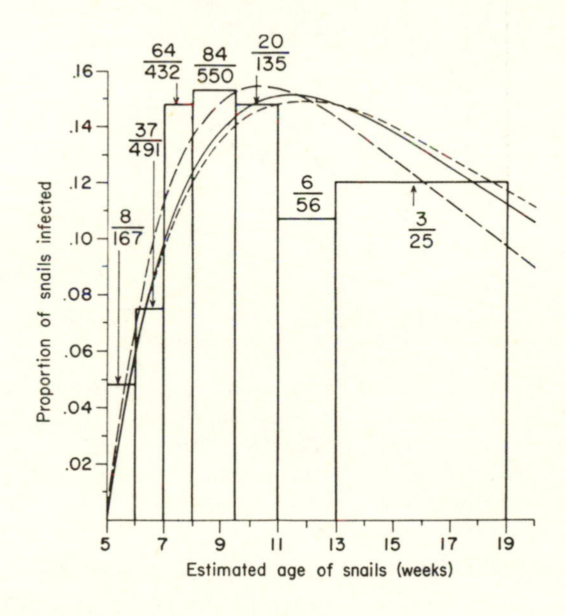

Fig. 13.4. Observed age prevalence (histogram) of infection in *Biomphalaria glabrata* snails in St Lucia, starting from the first week when snails shed cercariae; and three predicted age prevalence curves: solid curve, eq. (13.9); long dashes, force of infection a adjusted for differential mortality of infected individuals; short dashes, loss of infection b adjusted for differential mortality. (From Cohen, 1973b, p. 210.)

13.4.2 *Differential mortality due to infection*

Infection with schistosomes is unhealthy for both people and snails. For people, the increment in the probability of death at any age due to infecting schistosomes has never been measured credibly.

Laboratory experiments of Pan (1965) with the snails which transmit *S. mansoni* show unmistakably (Fig. 13.5) an increase in the death rate of snails shedding cercariae compared with uninfected snails.

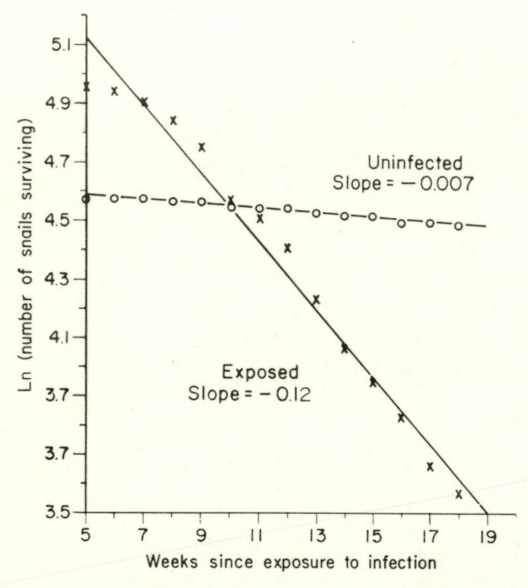

Fig. 13.5. Natural logarithms of the numbers of survivors in a group of snails exposed to *S. mansoni* (x) and in an unexposed group (open circles), based on data of C. T. Pan, and fitted straight lines; starting from first week when snails shed cercariae. The difference between the two slopes estimates ϵ in eq. (13.11). (From Cohen, 1973b, p. 208.)

A slight generalization (Cohen, 1973b) of the models above allows for differential mortality due to current infection. If μ is the constant mortality or emigration rate of individuals not currently shedding eggs (in the case of humans) or cercariae (snails), and $\mu+\epsilon$ is the increased mortality or emigration rate of individuals currently shedding, then suppose

$$(1/N)\, d(Nx)/dt = -ax-\mu x, \qquad x(0) = 1, \qquad (13.10)$$

$$(1/N)\, d(Ny)/dt = +ax-by-(\mu+\epsilon)\, y, \qquad y(0) = 0, \qquad (13.11)$$

$$(1/N)\, d(Nz)/dt = by - \mu z, \qquad z(0) = 0. \qquad (13.12)$$

The explicit solution for $y(t)$ satisfying eqs. (13.10) to (13.12) appears in Cohen (1973b, p. 200). When $\epsilon = 0$, putting $N(t) = N(0)e^{-\mu t}$ leads back to the eqs. (13.6) to (13.8) and resulting eq. (13.9) of section 13.4.1.

If $y(t)$ obtained from eqs. (13.10) to (13.12) is a better approximation to reality than eq. (13.9), but a curve of the form of eq. (13.9) is fitted to data in ignorance of ϵ, then the resulting estimates of the parameters a and b may be biased. The alternate two theoretical curves in Fig. 13.4 show the result of assuming that all the bias is absorbed either by a or by b. The differences among the predicted age prevalence curves are small, although the possible bias in the parameter estimates is not.

If the differential mortality due to infection is measured in the same population of snails whose age prevalence curve is being studied, even the possible bias in the parameter estimates is small (Sturrock *et al.*, 1975). For estimating rates of gain and loss of infection and describing the age prevalence distribution in the snail populations studied by Sturrock, a little verisimilitude is gained by allowing for differential mortality due to infection, but not much.

This example illustrates a kind of sensitivity analysis which ought, I think, to accompany the study of very simple ecological models. The model is eqs. (13.10) to (13.12) is literally more realistic than the model in eqs. (13.6) to (13.8) because it includes a representation of a real phenomenon known to occur. It is a bit more complicated to study mathematically. It does not cause major alterations in how data studied with eqs. (13.6) to (13.8) are understood. Hence, for rough purposes, one can be more assured of the adequacy of eqs. (13.6) to (13.8); for finer purposes, one has a more refined tool, eqs. (13.10) to (13.12).

13.4.3 *Immunity*

Linhart (1968) analyzes three different stochastic models leading to age-prevalence curves, based on three different formalizations of how an infected (human) individual might develop immunity to infection. Bradley (1974) emphasizes the probable importance of human immunity in the epidemiology of human schistosomiasis, though Warren (1973) favours another view; but so far Linhart's are the only explicit mathematical models incorporating immunity. It is unfortunate that some of Linhart's predicted age-prevalence curves have never been tested against observations.

An important task in the modeling of schistosomiasis is to see what the burgeoning biological information about the immunology of schistosomiasis (Butterworth, in press) implies for the validity of these and new models, and to translate that information into mathematically explicit, empirically testable, and epidemiologically useful form.

13.5 Snail population dynamics

Biomphalaria glabrata is the snail principally responsible for the transmission of *S. mansoni* in the New World. Jobin and Michelson (1967) raised laboratory populations of these snails with varying amounts F of food (measured in g of watercress), numbers N of snails (each 15 mm in diameter), and volumes V of water (measured in litres), at 25°C. For each such population they measured the fecundity (E) by the numbers of eggs laid per snail per day. Their results (Fig. 13.6) are conveniently summarized by

$$E = kF/(NV). \tag{13.13}$$

It is plausible that, over a certain range at least, fecundity should

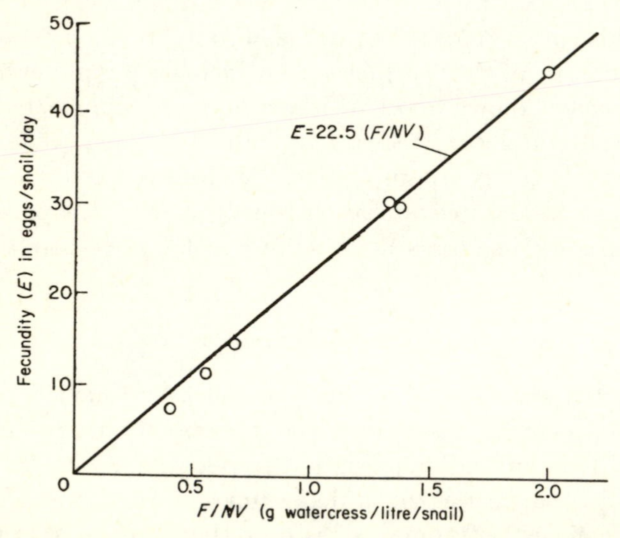

Fig. 13.6. Fecundity of *Biomphalaria glabrata* snails as a function of food F, number N of snails, and water volume V. (From Jobin and Michelson, 1967, p. 659.)

increase with food and decrease with the number of snails competing for that food. What is counterintuitive about eq. (13.13) is that a larger volume of water *decreases* fecundity. The reason is that the snails have a harder time finding the food.

Since eq. (13.13) predicts that snails in a vanishingly small volume of water have an infinite fecundity, eq. (13.13) obviously holds over only a limited range of V; in fact the experiments had jars only of 4·5 litres and 7·6 litres.

In very large volumes of water, such as lakes, which are not crowded with snails, the addition of one more snail has no effect on the fecundity of the other snails present. So Jobin and Michelson (1967) assume that the inverse dependence on N in eq. (13.13) disappears whenever the volume of water per snail exceeds a threshold called the 'crowding zone'.

These limitations on eq. (13.13) for small V and for large V show that the direct and inverse proportionalities in eq. (13.13) are deceptively simple. Eq. (13.13) should be regarded as a linearization or tangent approximation to a nonlinear function. This approximation is useful over the range of variables used in one set of experiments.

In a simulation based on eq. (13.13), Jobin and Michelson (1967)

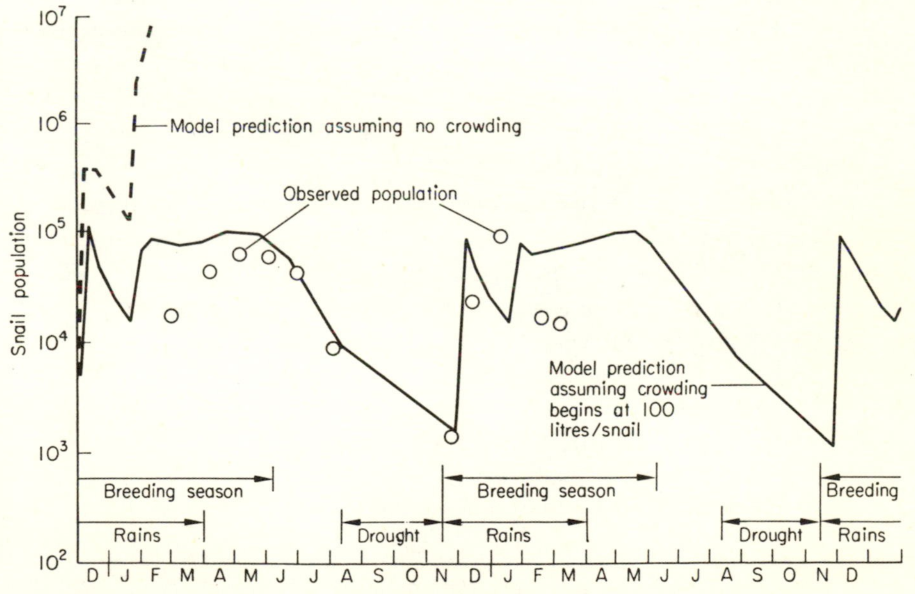

Fig. 13.7. Predictions (solid and dashed lines) of model for population of *Bulinus globosus* snails compared with field observations (open circles) from Foyle Pond. (From Jobin and Michelson, 1967, p. 661.)

treat the proportionality constant k as a function of the species and age of the snails, the nutritional value of food, and water temperature. Mortality is assumed to depend on demographic and ecological factors. The fecundity and mortality submodels are combined into difference equations whose trajectory is computed. The parameters are estimated from observations by another investigator of *Bulinus globosus*, a Rhodesian snail carrying *S. haematobium*. The model's predictions of snail populations agree roughly with censuses (Fig. 13.7). This study is an encouraging attempt to combine laboratory and field data (albeit from different snail species, since no one species has been studied comprehensively enough).

Sensitivity analysis of the sort discussed in section 13.4.2, with respect to both parameter values and the form of entire components of the model, has not been reported in this case. There is room for much useful work here.

13.6 Sex among the schistosomes

The preceding models describe either the human or the snail half of the schistosome life cycle, never both. A minimally realistic model of the entire life cycle must explicitly recognize that the worms reproduce sexually in people. The effect on egg output, and hence on transmission of infection, of one additional female worm in a person depends on the numbers of males and females already there and upon the mating system and success of worms in people.

Unfortunately, the sexual stage of schistosomes has yet to receive its Masters and Johnson. Mathematical models of mating functions quantify possibilities within the range of present ignorance. Three recent studies which cite earlier work are due to Leyton (1968), Nasell and Hirsch (1973), and Dietz (1975).

13.7 Life cycle models

A few models attempt to comprehend the entire schistosome life cycle.

13.7.1 *A life table model*

In Palo in the Philippines (Hairston, 1962, 1965a), the proportion of people infected with *S. japonicum* at each age changed very little from

1945 to 1959. Hence one suspects that the per capita risk of infection also changed very little over that period. Barring large changes in the human population, it follows that the total worm population must have been stationary (constant in total numbers and age structure). It is therefore appropriate to use a life table, the demographic model of a stationary population, to describe the population of *S. japonicum* there.

In this model, the net rate of reproduction R_0 of the worm, previously defined in eq. (2.14), may be written as the product of four factors reflecting four major stages in the worm's life cycle:

Worm's R_0 = (Larval R_0 in snail) × (Cercariae's probability of infecting a mammal) × (Adult worms' R_0 in mammal) × (Eggs' probability of infecting a snail). (13.14)

Hairston (1962) further decomposes each of these factors. For example, he represents the probability of an egg's infecting a snail (1962, p. 39) as the product of the probability that an egg is able to hatch times the probability that the egg is deposited near snails times the probability of penetrating a snail times the probability of establishing an infection in the snail after penetration.

Even the very extensive field data gathered in Palo are insufficient to estimate all the factors in eq. (13.14) directly. Hairston supplements the data with guesses (of the contribution of swamp rats to transmission, for example), assumptions [of a Poisson distribution of worms per person, for example, though a negative binomial distribution is more likely (Hairston, 1965a, p. 60)], and ingenuity.

For *S. japonicum*, Hairston's (1965a, p. 60) final estimate of the right side of eq. (13.14) is 0·6. Based on less reliable data from Egypt, his estimate for *S. mansoni* is 1·9 and for *S. haematobium* is 2·8. If Hairston's model and observations were correct, these estimates should be 1. That the estimates do not differ from 1 by orders of magnitude is an indication of the coherence of the observations and the approximate correctness of the model, under the given conditions.

Still, some qualifications deserve note.

First, although the age prevalence distribution remained nearly constant from 1945 to 1959, and hence probably also the per capita exposure, if the total human population increased over that period, the worm population may have also. But the rate of increase of the human population is likely to have been low, on the order of a few percent per year. So the deviation of the R_0 of the worms from 1 was probably much less than the uncertainty of the data.

Second, the decomposition of the probability of transmission around the life cycle into the product of the probabilities of the elementary events which make up the life cycle implicitly assumes that the simple events may be treated as independent. Independence is so unlikely that Hairston (1962, 1965a) avoids making the assumption in practice by estimating from his data, not the elementary probabilities which appear in some of his formulas, but clusters of these factors representing compound events.

Third, since 'the parasite population is able to come into equilibrium at different rates of transmission' in different ecological settings, 'net reproduction in one or both of the hosts must be curtailed with increasing transmission and enhanced with decreasing transmission', and hence (Hairston, 1965a, pp. 46–47) 'there is a range of transmission rates over which compensatory mechanisms operate to keep the parasite population in equilibrium'. This means that if an intervention programme reduces one of the four factors on the right of eq. (13.14), over at least some range, the other factors on the right will not remain constant but will increase to keep the product near 1.

For example, increasing the number of female worms of *S. japonicum* present in a person of specified age decreases the average daily number of eggs in faeces per female worm (Fig. 13.8; Hairston, 1962). The egg counts are based on direct measurement. The numbers of female worms are estimated from the rates of acquisition of worms and from the death rates of female worms. [See Hairston (1965a, p. 54) for a minor correction of Hairston (1962, p. 54).] These rates are in turn estimated from age prevalence distributions. For each age group in Fig. 13.8, the point with the higher worm load is from data on Palo, while the point with the lower worm load is from a different, urban or coastal area with lower transmission. Since the abscissae are obtained so indirectly from different ecologies, there is much room for discussion of the exact slopes of the lines in Fig. 13.8. But the qualitative point is consistent with a variety of experimental evidence that increasing parasite loads decrease reproductive output per parasite (Kennedy, 1975).

This example illustrates one of several general compensatory mechanisms which parasites have evolved (Bradley, 1974) to regulate population numbers. As a result of such mechanisms, one cannot use Hairston's calculated values of the factors in eq. (13.14) in a simple-minded way when evaluating a control programme which affects the values of some of those factors.

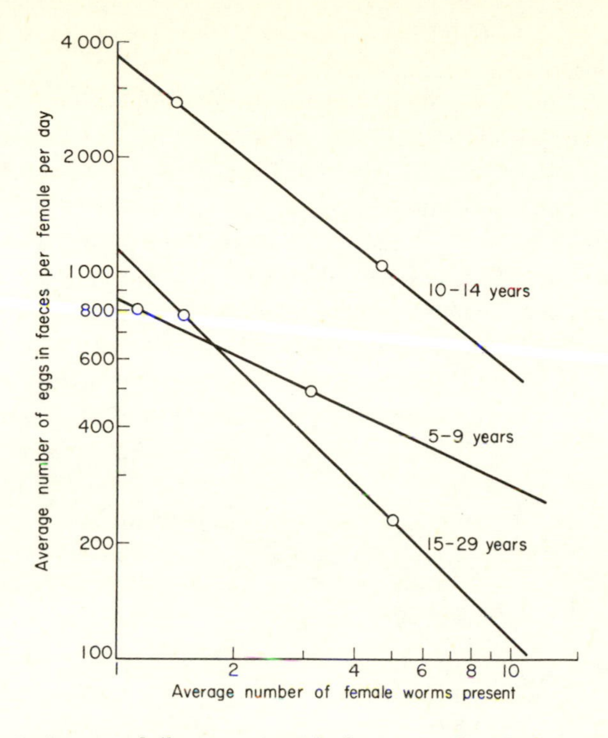

Fig. 13.8. Average daily egg output in faeces per female *S. japonicum* as a function of mean female worm load per person, in three age groups of people. Open circles estimated from data (see text); straight lines are simplest functions through two points. (From Hairston, 1962, p. 52.)

13.7.2 *Dynamic models*

Dynamic models, intended to describe what will happen when the life cycle is perturbed, attempt to represent these regulatory mechanisms and are therefore necessarily nonlinear. One such model (Macdonald, 1965) emphasizes the nonlinearity introduced by supposed monogamous mating in the sexual stage of the worms, and borrows other nonlinear bits from existing models of malaria.

Dynamic models examine the transient and asymptotic behaviour of indicators such as the mean worm load m in people or the fraction of people infected as functions of various parameters, like the probability that an egg reaches and infects a snail, the snail population, and the probability that a cercaria reaches and infects a person. Numerical analysis of one simple model (Macdonald, 1965) suggests the existence of a threshold in m. Once m is below this threshold, transmission of infection disappears in a few years; once above it, infection remains endemic indefinitely.

All but the most sophisticated dynamic models (including Nasell and Hirsch, 1973 and Nasell, unpublished a) posit a single isolated endemic focus, closed against infection from without. When an external source of infection is considered (Nasell, unpublished b) the asymptotic or equilibrium behaviour of the mean worm load m in people changes qualitatively (see Fig. 13.9). Whereas previously the threshold switched m between a positive level of infection and $m = 0$, Fig. 13.9 shows that certain combinations of parameter values make possible a threshold which switches m between two positive levels; moreover, this threshold arises discontinuously as a function of other parameters.

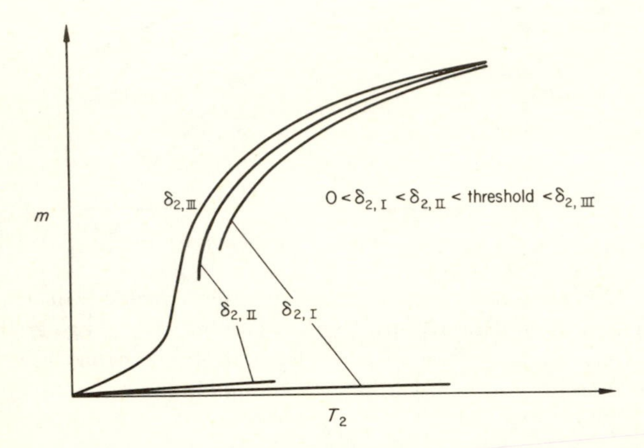

Fig. 13.9. Asymptotic or equilibrium mean worm load m in humans as a function of the amount of transmission T_2 of eggs from people to snails, for various external egg inputs δ_2 and fixed cercarial transmission capacity and fixed external input of cercariae. (Adapted from Nasell, unpublished b, p. 90.)

All the curves in Fig. 13.9 correspond to the same, fixed transmission capacity of cercariae from snails to people and the same, fixed external input of cercariae. The three curves correspond to different levels of external input δ_2 of eggs. The highest level of external egg input $\delta_{2,III}$ leads to a unique worm load as a function of the abscissa T_2, which measures the amount of transmission of eggs from people to snails (T_2 is proportional to the number N_1 of people). In this case the only way to lower m is to lower T_2, and the equilibrium m varies smoothly as T_2 varies. But for smaller values like $\delta_{2,I}$, $\delta_{2,II}$ of external egg input which are below a certain threshold, m may fall either on the lower or on the upper curve. If m falls on the upper curve, then one strategy of control, as before, is to reduce T_2 and obtain a corresponding smooth reduction in m. A more dramatic possibility is to shift the initial conditions by coordinated programmes of human case finding and treatment so that m shifts to the lower line. Then, if it is still desirable, T_2 can be reduced further. Unfortunately, this possibility leaves the community vulnerable to transfer back to a higher m if an uncontrolled influx of infection effectively reverses the change in initial conditions.

A more detailed review of all these mathematical models of schistosomiasis appears in Cohen (1976). See also Hairston (1973).

13.8 A broader view

"N. A. Croll: Are definitive hosts always idiots?

J. C. Holmes: Definitive hosts are not idiots. Quite the contrary." (Canning and Wright, 1972, pp. 148–49)

13.8.1 *The ecology of human host-parasite systems*

For concreteness, this chapter focuses on schistosomiasis. Other diseases, described by Crosby (1972), May (1958, 1961), Cockburn (1963), Van Oye (1964), Ford (1971), and, magnificently, Burnet and White (1972), deserve the same ecological approach. When descriptions revel mystically in the incomprehensible subtlety and unmanageable complexity of disease ecology, the success of simple models in schistosomiasis should not be forgotten.

Some attempts to generalize about human host-parasite systems immediately formalize an image of the transmission of infection, without (at least until recently) overmuch attention to biological mechanisms (Bailey, 1957, whose new edition has recently appeared: 1975; Dietz, 1967; Crofton, 1971a, 1971b). Other attempts to elucidate classes of biological mechanisms (Bradley, 1974) are ripe for, but have yet to receive, adequate mathematical formalization; hence many implications of such attempts remain to be worked out.

In view of the generally primitive level of theoretical ecological understanding of human host-parasite systems, policies for public health which follow from models should probably be presented, not as unconditional recommendations, but as proposals for experimental tests using the best field controls possible.

13.8.2 *Broader aspects of disease*

The models in this chapter describe the dynamics of infection. Infection is not tantamount to disease. Coping with disease requires, in addition to purely biomedical understanding, a grasp of its significant economic and social aspects.

For example, using quantitative models from economics and demography with extensive observations from St Lucia, a Caribbean island, Weisbrod *et al.* (1973) ask: Does infection cause disease? They find it impossible to evaluate the impact of *S. mansoni* on birth and death rates, the achievement of school children, and the labour productivity of adults without considering four intestinal nematodes of man which are also very widespread there. Throughout the poor countries of the world, most parasitized people are miniature ecosystems of multiple parasites which interact (see also Lehman *et al.*, 1970). Moreover, ecological alterations such as the damming of rivers, the construction of irrigation canals, and the resettlement of human populations into new lands have greatly increased opportunities in some regions for schistosome bearing snails to propagate and to contact people. Other quantitative approaches to evaluating economic and demographic effects of schistosomiasis are presented and cited by Cohen (1974, 1975) and Ansari and Junker (1969).

Because the real world seems painfully negligent of the boundaries between academic departments in universities, models which will be practically useful in coping with that world require an integration of ecology and of the biology of individuals with economics and sociology. The final chapter of this book explores such models.

14
Man Versus Pests

GORDON CONWAY

14.1 Introduction

Pest control constitutes an ancient war, waged by man for 4000 years or more against a great variety of often small and remarkably persistent enemies. Surprisingly, although the war is old, its dynamics and the nature of the principal protagonists seem poorly understood. Even the objectives, at least of man, are ill-defined. It is as though man, in the heat of the battle, has not had the time to analyse in any sophisticated fashion the conflict in which he finds himself. Battles have been won and lost but lessons have been learned slowly and painfully. It is only in recent years that people have begun to ask the fundamental questions of principle and to raise doubts about implicit beliefs and objectives.

The term pest connotes a value judgement. A pest is a living organism which causes damage or illness to man or his possessions or is otherwise, in some sense, 'not wanted'. Thus, in the first instance, pest control is a problem for the social sciences and in particular for applied economics. Simply stated, the problem is to assess whether the damage or illness caused by a pest can be reduced in a manner which is profitable or desirable. Formal economic analysis requires that the damage or illness be quantified, usually in monetary terms, so that the potential benefits ensuing from control of the pest can be compared with the costs involved.

The classical economist's tool for tackling this kind of problem is marginal analysis (Fig. 14.1). A control action is termed rational when its cost is less than or equal to the net increase in revenue it produces and is economically most efficient (i.e. most profitable) when the marginal cost of control equals the marginal revenue produced (Southwood and Norton, 1973). If we consider an agricultural crop, then the object of pest control is to maximise the quantity

$$Y[A(S)]\,P[A(S)]-C(S). \qquad (14.1)$$

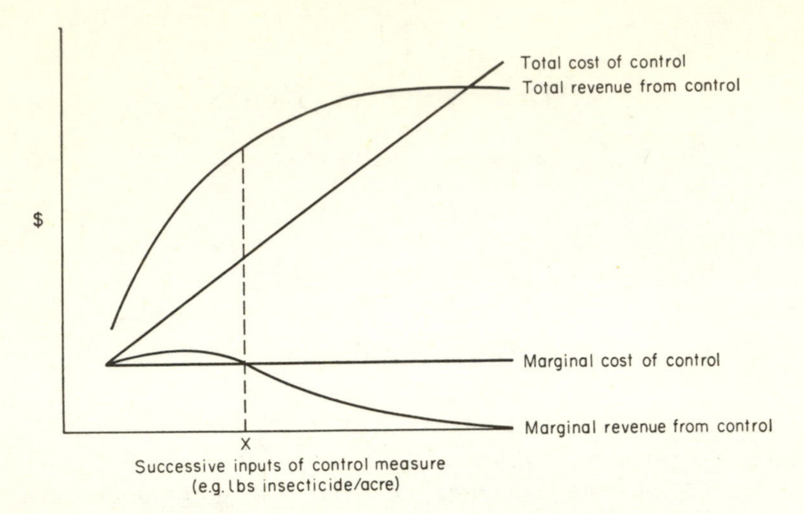

Fig. 14.1. Cost and revenue curves for pest control. The marginal curves give the increment in total cost or revenue for each successive input of control. Profit is greatest when marginal revenue equals marginal cost. (After Southwood and Norton, 1973.)

This expression, which relates cost of control to revenue, has within it four important functions:

(i) the cost of control function, $C(S)$, which relates the cost of control to the control strategy S;

(ii) the control function, $A(S)$, which relates the level of pest attack to the control strategy S;

(iii) the quantity damage function, $Y[A(S)]$, which relates the yield to the level of attack $A(S)$;

(iv) the quality damage function, $P[A(S)]$, which relates the price to the level of attack $A(S)$.

Figure 14.2 illustrates how the functions are linked.

To this point analysis of the problem has been purely in economic terms. But, by definition, pests are also living organisms and as we dissect the constituent functions and investigate their properties biological, and in particular ecological, knowledge dominate the analysis. Populations of organisms—as the early chapters of this book make clear—respond, often in a complex and dynamic fashion, to outside interference. Only rarely will the level of pest attack be a simple function of control strategy. For example, a high kill produced by a pesticide may evoke a density dependent response, so that the pest population rebounds, sometimes to a new and higher level. In the longer term there may be genetic responses. Pest populations may develop resistance to control, notably to chemical pesticides. Moreover, the damage function may contain complex ecological components since the object of pest attack—usually a crop plant, domestic animal or man himself—is also a living organism. The outcome of attack is thus the product of the dynamic interaction of two populations. For example, crop plants are frequently able to tolerate or compensate for low levels of pest attack. Similarly, where pests are vectors of disease the ensuing incidence of illness in the animal or human population will depend on such features as the age structure and the level of immunity present.

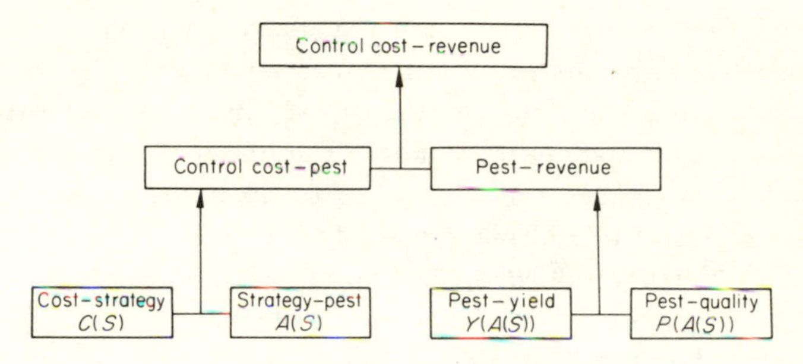

Fig. 14.2. Constituent functions of the relationship between revenue and cost of control. (After Southwood and Norton, 1973.)

In this chapter I shall discuss more fully how modern ecological and economic analysis illuminates the problem of pest control. I describe, in the next section, how a control strategy can be chosen which is appropriate to the ecological characteristics of the pest population. I then show how knowledge of the damage function modifies the choice of strategy. Finally I discuss the objectives of the pest control decision-maker and the broader context within which the choice has to be made.

14.2 The control function

14.2.1 *The strategy of pests*

In traditional text books, pests are classified and dealt with in taxonomic order or according to the crop or other object of attack. A potentially more illuminating classification is according to the spectrum of ecological strategies which are described by Southwood in chapter 3. On this basis we can then speak of *r*-pests and *K*-pests with, in between, 'intermediate' pests.

r-Pests are the classic pests. They fit most closely the sense of the latin *pestis* for plague, from which the term pest is derived. Rats, locusts, aphids, mosquitoes and fleas, the organisms which the term immediately conjures up in the mind, all show *r*-strategy features of high potential rates of increase, strong dispersal and host finding ability and, relative to other members of the same taxa, small size.

An illuminating example of an *r*-pest is the plague caterpillar, *Tiracola plagiata*, of South East Asia. This belongs to the moth family Noctuidae which contains many other *r*-pests, including the army worm and the cutworm. The adult moths are strong fliers and capable of laying batches of a 1000 eggs or more. The life cycle through the egg, larval and pupal stages takes only 30 to 40 days. Conway (1971a) describes an outbreak of this species in an area of cleared primary rainforest in Northern Borneo. A strip of forest about a quarter of a mile long had been cleared some five years earlier, but had been abandoned and was in the process of passing back through the long succession to primary forest. At the time of the outbreak the vegetation consisted of a variety of secondary forest trees standing six to fifteen feet high. When first seen the outbreak, a huge army of caterpillars comprising several million individuals, had progressed half way along the strip, removing the leaves from virtually every plant. Since the land was abandoned, with no value attached to the vegetation, the caterpillars were not pests. However, at the end of the strip was a young plantation of cocoa which had also been planted on cleared forest land. The population was not sprayed and two weeks later all the caterpillars pupated in the soil close to the cocoa plantation. Only a few adult moths emerged from the pupae, the majority dying from a viral or other infection, and the cocoa was spared. But in other countries in South East Asia the plague caterpillar has become a serious pest of cocoa.

The species is indigenous to the region, living within the boundaries of the primary rainforest, but adopting an *r*-strategy so as to exploit the scattered ephemeral clearings in the forest which, in the absence of man, are created by lightning strikes or by the overflowing of rivers. It is a species which is catholic in its food preferences and hence is, in every sense, ideally preadapted to the role of agricultural pest. With the planting of annual and perennial crops it continues its life strategy with the difference that is no longer confined to temporary habitats.

We can speculate that the pattern described here for the emergence of *r*-pests in the tropical rainforest ecosystem is common to other ecosystems in other climates. For example, the complicated and unique life cycle of the locust is adapted to the exploitation of the ephemerally favourable habitats of the semi-arid regions of the world. The existence of cropping has served to extend and prolong their periodic outbreaks. In the temperate regions of the world, perhaps the majority of pests are *r*-pests, consisting of species which have adopted this strategy in order to exploit the essentially short-lived plants which depend upon each annual growing season. The aphid *Myzus persicae* is a good example. It feeds on over 200 species of wild plants and is now a major pest of such crops as sugar-beet, potatoes, peaches, etc.

Table 14.1 lists a number of characteristic *r*-pests, with their biological features and the damage they cause.

Table 14.1. Some important *r*-pests and their characteristics.

Species	Fecundity per female (approx.)	Generation time (approx.)	Pest features
Desert locust (*Schistocerca gregaria*)	400	1–2 months	Periodic outbreaks which defoliate any crop
Black bean aphid (*Aphis fabae*)	100	1–2 weeks	Wide range of crop plants, including beans and beet
House fly (*Musca domestica*)	500	2–3 weeks	Feeds on organic waste
Black cutworm (*Agrotis ipsilon*)	1500	1–1½ months	Attacks seedlings of most crops

K-pests, by contrast, have low rates of potential increase, greater competitive ability, more specialized food preferences and, by comparison with other members of the same taxa, greater size. Characteristic *K*-pests are the sheep ked, codling moth, the rhinoceros beetle and the tsetse fly. A good example is another Southeast Asian pest, the bark borer, *Endoclita hosei*, which is a member of the Hepialidae, a family of large, slow flying moths with low fecundity and long life cycles. In nature the larvae bore in the trunk of a secondary forest tree, where they seem to cause little damage. But they also attack cocoa and there a single larva boring in a cocoa tree will cause its death. Thus even small populations can cause considerable damage.

Most organisms which show extreme *K*-strategies do not become pests; their specialised niche is not of interest to man. However, if man does expand the niche or provide a new niche of very similar composition then *K*-strategists can become important pests, causing damage less because of their numbers and more because of the nature of their feeding habits. Thus the codling moth whose damage to fruit would be tolerable to wild trees becomes a major pest for man who is seeking high quality, unblemished apples.

Table 14.2 lists a number of characteristic *K*-pests with their biological characteristics and the damage they cause.

Table 14.2. Some important *K*-pests and their characteristics.

Species	Fecundity per female (approx.)	Generation time (approx.)	Pest features
Rhinoceros beetle (*Oryctes rhinoceros*)	50	3–4 months	Adult feeds on apical growing point of coconuts, often killing palms
Tsetse fly (*Glossina* spp.)	10	2–3 months	Vector of trypanosomiasis affecting man and cattle
Codling moth (*Cydia pomonella*)	40	2–6 months	Larvae damage apples and other fruit
Sheep ked (*Melophagus ovinus*)	15	1–2 months	External parasite on sheep

Probably the great majority of pest species lie between these two extremes and can be classified as intermediate pests. As Southwood points out in chapter 3 there is a continuum, in nature, from *r*-strategist to *K*-strategist and a sharp demarcation between the categories is unrealistic. Intermediate pests may show any mixture of features, but probably their most important characteristic from the viewpoint of control is the high degree to which they are normally regulated by natural enemies, such as parasites and predators (see Fig. 3.6). On this criterion, but also because of their intermediate rates of reproduction and medium length life cycles, some important pests in this category are the spruce bud-worm, the cottony cushion scale and the grape leafhopper.

14.2.2 *Strategies of control*

Such a categorisation of pests on the basis of ecological strategy has major implications for control. *r*-Pests occur in the form of massive outbreaks, either at frequent or infrequent intervals. Where crops are attacked it is usually the leaves and foliage that are damaged. *r*-Pests may be subject to natural enemies and to disease, but these rarely act before considerable damage has been done. Most important, *r*-pests are resilient to disturbance and can rebound even after heavy mortalities. At the other extreme *K*-pests are a constant problem. Often present in low numbers, they are nevertheless pests since on crops they attack the products, such as the fruit, rather than the leaves or roots. They have few natural enemies and bounce back from low mortalities, but can be driven to extinction if mortality is high enough. Intermediate pests feed either on roots and foliage or on the products but can be well regulated by natural enemies. Particularly if roots or foliage are being damaged this degree of regulation may represent a satisfactory form of control for man, as for example in the case of forest tree pests.

Five major techniques of pest control are currently practised:

(i) *Pesticide control:* the use of chemical compounds directly to kill pests.

(ii) *Biological control:* the use of natural enemies, viruses, bacteria or fungi either by augmenting those already present or by introducing species from other regions and countries.

(iii) *Cultural control:* the use of agricultural or other practices to change the habitat of the pest.

(iv) *Plant and animal resistance:* the breeding of animals and crop plants for resistance to pests.

(v) *Sterile mating control:* sterilisation of pest populations by various techniques to reduce the rate of reproduction.

Unfortunately the history of pest control has been dominated by the search for panaceas. With the advent of organochlorine and organophosphorus insecticides at the end of the second world war, it was believed that most pest problems would be quickly solved. But within fifteen years the limitations of pesticides had become widely apparent. Control programmes failed or sometimes created worse problems than they solved because of the effects of pesticides on natural enemies, or the development of resistance. Moreover evidence grew that these pesticides posed serious hazards to wildlife and in some cases to man himself. Biological control became the new panacea. When it, too, was seen to be limited in scope, attention shifted to sterile mating techniques and most recently has focussed on the development of animal and crop resistance. An attempt at a more rational approach began with the concept of integrated control which sought to combine biological and chemical control. Latterly the term has been used to denote the mix of all appropriate techniques in a given situation (Smith and Reynolds, 1966). However, as Way (1973) points out, it has been largely an empirical approach; successful integrated control programmes have depended on a combination of insight and trial and error. There has been little attempt, so far, to define a theoretical basis for choosing appropriate strategies. Recognition of the *r-K* categorisation, I believe, provides a step in this direction.

For instance, it makes clear that an eradication strategy is foredoomed to failure, as the history of malaria mosquito control demonstrates (Harrison, 1977). Although adoption by the World Health Organization in 1956 of a policy of malaria eradication did not imply extermination of the vectors, it did require laying down and maintaining over many years very extensive and virtually perfect chemical barriers to prevent contact between the vectors and man. In much of the world, most notably in sub-Saharan Africa, that proved an unequal struggle in which the *r*-pest had insuperable advantages. The policy has now been changed to one of containment. *r*-Pests—mosquitoes again are a good example—are also unlikely to be satisfactorily controlled by natural enemies, although biological control may play a subsidiary role. Thus, despite all their inherent drawbacks, pesticides are likely to remain the main counter to *r*-pests; it is the only technique which has the speed and flexibility required to respond to the outbreak situation characteristic of *r*-pests.

By contrast, *K*-pests can be eradicated or at least forced to low non-damaging population levels, and for this reason are the most suitable targets for the sterile mating technique. Pesticides may be appropriate, particularly if very small populations cause high losses, for example when fruit is blemished. The most appropriate strategies for *K*-pests, however, will be based on cultural control and the development of plant and animal resistance, since both of these are directly aimed at reducing the effective size of the pest niche.

Biological control has its greatest pay-off against intermediate pests and by corollary it is against these pests that the use of pesticides is likely to be counter-productive unless carefully integrated.

Table 14.3 lists the principle control strategies associated with each category of pest.

Table 14.3. Principal control techniques appropriate for use against different pest strategies.

r-pests	intermediate pests	K-pests
Pesticides		Pesticides
	Biological control	
	← Cultural	control →
		Plant and animal resistance
		Sterile mating techniques

14.3 The damage function

It is common to find a linear relationship between the size of a pest population and the direct, immediate injury that it typically causes. For example, the leaf area of a crop consumed is roughly proportional to the number of pest individuals present in the crop, and the numbers of humans bitten by mosquitoes is proportional to the size of the man-biting mosquito population. The relationship becomes more complex, however, when we consider the yield of the crop or the ensuing incidence of illness.

14.3.1 *Crop compensation*

The basic form of the relationship between crop yield and pest population size is sigmoid (Fig. 14.3) but, as Southwood and Norton (1973) point out, only a portion of this curve is usually exhibited. Thus for pests of foliage or roots, the plateau of high yields indicated by the left hand portion of the curve is extended; severe losses are only caused by high populations. For pests attacking fruit, on the other hand, the relationship is typified by a rapid decline in yields, as demonstrated in the right hand portion of the curve; losses can be severe from low populations. In consequence K-pests typically are pests of fruit while foliage and leaf attacking pests tend to be r-strategists.

The plateau of high yields at low pest population densities, which is characteristic of foliage and root attacking pests, results from the ability of many crops to compensate for pest injury. Up to a certain level of attack a crop is able to replace the photosynthetic tissues removed by the pest population. The individual plant may respond by growing new leaves or shoots or by extending the photosynthetic area of existing leaves. Alternatively the compensation may occur at the level of the crop population, neighbours expanding to fill the gaps caused by dead or damaged plants.

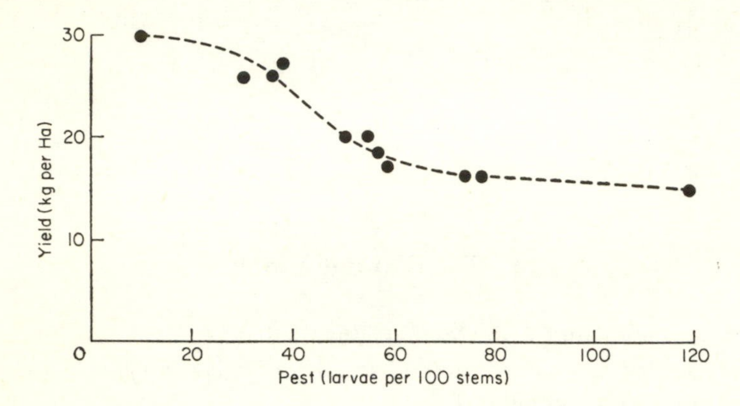

Fig. 14.3. The yield-pest relationship as shown by the African white rice borer, *Maliorpha separatella*, on rice. (After Brénière and Walker, 1970.)

Salazar (1976) has recently described the way in which the field bean (*Vicia faba*) compensates for damage by the aphid *Aphis fabae*. Large populations of aphids can produce a total loss of crop, but medium or light infestations may result in a final yield of beans which does not greatly differ from the yield produced by uninfested plants. Salazar showed that the bean plant is shorter as a result of aphid injury but also reacts by increasing leaf area in the region of the plant where the bean pods are produced (Fig. 14.4). In effect the bean plant diverts energy and resources to ensuring, as far as possible, that the yield is maintained. In some reported instances the compensatory mechanism is so powerful that low infestations early in the growth of a crop have actually led to a higher yield than would have occurred in the absence of the pest (Gough, 1946; Kincade *et al.*, 1970).

From these studies the implication is that the presence in a crop of a population of pests known to cause injury is not *per se* a justification for control. In addition to measuring the density of the pest population, an assessment must be made of where the crop is in its phenological development, and this knowledge related to what is known of the compensatory mechanism, before a rational control decision can be taken. This is not easy and a common consequence of poor knowledge is that control measures are taken when they are not strictly necessary. Profitability is reduced and where pesticide control is decided upon there is a greater likelihood of selecting for pesticide resistance or of incurring environmental contamination.

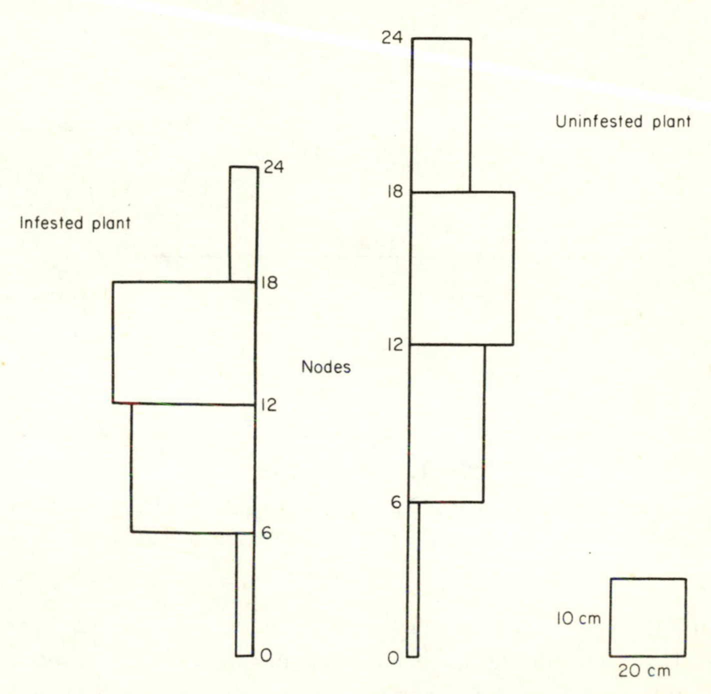

Fig. 14.4. Compensatory growth in field bean plants infested by *Aphis fabae*. The diagrams indicate the height and leaf area in different regions of an uninfested and moderately infested plant. On the infested plant leaf area is increased in the region where the bean pods are carried. (After Salazar, 1976.)

In an attempt to minimise insecticide application in a situation of this kind Wilson *et al.* (1972) built a computer model describing the effects of the bollworm *Heliothis* on cotton in Western Australia. Their model was based on three observations. First, there is a definite maximum crop of bolls that can ripen. Second, the bolls require a minimum number of day-degrees to mature and hence have to be protected for this period in order for the maximum crop to be obtained. Third, the day-degree requirement to set the full crop of bolls depends on the size of the plant. The bigger the plant the more quickly the crop is set. As the model clearly demonstrated (Fig. 14.5), if treatment is delayed then the period of protection can be shorter. Less insecticide is used and the control is cheaper and more efficient.

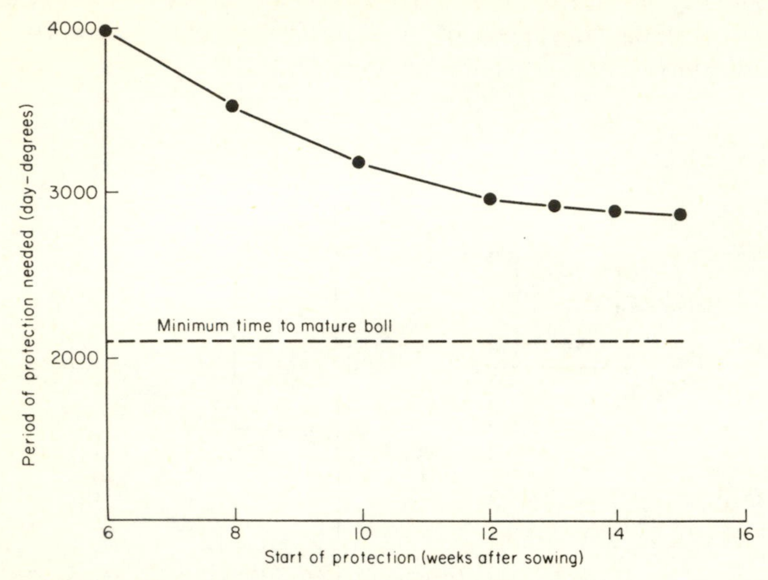

Fig. 14.5. Minimum period of time for which cotton in Western Australia needs to be protected from bollworm attack so as to ensure a maximum crop of bolls. (After Wilson *et al.*, 1972.)

Unfortunately there is very little information on the pest-yield relationship, even for the most common pests, and only rarely is there a good understanding of the compensatory mechanism, where present. If more were know, it would be possible not only to minimise wasteful control actions but also to enhance the degree of compensation which naturally occurs, using cultural techniques or plant breeding.

14.3.2 *Illness*

Where pests carry disease the relationship is a product of the dynamics of three populations: that of the pest or vector (e.g. a mosquito, *Anopheles* spp.), the host (e.g. man) and the disease organisms (e.g. the malaria parasite, *Plasmodium* spp.). Macdonald (1957) developed a simple and elegant mathematical model to describe this relationship for malaria. The basic model depends on six key parameters:

m = the density of anopheline mosquitoes in relation to man;

a = the average number of men bitten by one mosquito in one day;

b = the proportion of anophelines with sporozoites in their glands which are actually infective;

p = the probability of a mosquito surviving through one day;

n = the time taken for completion of the extrinsic cycle of the malaria parasite; and,

r = the reciprocal of the period of infectivity in man.

When combined, these parameters give a measure of the rate of production of new infections arising from a single primary infection in a community. This is called the basic reproduction rate of malaria and equals

$$\frac{m\,a^2\,b\,p^n}{r \ln(1/p)} \tag{14.2}$$

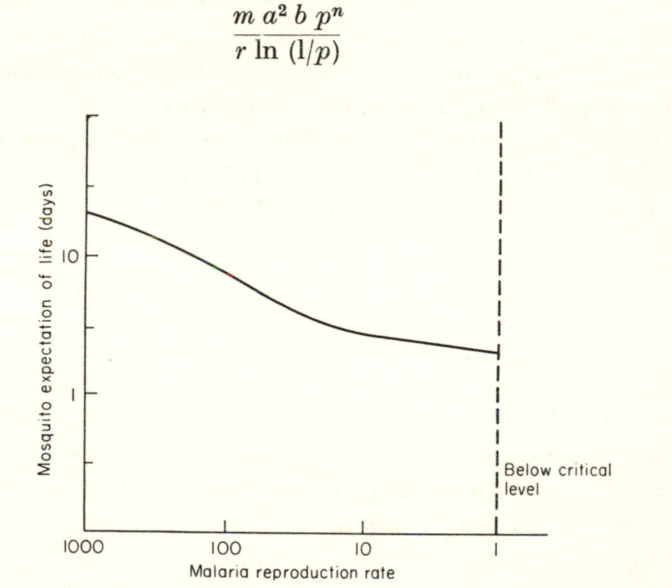

Fig. 14.6. Effect of change in the mosquito's expectation of life on the basic reproduction rate of malaria. (After Macdonald, 1961.)

In contrast to crop damage the relationship between illness and pest density (m) is simple and linear. The critical population parameter is the survival probability (p) which is contained in eq. (14.2) both raised to a power and as a logarithm. In consequence small changes in survival probability produce large changes in the malaria reproduction rate. As Fig. 14.6 shows, a decrease in the mosquito's mean expectation of life from 20 days ($p \sim \cdot 95$) to under 3 days ($p \sim \cdot 7$) is sufficient to push the reproduction rate below unity, and hence to cause extinction of disease transmission. This explains why pesticides such as DDT, which reduce survival probability, have been so heavily relied upon in malaria control campaigns. In practice, however, it has proved very difficult to achieve eradication. The rapid rates of increase of mosquito populations, their powers of host finding, dispersal and adaptation to a wide variety of habitats, and their ability to evolve resistance, have limited the effective use of pesticides. Few nations have been able to provide the time, organisation and funds which control of so powerful a pest requires.

Fortunately, the existence of three populations in the illness-pest relationship means that pest populations are no longer the sole target for control strategies. The disease pathogen populations can themselves be the subject of control, either through immunization or through chemotherapy. Macdonald *et al.* (1968) have suggested by means of a computer model how chemotherapy, based on drugs, and pesticides could be utilised in an integrated fashion for malaria control. The search is now on for a vaccine which will confer immunity against malaria. Other examples of the ecology of human diseases were discussed in chapter 13.

14.4 The decision-maker

Knowledge of the dynamics of the pest and of the pest-damage relationship can provide a guide to those control strategies which, in ecological terms, are most likely to succeed. But the final choice of strategy depends on the objectives of the decision-maker and on the context within which he makes his decisions.

14.4.1 *Objectives*

At the beginning of the chapter I assumed that the objective of pest control was profit maximisation, i.e. the farmer or other decision-maker wished to maximise the difference between the monetary return from control and the cost.

This is an objective which is probably pursued by most large scale farming operations. A good example occurs on the West Indian island of Trinidad, where sugar-cane is grown over several thousand acres by a single company. The cane is attacked by a froghopper, *Aeneolamia varia saccharina*, the adult of which causes a loss in the quantity of cane produced. Conway *et al.* (1975) have produced, for this case, a procedure for deciding on the profit maximising strategy. They assume, initially, a single brood (generation) of froghoppers whose population growth can be described by a simple mathematical function (Fig. 14.7). From field experiments it appears that the loss of sugar-cane yield is then a simple linear function of the number of adult days, i.e. of the area underneath the population curve. Two kinds of insecticide are available: (1) a cheaper non-residual compound which only acts on the day it is applied, and (2) a more expensive residual compound which persists for five days. These can be applied at any time and in any permutation. If a limit is set of three applications per brood, ten dominant strategies are possible (five further strategies exist but these eliminate fewer adult days than other permutations of the same combination of insecticides). Table 14.4 gives the optimal spraying times for these strategies.

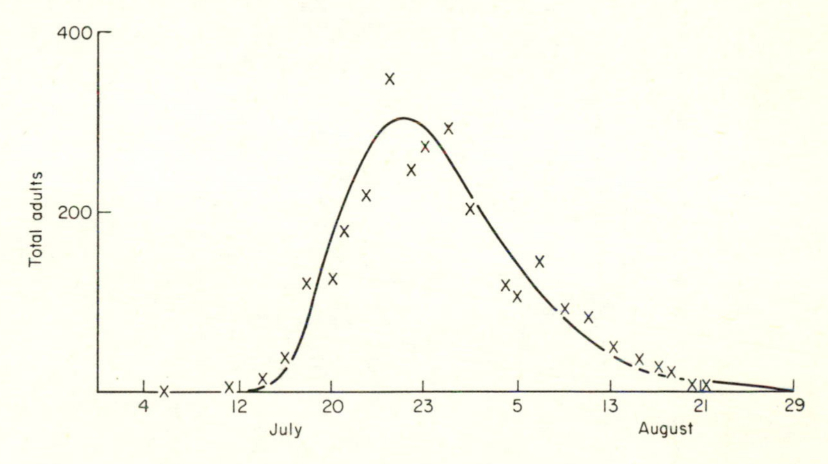

Fig. 14.7. Field records of a single brood of froghopper on sugar-cane in Trinidad. The fitted line is a simple beta function. (After Conway *et al.*, 1975.)

They then show how to arrive at a set of net revenue lines for different sizes of pest population, using information from the cost and damage function (Fig. 14.8). Each spraying strategy is rational when its revenue line is above that of the no-spraying strategy (1) and the most efficient strategy at each population level is the highest revenue line. The optimal control policy for the range of possible population sizes is given by the curve made up of the segments of revenue lines representing the most efficient strategies.

Table 14.4. Optimal application times for ten spraying strategies directed against a single froghopper brood. (From Conway *et al.*, 1975.)

Strategy	Time from first adult emergence (days)																		Percentage of adult-days removed
	1	2	3	4	5	6	7	8	9	10	11	12	13	14	15	16	17	18	
1																			0·0
2											NR								19·3
3									NR					NR					32·7
4										R									37·6
5							NR				NR					NR			42·8
6								NR				R							49·0
7							NR			NR			R						57·2
8							R							R					60·9
9						NR		R							R				67·4
10				R						R						R			74·9

R = residual spray; NR = non-residual spray.

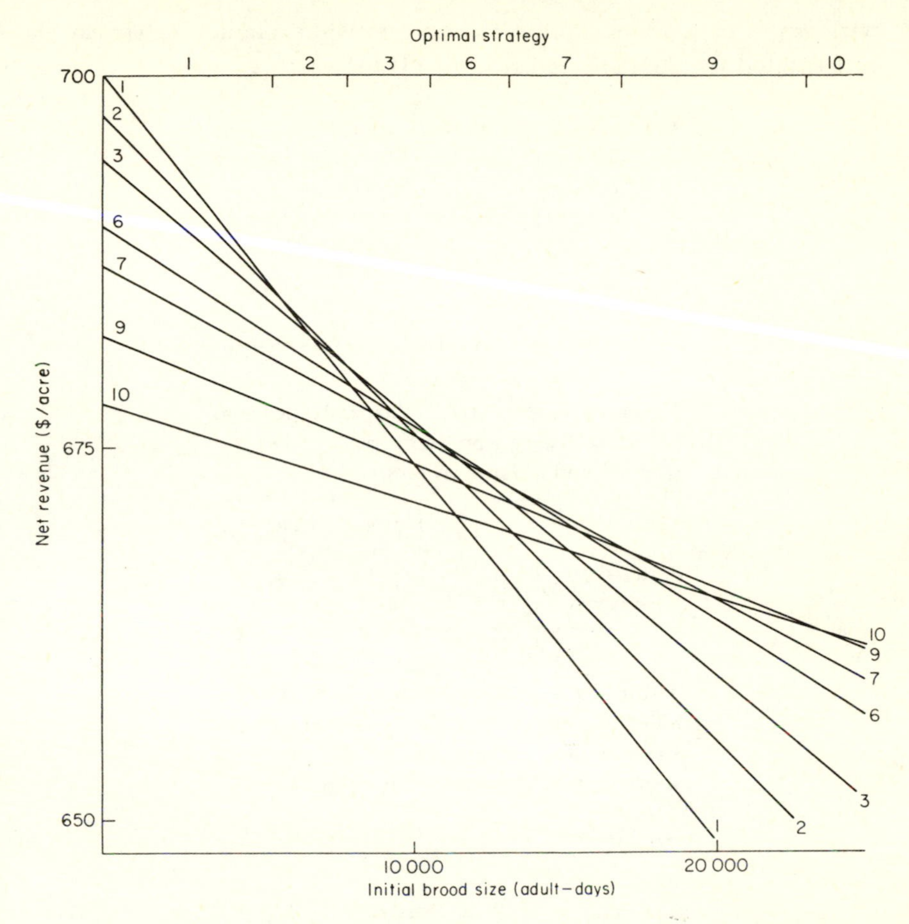

Fig. 14.8. Optimal spraying strategies (see Table 14.4) against a single brood of froghoppers. Net revenue is measured in Trinidad and Tobago dollars. (After Conway *et al.*, 1975.)

This procedure assumes that the eventual size of a pest population can be accurately estimated. But in most situations one knows only the probability of an approximate level of pest attack and the problem is best approached by use of a Bayesian decision matrix (Halter and Dean, 1972; Norton, 1976a). Let us assume an imaginary crop which can be either heavily or lightly attacked by a pest; further we assume that the probabilities of attack are different for two regions, the North and South of the country (see Table 14.5).

On the basis of past experience we can estimate the yields at different levels of attack for an unsprayed and sprayed crop (Table 14.6). Then with knowledge of current costs and prices a similar table can be constructed for the expected returns (Table 14.7).

Table 14.5. Probabilities of heavy and light pest attack for two different regions of a country.

	Heavy	Light
North	·4	·6
South	·1	·9

Table 14.6. Expected yields in lbs per acre following heavy and light pest attack on sprayed and unsprayed crops.

	Heavy	Light
Sprayed	85	90
Unsprayed	20	90

Table 14.7. Expected return in dollars following heavy and light pest attack on sprayed and unsprayed crops. Cost of spraying = $100/acre; value of crop = $10/lb.

	Heavy	Light
Sprayed	750	800
Unsprayed	200	900

The farmer in the North who sprays every year will thus expect an average annual monetary outcome of $\cdot 4 \times 750 + \cdot 6 \times 800 = \$780$ and the farmer who does not spray will expect $\cdot 4 \times 200 + \cdot 6 \times 900 = \$620$. In the South on the other hand the farmer who sprays expects $\cdot 1 \times 750 + \cdot 9 \times 800 = \$795$ and the farmer who does not spray expects $\cdot 1 \times 200 + \cdot 9 \times 900 = \$830$. It is rational, in terms of average long run profitability, for farmers to spray in the North but not in the South.

The Bayesian approach can be particularly illuminating in showing the potential value of forecasting pest attack. For example, if it is possible to forecast the level of infestation with 70 per cent accuracy then it becomes profitable to spray in the South when heavy infestations are forecast. The expected monetary outcome is then $(\cdot 7 \times \cdot 1 \times 750) + (\cdot 3 \times \cdot 1 \times 200) + (\cdot 3 \times \cdot 9 \times 800) + (\cdot 7 \times \cdot 9 \times 900) = \$841 \cdot 50$, compared to \$830 for a strategy of no spraying at all. Often a great deal of effort and money is put into developing pest forecasting schemes on the assumption that any forecast is an improvement on no forecast at all. Use of a Bayesian approach can provide the ecologist with the minimum level of accuracy of forecasting which is required and hence provide a meaningful assessment of his work.

Although profit maximisation may be taken as the objective among large farmers, pest control may be viewed in a significantly different fashion by smaller farmers. The goal of profit maximisation assumes that the decision-maker is risk neutral; in other words, he values each successive increment in income in the same way. But many farmers can be classified as risk averse (Norton, 1976b). They value initial increases in income far higher than later increases. For example, a farmer who has just started farming and has little capital available will set his primary goal as the achievement of a certain minimum level of income. This situation will be particularly true of subsistence farmers in developing countries who will put a much higher value on achieving, each year, the necessary level of subsistence than on each increment of income above that level. To illustrate the point let us take the Bayesian example above and assume the risk averse farmer lives in the South region and no forecast is available. As we have seen the rational strategy in terms of the highest expected monetary value will be not to spray. However, if we look back at Table 14.5 we see that unsprayed crops experience a much greater variance in outcome than sprayed crops. Although not spraying gives the highest overall profit, one year in ten it will produce an extremely low return and the risk-averse farmer may decide to spray every year to prevent this one in ten occurrence. Norton (1976b) illustrates this point with a detailed analysis of the control of potato blight in the United Kingdom.

14.4.2 *Time*

So far I have assumed that the control decision is made before the pest attack occurs; the decision is prophylactic in nature (Fig. 14.9). The farmer who is risk averse will tend to rely on prophylactic control. In many cases the measures will be strictly pre-emptive, i.e. taken before the attack occurs. For example, the farmer may use a resistant crop or he may dress his seed before sowing with a fungicide or insecticide. Most control measures against K-pests are likely to be prophylactic, since the strategy is to reduce the size of the pest niche and, in practice, this can only be done ahead of the pest attack. Spraying according to some fixed calendar schedule is also a form of prophylactic control since, again, the decision is made in advance of knowledge of the actual level of pest attack.

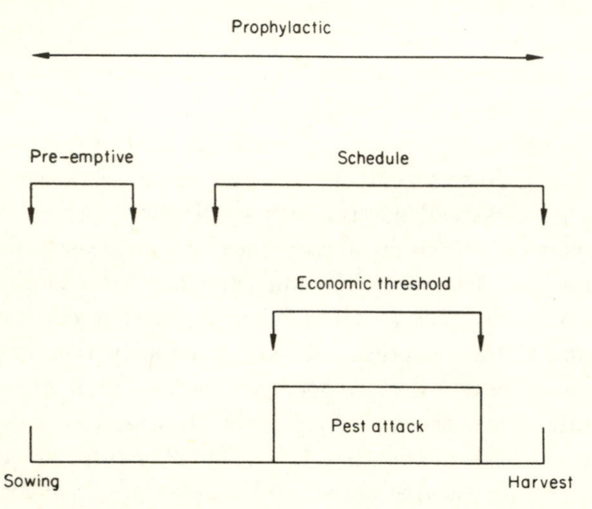

Fig. 14.9. Relationship between pest control strategies and the time of pest attack. (After Norton, 1977.)

The alternative to prophylactic control is to utilise the concept of economic threshold (Stern, 1966). This is defined as the density of the pest population at which a control action becomes profitable. Clearly it depends on the state of the crop, the kind of control action contemplated and many other factors. The advantage of the economic threshold approach is that it enables pest control measures to be more precisely tuned to the pest population and the damage that is being caused. Control is potentially more efficient and hence can serve more closely the goal of profit maximisation. Where pesticides are used there is the added benefit that the amounts applied are likely to be less than under a prophylactic or schedule spraying programme and hence the degree of environmental contamination is reduced. One estimate suggests that pesticide usage in the U.S.A. could be cut by over 50 per cent if pesticides were applied only when and where they were really required (Pimentel, 1973).

But use of the economic threshold approach implies that the farmer has to make a sequence of decisions. He also has to have a greater knowledge of the dynamics of the pest population and its interaction with the crop. The problem now becomes complex. Fortunately a number of techniques stemming from operations research are available to handle problems of this kind. One of these, dynamic programming, was first used in a pest control problem by Watt in 1963 and has been explored in more detail by Shoemaker (1973). One advantage of the technique is the relative ease with which the dynamics of the pest population can be allowed for in determining optimal control strategies.

Conway *et al.* (1975) have illustrated its use in their study of the sugar-cane froghopper. In the example discussed above only a single brood was considered, but in actuality the sugar-cane is attacked by four broods in a year. Clearly if there is no density dependent relationship between the broods then the problem of control is relatively straightforward: the size of all four broods is a function of the level of control of the first and the procedure, described above, for determining the best strategy is adequate. However, it is reasonable to expect that all pests, whether r or K-strategists, are subject to a degree of density dependence. In the case of the froghopper we have assumed that this is described by eq. (3.3), that is

$$A_{t+1} = \lambda A_t^{1-b}. \qquad (14.3)$$

Here A_{t+1}, A_t are the numbers of adult days at time t and $t+1$, and (as discussed in chapters 2 and 3) λ is the net rate of increase, and b is a measure of density dependence (with the characteristic return time $T_R = 1/b$). For the sake of illustration, we assumed the ten possible control strategies described in Table 14.4, and determined the optimal spraying policy for different values of the characteristic return rate b. Table 14.8 shows the result. As can be seen, with b equal to 0.5 (i.e. undercompensation) the spraying becomes concentrated in the second brood and with $b = 1{\cdot}25$ (i.e. overcompensation) it becomes concentrated in the second and third broods.

As the time horizon is extended a further complexity arises from the probability of insecticide resistance developing in the pest population. This is a phenomenon which has become increasingly common in recent years: there are now over 200 species of pest which are resistant to one or more chemical compounds. There are, as yet, no predictive models useful to the decision-maker in designing a control strategy

Table 14.8. Optimal spraying strategies (see Table 14.4) for four froghopper broods with different degrees of density dependence between the broods (Revenue in Trinidad and Tobago dollars) (From Conway *et al.*, 1975.)

Initial size of 1st brood	Density dependence (value of b)														
	$b = 0$					$b = 0 \cdot 5$					$b = 1 \cdot 25$				
	Brood strategy 1st	2nd	3rd	4th	Net revenue ($/acre)	Brood strategy 1st	2nd	3rd	4th	Net revenue ($/acre)	Brood strategy 1st	2nd	3rd	4th	Net revenue ($/acre)
0	1	1	1	1	700	1	1	1	1	700	1	1	1	1	700
1000	2	2	1	1	675	1	9	2	1	649	1	10	10	1	608
2000	7	3	1	1	664	1	10	3	1	641	1	10	10	1	608
3000	10	2	1	1	657	2	10	3	1	635	1	9	10	1	607
4000	9	6	1	1	652	3	10	3	1	630	1	9	10	1	606
5000	10	5	1	1	648	6	10	3	1	627	1	9	10	1	604
6000	10	7	1	1	644	7	10	3	1	624	1	9	10	1	602
7000	10	7	1	1	640	9	10	3	1	621	1	9	10	1	500
8000	10	9	1	1	638	9	10	3	1	619	2	9	10	1	598
9000	10	9	1	1	635	9	10	3	1	617	2	9	10	1	597
10000	10	10	1	1	633	10	10	3	1	615	2	9	10	1	595

which will reduce the probability of resistance occurring. Comins (1976) has, however, produced a theoretical model in which the rates of migration between sprayed and unsprayed populations, the degree of density dependence in the sprayed population, and the rate at which insecticide resistance develops are related to one another. The model shows very clearly that there is a critical migration rate above which resistance is greatly retarded. It also shows that density dependence is an important factor. Comins found that for perfect density dependence a very highly effective insecticide treatment will tend to increase the rate of development of resistance, but where density dependence is undercompensating the rate is reduced at both high and low kills; intermediate kills provide the fastest resistance (Fig. 14.10).

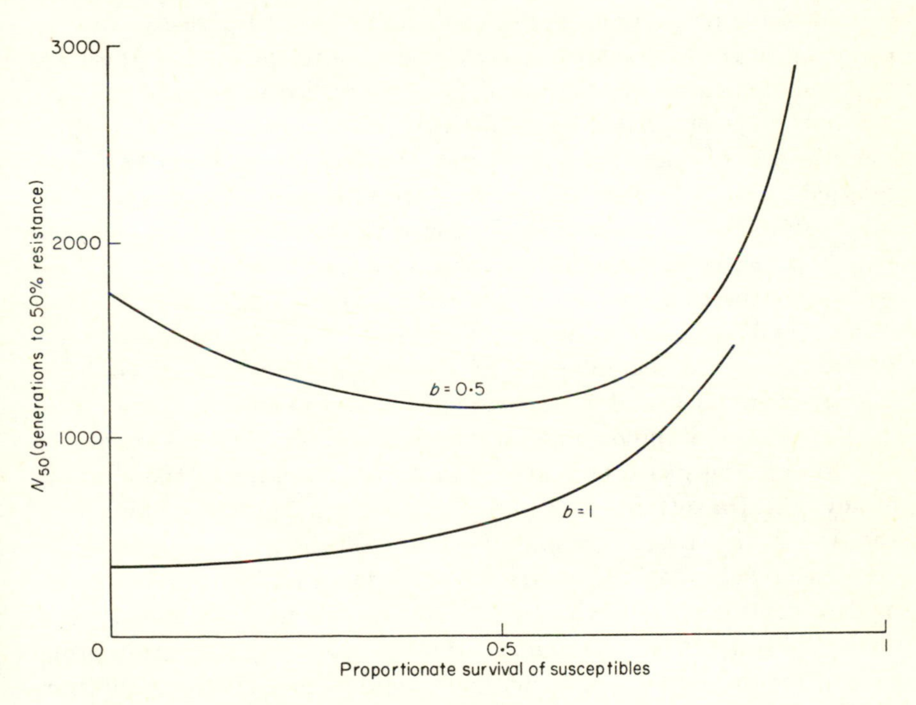

Fig. 14.10. Effects of density dependence on the development of insecticide resistance. (After Comins, 1976.)

14.4.3 *Space*

To this point I have considered a single pest on a single crop; but most farmers have to cope with an assortment of pests on a variety of crops. The control techniques which seem, on the previous analysis, to be most appropriate for a particular pest now have to be modified in the light of the other pest problems.

Many of the pioneer integrated control programmes were expressly designed to cope with situations of this kind (Wildbolz and Meier, 1973; Wood, 1971). For instance in the early 1960's, in northern Borneo, an integrated control programme was designed for a complex of half a dozen pests on cocoa (Conway, 1971b). Most of these pests were intermediate pests, although close to the *r* end of the spectrum. In their natural habitat of primary or secondary forest, they were regulated by a variety of natural enemies, but heavy spraying with organochlorine insecticides had apparently destroyed these enemies resulting in severe pest outbreaks on the cocoa. As a first step the use of organochlorine insecticides was stopped and within a few months several of the major pest species came under natural parasite control, reverting to levels which caused little damage to the cocoa. However, two pests remained and these had to be controlled in a manner which did not upset the balance, so recently re-established for the other species. One of these two pests was the tree borer *Endoclita hosei* referred to earlier in the chapter as a typical *K*-pest. This was successfully controlled by combining injection with insecticide of any borings found, and the destruction of the secondary forest tree species which was the primary host of the borer in the vicinity of the cocoa. The second pest, a bag-worm (Psychidae), which showed a high reproductive rate characteristic of *r* pests, was knocked down by a selective insecticide and then came under parasite control.

Experience of producing strategies for dealing with complexes of pests on crops such as cotton and apples has lead to the concept of key pest. On cotton, for example, over 80 per cent of the damage in the U.S.A. is caused by boll weevils, bollworms and lygus bugs (Reynolds *et al.*, 1975). These are the prime targets of control, but if the control measures are ill-chosen the key species may be joined by a variety of new, 'upset' pests. Often attention has diverted to attempting to control these upset pests without appreciating that the solution to the problem lies in more appropriate control of the key pests.

Although the growing of more crops will tend to increase the number of pest problems with which a farmer has to cope, the growing of different crops in close proximity may tend to reduce the overall pest problem. As we saw in chapter 5, spatial heterogeneity tends to promote stability in animal and plant populations. It has been frequently remarked that the highly variegated cropping systems of peasant farmers in the tropics are often characterised by a lack of pest disease and weed problems (Conway, 1973; Norton, 1975). As we have seen, such farmers are risk averse and prefer a multiple cropping pattern because this gives a better guarantee of minimal yield in the face of a variety of hazards such as pests, floods, drought, illness etc. Satisfactory pest control is thus only one of the ends served by the overall cropping strategy.

As the spatial horizon of the decision-maker extends, regional and finally national priorities become important in pest control. The farmer, for example, may be less concerned with the problem of insecticide resistance than is the regional or national policy-maker faced with the broader implications of having to provide new pesticide compounds every few years. This may entail heavy research and development costs or, in the case of the developing countries, a drain on foreign exchange. Although such elements frequently intrude in the development of specific pest control strategies they are rarely explicitly discussed.

References

Abrahamson, W.G. and Gadgil, M. 1973. Growth form and reproductive effort in Goldenrods (*Solidago*, Compositae). *Am. Nat.* 107, 651–61.

Abrams, P. 1976a. Limiting similarity and the form of the competition coefficient. *Theoret. Pop. Biol.*, 8, 356–75.

Abrams, P. 1976b. Environmental variability and niche overlap. *Math. Biosci.*, (in press).

Akinlosotu, T.A. 1973. *The Role of* Diaeretiella rapae (*McIntosh*) *in the Control of the Cabbage Aphid*. Unpublished Ph.D. thesis, University of London.

Alexander, R.D. 1974. The evolution of social behaviour. *A. Rev. Ecol. Syst.*, 5, 325–83.

Anderson, M.C. 1966. Ecological groupings of plants. *Nature*, (*Lond.*), 212, 54–6.

Andrewartha, H.G. and Birch, L.C. 1954. *The Distribution and Abundance of Animals*. Chicago, University of Chicago Press.

Anonymous. 1969. What will happen to geology? *Nature*, (*Lond.*), 221, 903.

Ansari, N., and Junker, J. 1969. Economic aspects of parasitic diseases. *Biotechnology and Bioengineering Symposia*, 1, 235–52.

Armstrong, N.E., Storrs, P.N., and Pearson, E.A. 1971. Development of a gross toxicity criterion in San Francisco Bay. In *Advances in Water Pollution Research*: *Proceedings of Vth International Conference*. Pp. III 1–15. Oxford, Pergamon Press.

Auslander, D., Oster, G., and Huffaker, C. 1974. Dynamics of interacting populations. *J. Franklin. Inst.*, 297, 345–76.

Austin, M.P., Aston, P.S., and Greig-Smith, P. 1972. The application of quantitative methods to vegetation survey. III. A re-examination of rain forest data from Brunei. *J. Ecol.*, 60, 305–24.

Axelrod, D.I., and Bailey, H.P. 1968. Cretaceous dinosaur extinction. *Evolution*, 22, 595–611.

Ayala, F.J., Gilpin, M.E., and Ehrenfeld, J.G. 1973. Competition between species: theoretical models and experimental tests. *Theoret. Pop. Biol.*, 4, 331–56.
[Referred to as Ayala *et al.*, 1973a.]

Ayala, F.J., Hedgecock, D., Zumwalt, G., and Valentine, J.W. 1973. Genetic variation in *Tridacna maxima*, an ecological analog of some unsuccessful evolutionary lineages. *Evolution*, 27, 177–91.
[Referred to as Ayala *et al.*, 1973b.]

Bailey, N.T.J. 1957. *The Mathematical Theory of Epidemics*. London, Charles Griffin & Company Limited.

Bakker, K., Bagchee, S.N., Van Zwet, W.R., and Meelis, E. 1967. Host discrimination in *Pseudeucoila bochei* (Hymenoptera: Cynipidae). *Entomologia exp. appl.*, 10, 295–311.

Bakker, R. 1974. Dinosaur bioenergetics: a reply to Bennett and Dalzell, and Feduccia. *Evolution*, 28, 497–503.

Bakker, R. 1975a. Experimental and fossil evidence for the evolution of tetrapod bioenergetics. In D.M. Gates and R.B. Schmerl (eds.), *Perspectives in Biophysical Ecology*. New York, Springer-Verlag.

Bakker, R. 1975b. Dinosaur renaissance. *Sci. Amer.*, 232(4), 58–78.

Baltensweiler, W. 1971. The relevance of changes in the composition of larch bud moth populations for the dynamics of its numbers. In P.J. den Boer and G.R. Gradwell (eds.), *Dynamics of Populations*, pp. 208–19. Wageningen, Center for Agricultural Publishing.

Barclay, H. 1975. Population strategies and random environments. *Can. J. Zool.* 53, 160–5.

Bazzaz, F.A. 1975. Plant species diversity in old-field successional ecosystems in southern Illinois. *Ecology*, 56, 485–8.

Beauchamp, R.S.A., and Ullyett, P. 1932. Competitive relationships between certain species of fresh-water triclads. *J. Ecol.*, 20, 200–8.

Beddington, J.R. 1974. Age distribution and the stability of simple discrete time population models. *J. Theor. Biol.*, 47, 65–74.

Beddington, J.R. 1975. Mutual interference between parasites or predators and its effect on searching efficiency. *J. Anim. Ecol.*, 44, 331–40.

Beddington, J.R., and May, R.M. 1975. Time delays are not necessarily destabilizing. *Math. Biosci.*, 27, 109–17.

Beddington, J.R., Free, C.A., and Lawton, J.H. 1975. Dynamic complexity in predator-prey models framed in difference equations. *Nature*, (*Lond.*), 255, 58–60.

Beddington, J.R., Free, C.A., and Lawton, J.H. 1976. Concepts of stability and resilience in predator-prey models. (In press.)
[Referred to as Beddington *et al.*, 1976a.]

Beddington, J.R., Hassell, M.P., and Lawton, J.H. 1976. The components of arthropod predation. II. The predator rate of increase. *J. Anim. Ecol.* 45, 165–85.
[Referred to as Beddington *et al.*, 1976b.]

Benson, J.F. 1973. *Laboratory studies of insect parasite behaviour in relation to population models*. Unpublished D.Phil. thesis, Oxford University.

Berkner, L.V., and Marshall, L.C. 1964. The history of oxygenic concentration in the earth's atmosphere. *Discuss. Faraday Soc.*, 37, 122–41.

Birch, L.C. 1957. The meanings of competition. *Am. Nat.*, 91, 5–18.

Birch, L.C. 1960. The genetic factor in population ecology. *Am. Nat.*, 94, 5–24.

Birch, L.C. 1971. The role of environmental heterogeneity and genetical heterogeneity in determining distribution and abundance. *Proc. Adv. Study Inst. Dynamics Numbers Popul.* (Oosterbeck, 1970), pp. 109–28.

Biswell, H.H. 1974. Effects of fire on chaparral. In T.T. Kozlowski and C.E. Ahlgren (eds.), *Fire and Ecosystems*, pp. 321–64. New York, Academic Press.

Biswell, H.H., Buchanan, H., and Gibbens, R.P. 1966. Ecology of the vegetation of a second-growth Sequoia forest. *Ecology*, 47, 630–4.

Blackith, R.E. 1974. Strategies and tactics in evolution. *Recherches Biologiques Contemporaines*, 427–35.

Blydenstein, J. 1967. Tropical savanna vegetation of the Llanos of Columbia. *Ecology*, 48, 1–15.

Bonner, J.T. 1965. *Size and Cycle: an essay on the structure of biology*. Princeton, Princeton University Press, pp. 219.

Boorman, S.A., and Levitt, P.R. 1973. Group selection on the boundary of a stable population. *Theoret. Pop. Biol.*, 4, 85–128.

Botkin, D.B., Janak, J.F., and Wallis, J.R. 1972. Some ecological consequences of a computer model of forest growth. *J. Ecol.*, 60, 849–72.

Bradley, D.J. 1974. Stability in host-parasite systems. In M.B. Usher and M.H. Williamson (eds.), *Ecological Stability*. London, Chapman and Hall.

Brénière, J., and Walker, P.T. 1970. Host: *Oryza sativa* (rice); Organism: Maliorpha *separatella* (African white rice borer). In *Crop Loss Assessment Methods*. Rome, Food and Agriculture Organization.

Bretsky, P.W., and Lorenz, D.M. 1970. Adaptive response to environmental stability: a unifying concept in paleoecology. *Proc. N. Amer. Paleont. Convention* (part E), 522–50.

Bretsky, P.W., and Bretsky, S.S. 1975. Succession and repetition of Late Ordovician fossil assemblages from the Nicolet River Valley, Quebec. *Paleobiology*, 1, 225–37.

Brian, M.V. 1952. Structure of a dense natural ant population. *J. Anim. Ecol.*, 21, 12–24.

Brougham, R.W. 1955. A study in rate of pasture growth. *Aust. J. Agric. Res.*, 6, 804–12.

Brown, E.S. 1962. *The African Army Worm, Spodoptera exampta [Walker] [Lep., Noctuidae]: A review of the literature*. Commonwealth Institute of Entomology, p. 57.

Brown, J.H. 1971. Mammals on mountaintops: Non-equilibrium insular biogeography. *Am. Nat.*, 105, 467–78.

Brown, J.H., and Lieberman, G.A. 1973. Resource utilization and coexistence of seed-eating desert rodents in sand dune habitats. *Ecology*, 54, 788–97.

Brown, W.L., and Wilson, E.O. 1956. Character displacement. *Syst. Zool.*, 5, 49–64.

Buell, M.F., Buell, H.F., and Small, J.A. 1954. Fire in the history of Mettler's Woods. *Bull. Torrey Bot. Club*, 81, 1–3.

Burnet, M., and White, D.O. 1972. *Natural History of Infectious Disease*. 4th Ed. London, Cambridge University Press.

Burnett, T. 1956. Effects of natural temperatures on oviposition of various numbers of an insect parasite (Hymenoptera, Chalcididae, Tenthredinidae). *Ann. ent. Soc. Am.*, 49, 55–9.

Burnett, T. 1958. Dispersal of an insect parasite over a small plot. *Can. Ent.*, 90, 279–83.

Butlin, N.G. 1962. Distribution of sheep population: preliminary statistical picture, 1860–1957. In A. Barnard (ed.), *The Simple Fleece*, pp. 281–307. Melbourne, Melbourne University Press.

Butterworth, A.E. (rapporteur). In press. *The Immunology of Schistosomiasis*. Report of a Meeting of Investigators held in Nairobi, 11–17 December 1974.

Caccamise, D.F. 1974. Competitive relationships of the common and lesser nighthawks. *Condor*, 76, 1–20.

Canning, E.U., and Wright, C.A. 1972. *Behavioural Aspects of Parasite Transmission*. London, Academic Press.

Cates, R.G., and Orians, G.H. 1975. Successional status and the palatability of plants to generalized herbivores. *Ecology*, 56, 410–18.

Caughley, G. 1970. Eruption of ungulate populations, with emphasis on Himalayan thar in New Zealand. *Ecology*, 51, 53–72.

Caughley, G. 1976. Wildlife management and the dynamics of ungulate populations. In T.H. Coaker (ed.), *Advances in Applied Biology*, Vol. 1, in press.

Cavalli-Sforza, L.L., and Bodmer, W.F. 1971. *The Genetics of Human Populations*. San Francisco, W.H. Freeman.

Cavalli-Sforza, L.L., and Feldman, M.W. 1973. Models for cultural inheritance: I, group mean and within group variation. *Theoret. Pop. Biol.*, 4, 42–55.

Charnov, E.L. 1973. *Optimal foraging: some theoretical explorations.* Ph.D. Thesis, Seattle, University of Washington.

Charnov, E.L. 1975. Optimal foraging: attack strategy of a mantid. *Am. Nat.*, 110, (in press).

Chitty, H. 1950. Canadian arctic wildlife enquiry, 1945–49, with a summary of results since 1933. *J. Anim. Ecol.*, 19, 180–93.

Chua, T.C. 1975. *Population studies on the cabbage aphid,* Brevicoryne brassicae (L.), *and its parasites, with special reference to synchronization.* Unpublished Ph.D. thesis, London University.

Clapham, W.B. 1973. *Natural Ecosystems.* New York, Collier-Macmillan.

Clements, F.E. 1916. Plant succession: an analysis of the development of vegetation. *Carnegie Inst. Wash. Publ.*, 242. pp. 512.

Cockburn, A. 1963. *The Evolution and Eradication of Infectious Diseases.* Baltimore, The Johns Hopkins Press.

Cody, M.L. 1966. A general theory of clutch size. *Evolution*, 20, 174–84.

Cody, M. 1968. On the methods of resource division in grassland bird communities. *Am. Nat.*, 102, 107–48.

Cody, M.L. 1969. Convergent characteristics in sympatric populations: A possible relation to interspecific territoriality. *Condor*, 71, 222–39.

Cody, M.L. 1974. *Competition and the Structure of Bird Communities.* Princeton, Princeton University Press.

Cody, M.L. 1975. Towards a theory of continental species diversities. In M.L. Cody and J.M. Diamond (eds.), *Ecology and Evolution of Communities.* Cambridge, Mass., Harvard University Press, pp. 214–57.

Cohen, J.E. 1972. When does a leaky compartment model appear to have no leaks? *Theoret. Pop. Biol.*, 3, 404–5.

Cohen, J.E. 1973a. Heterologous immunity in human malaria. *Quart. Rev. Biol.*, 48, 467–89.

Cohen, J.E. 1973b. Selective host mortality in a catalytic model applied to schistosomiasis. *Am. Nat.*, 107, 199–212.

Cohen, J.E. 1974. Some potential economic benefits of eliminating mortality due to schistosomiasis in Zanzibar. *Social Science and Medicine*, 8, 383–98.

Cohen, J.E. 1975. Livelihood benefits of small improvements in the life table. *Health Services Research*, Spring, 82–96.

Cohen, J.E. 1976. Mathematical models of schistosomiasis: a critical review. (In press).

Cole, L.C. 1960. Competitive exclusion. *Science*, 312, 348–9.

Colinvaux, P.A. 1973. *Introduction to Ecology.* New York, Wiley.

Colwell, R.K. 1973. Competition and coexistence in a simple tropical community. *Am. Nat.*, 107, 737–60.

Colwell, R.K., and Futuyma, D.J. 1971. On the measurement of niche breadth and overlap. *Ecology*, 52, 567–76.

Colwell, R.K. and Fuentes, E.R. 1975. Experimental studies of the niche. *A. Rev. Ecol. Syst.*, 6, 281–310.

Comins, H. 1976. The development of insecticide resistance in the presence of migration. *J. Theor. Biol.*, (in press).

Connell, J.H. 1961. The influence of interspecific competition and other factors on the distribution of the barnacle *Chthamalus stellatus. Ecology*, 42, 710–23.

Connell, J.H. 1972. Community interactions on marine rocky intertidal shores. *A. Rev. Ecol. Syst.*, 3, 169–92.

Connell, J.H. 1974. Field experiments in marine ecology. In R. Mariscal (ed.), *Experimental Marine Biology*. New York, Academic Press.

Connell, J.H. 1975. Some mechanisms producing structure in natural communities: a model and evidence from field experiments. In M.L. Cody and J.M. Diamond (eds.), *Ecology and Evolution of Communities*, pp. 460–90. Cambridge, Harvard University Press.

Conway, G.R. 1971. *Pests of Cocoa in Sabah and their Control*. Ministry of Agriculture and Fisheries, Sabah, Malaysia. Pp. 125.

Conway, G.R. 1972. Ecological aspects of pest control in Malaysia. In M.T. Farvar, and J.P. Milton (eds.), *The Careless Technology*. New York, The Natural History Press.

Conway, G.R. 1973. Aftermath of the Green Revolution. In N. Calder (ed.), *Nature in the Round*, pp. 226–35. London, Weidenfeld and Nicolson.

Conway, G.R., Norton, G.A., Small, N.J., and King, A.B.S. 1975. A systems approach to the control of the sugar cane froghopper. In G.E. Dalton (ed.), *Study of Agricultural Systems*, pp. 193–229. London, Applied Science Publishers.

Cooper, C.F. 1960. Changes in vegetation, structure and growth of southwestern pine forests since white settlement. *Ecol. Monogr.*, 30, 129–64.

Crocker, R.L., and Major, J. 1955. Soil development in relation to vegetation and surface age at Glacier Bay, Alaska. *J. Ecol.*, 42, 427–48.

Crofton, H.D. 1971a. A quantitative approach to parasitism. *Parasitology*, 62, 179–93.

Crofton, H.D. 1971b. A model of host-parasite relationships. *Parasitology*, 63, 343–64.

Crombie, A.C. 1947. Interspecific competition. *J. Anim. Ecol.*, 16, 44–73.

Crosby, A.W. 1972. *The Columbian Exchange: Biological and Cultural Consequences of* 1492. Westport, Conn., Greenwood Press.

Crow, J.F., and Kimura, M. 1970. *An Introduction to Population Genetics Theory*. New York, Harper & Row.

Crowell, K.L. 1962. Reduced interspecific competition among the birds of Bermuda. *Ecology*, 43, 75–88.

Culver, D.C. 1970. Analysis of simple cave communities: niche separation and species packing. *Ecology*, 51, 949–58.

Dahlsten, D.L. 1967. Preliminary life tables for pine sawflies in the *Neodiprion fulviceps* complex (Hymenoptera: Diprionidae). *Ecology*, 48, 275–89.

Dajoz, R. 1974. *Dynamique des Populations*. Paris, Masson et cie.

Dammerman, K.W. 1948. The fauna of Krakatau 1883–1933. *Verhandel. Koninkl. Ned. Akad. Wetenschap. Afdel. Natuurk.*, 44(2), 1–594.

Davidson, J.L., and Donald, C.M. 1958. The growth of swards of subterranean clover with particular reference to leaf area. *Aust. J. Agric. Res.*, 9, 53–72.

Davis, J. 1973. Habitat preferences and competition of wintering juncos and golden-crowned sparrows. *Ecology*, 54, 174–80.

Dayton, P.K. 1971. Competition, disturbance, and community organization: the provision and subsequent utilization of space in a rocky interidal community. *Ecol. Monogr.*, 41, 351–89.

Dayton, P.K. 1975. Experimental evaluation of ecological dominance in a rocky intertidal algal community. *Ecol. Monogr.*, 45, 137–59.

DeBach, P. 1966. The competitive displacement and coexistence principles. *A. Rev. Ent.*, 11, 183–212.

DeBach, P. 1974. *Biological Control by Natural Enemies*. Cambridge, Cambridge University Press.

DeBach, P., and Smith, H.S. 1941. The effect of host density on the rate of reproduction of entomophagous parasites. *J. econ. Ent.*, 34, 741–5.

De Beer, G. 1958. *Embryos and Ancestors*. Oxford, Clarendon Press.

Deevey, E.S. 1965. Environments of the geologic past. *Science*, 147, 592–4.

DeLong, K.T. 1966. Population ecology of feral house mice: interference by *Microtus*. *Ecology*, 47, 481–4.

Den Boer, P.J. 1968. Spreading of risk and stabilization of animal numbers. *Acta. Biotheor.* 18, 165–94.

Diamond, J.M. 1969. Avifaunal equilibria and species turnover rates on the Channel Islands of California. *Proc. Natl. Acad. Sci. US.*, 64, 57–63.

Diamond, J.M. 1971. Comparison of faunal equilibrium turnover rates on a tropical island and a temperate island. *Proc. Natl. Acad. Sci. U.S.*, 68, 2742–5.

Diamond, J.M. 1972. Biogeographic kinetics: Estimation of relaxation times for avifaunas of Southwest Pacific Islands. *Proc. Natl. Acad. Sci. U.S.*, 69, 3199–203.

Diamond, J.M. 1973. Distributional ecology of New Guinea birds. *Science*, 179, 759–69.

Diamond, J.M. 1974. Colonization of exploded volcanic islands by birds: the Supertramp strategy. *Science*, 184, 803–6.

Diamond, J.M. 1975a. The island dilemma: Lessons of modern biogeographic studies for the design of natural reserves. *Biol. Conservation*, 7, 129–46.

Diamond, J.M. 1975b. Assembly of species communities. In M.L. Cody and J.M. Diamond (eds.), *Ecology and Evolution of Communities*. *op. cit.* pp. 342–444.

Diamond, J.M., and Jones H.L. 1976. *Dynamic Island Biogeography*. Princeton. Princeton University Press, (in press).

Diamond, J.M., and May, R.M. 1976. Species turnover rates on islands: dependence on census interval, (in press).

Diamond, J.M., and Mayr, E. 1976. The species-area relation for birds of the Solomon Archipelago. *Proc. Natl. Acad. Sci. U.S.*, 73, 262–6.

Dietz, K. 1967. Epidemics and rumours: a survey. *J. Royal Statist. Soc. A*, 130, 505–28.

Dietz, K. 1975. A pairing process. *Theoret. Pop. Biol.*, 8, 81–6.

Dingle, H. 1974. The experimental analysis of migration and life history strategies in insects. In L. Barton-Browne (ed.), *Experimental Analysis of Insect Behaviour*, pp. 329–42. Berlin, Springer-Verlag.

Dixon, A.F.G. 1959. An experimental study of the searching behaviour of the predatory coccinellid beetle *Adalia decempunctata* (L.). *J. Anim. Ecol.*, 28, 259–81.

van Dobben, W.H., and Lowe-McConnell, R.H. (eds.) 1975. *Unifying Concepts in Ecology*. The Hague, W. Junk.

Docters van Leeuwen, W.M. 1936. Krakatau, 1883–1933. *Ann. Jard. Botan. Buitenzorg*, 56–7, 1–506.

Dodd, A.P. 1940. *The Biological Campaign against Prickly-Pear*. Brisbane, Government Printer.

Dransfield, R. 1975. *The Ecology of Grassland and Cereal Aphids*. London, unpublished Ph.D. thesis.

Drury, W.H., and Nisbet, I.C.T. 1971. Inter-relations between developmental models in geomorphology, plant ecology, and animal ecology. *Gen. Syst.*, 16, 57–68.

Drury, W.H., and Nisbet, I.C.T. 1973. Succession. *J. Arnold Arboretum Harvard University*. 54, 331–68.

Eberhard, Mary Jane West. 1975. The evolution of social behaviour by kin selection. *Quart. Rev. Biol.*, 50, 1–33.

Ehler, L.E., and van den Bosch, R. 1974. An analysis of the natural biological control of *Trichoplusia ni* (Lepidoptera: Noctuidae) on cotton in California. *Can. Ent.*, 106, 1067–73.

Ehrlich, P., and Gilbert, L.E. 1973. Population structure and dynamics of the tropical butterfly, *Heliconius ethilla. Biotropica*, 5, 69–82.

Ehrlich, P.R., White, R.R., Singer, M.C., McKechnie, S.W., and Gilbert, L.E. 1975. Checkerspot butterflies: A historical perspective. *Science*, 188, 221–8.

Eldredge, N., and Gould, S.J. 1972. Punctuated equilibria: an alternative to phyletic gradualism. In T.J.M. Schopf (ed.), *Models in Paleobiology*. San Francisco, Freeman, Cooper and Company. pp. 82–115.

Elliott, H. 1971. Wandering albatross. In J. Gooders (ed.), *Birds of the World*. 1, pp. 29–31. London, I.P.C.

Elliott, P.F. 1975. Longevity and the evolution of polygamy. *Am. Nat.* 109, 281–7.

Elton, C. 1927. *Animal Ecology*. London, Sidgwick and Jackson.

Elton, C. 1942. *Voles, Mice and Lemmings: problems in population dynamics*. Oxford, Oxford University Press.

Elton, C. S. 1958. *The Ecology of Invasions by Animals and Plants*. London, Methuen.

Emlen, J.M. 1966. The role of time and energy in food preference. *Am. Nat.*, 100, 611–17.

Emlen, J.M. 1968. Optimal choice in animals. *Am. Nat.*, 102, 385–90.

Emlen, J.M. 1973. *Ecology: an evolutionary approach*. Addison-Wesley.

Eshel, I. 1972. On the neighbor effect and the evolution of altruistic traits. *Theoret. Pop. Biol.*, 3, 258–77.

Evans, H.F. 1973. *A study of the predatory habits of* Anthocoris *species* (*Hemiptera-Heteroptera*). Unpublished D. Phil. Thesis, Oxford University.

Faaborg, J. 1976. Patterns in the structure of West Indian bird communities, (in press).

Fager, E.W. 1968. The community of invertebrates in decaying oak wood. *J. Anim. Ecol.*, 37, 121–42.

Farnworth, E.G., and Golley, F.B. (eds.). 1974. *Fragile Ecosystems: evaluation of research and applications in the neotropics*. New York, Springer Verlag.

Feldman, M., and Roughgarden, J. 1975. A population's stationary distribution and chance of extinction in a stochastic environment with remarks on the theory of species packing. *Theoret. Pop. Biol.*, 7, 197–207.

Fenchel, T. 1974. Intrinsic rate of natural increase: the relationship with body size. *Oecologia*, 14, 317–26.

Fenchel, T. 1975. Character displacement and coexistence in mud snails (Hydrobiidae). *Oecologia*, 20, 19–32.

Fenchel, T., and Christiansen, F.B. 1976. *Theories of Biological Communities*. New York, Springer-Verlag.

Ford, J. 1971. *The Role of the Trypanosomiases in African Ecology. A Study of the Tsetsefly Problem*. Oxford, Clarendon Press.

Forsyth, A.B., and Robertson, R.J. 1975. K-reproductive strategy and larval behaviour of the pitcher plant sarcophagid fly, *Blaesoxipha fletcheri. Can. J. Zool.*, 53, 174–9.

Fried, M.H. 1975. The myth of tribe. *Nat. Hist.*, 84, 12–20.

Fujii, K. 1968. Studies on interspecies competition between the azuki bean weevil and the southern cowpea weevil: III, some characteristics of strains of two species. *Res. Popul. Ecol.*, 10, 87–98.

Gadgil, M. 1975. Evolution of social behaviour through interpopulation selection. *Proc. Natl. Acad. Sci., U.S.*, 72, 1199–201.

Gadgil, M., and Solbrig, O.T. 1972. The concept of r- & K-selection: evidence from wild flowers and some theoretical considerations. *Am. Nat.*, 106, 14–31.

Gardner, M.R., and Ashby, W.R. 1970. Connectance of large dynamical (cybernetic) systems: critical values for stability. *Nature*, (*Lond.*), 228, 784.

Gates, D.M., and Schmerl, R.B., (eds.) 1975. *Perspectives in Biophysical Ecology*. New York, Springer-Verlag.

Gause, G.F. 1934. *The Struggle for Existence*. Baltimore, Williams and Wilkins. Reprinted, 1964, by Hafner, New York.

Geisel, T.S. 1955. *On Beyond Zebra*. New York, Random House.

Gilbert, L.E. 1975. Ecological consequences of a coevolved mutualism between butterflies and plants. In L.E. Gilbert and P.H. Raven (eds.), *Coevolution of Animals and Plants*. Austin, Texas, University of Texas Press.

Gill, F.B., and Wolf, L.L. 1975a. Economics of Feeding Territoriality in the Golden-Winged Sunbird. *Ecology*, 56, 333–45.

Gill, F.B., and Wolf, L.L. 1975b. Foraging strategies and energetics of east African sunbirds at mistletoe flowers. *Am. Nat.*, 109, 491–510.

Gilpin, M.E. 1972. Enriched prey-predator systems: theoretical stability. *Science*, 177, 902–4.

Gilpin, M.E. 1974. A Liapunov function for competition communities. *J. Theoret. Biol.*, 44, 35–48.

Gilpin, M.E. 1975a. Limit cycles in competition communities. *Am. Nat.*, 109, 51–60.

Gilpin, M.E. 1975b. *Group Selection in Predator-prey Communities*. Princeton, Princeton University Press.

Gilpin, M.E., and Justice, K.E. 1972. Reinterpretation of the invalidation of the principle of competitive exclusion. *Nature*, (*Lond.*), 236, 273–4 and 299–301.

Gilpin, M.E., and Ayala, F.J. 1973. Global models of growth and competition. *Proc. Natl. Acad. Sci., U.S.*, 70, 3590–3.

Gilpin, M.E., and Diamond, J.M. 1976. Calculation of immigration and extinction curves from the species-area-distance relation. *Proc. Natl. Acad. Sci. U.S.*, (in press).

Gleason, H.A. 1926. The individualistic concept of the plant asscoiation. *Bull. Torrey Bot. Club*, 53, 331–68.

Gleason, H.A. 1927. Further views of the succession concept. *Ecology*, 8, 299–326.

Golley, F.B. 1968. Secondary productivity in terrestrial ecosystems. *Am. Zool.*, 8, 53–62.

Goodman, D. 1974. Natural selection and a cost ceiling on reproductive effort. *Am. Nat.*, 108, 247–68.

Gough, H.C. 1946. Studies on wheat bulb fly *Leptohylemia coarctata* Fall. II Numbers in relation to crop damage *Bull. ent. Res.*, 37, 439–54.

Gould, S.J. 1974. The great dying. *Nat. Hist.*, 83, No. 8, 22–7.

Gould, S.J. 1975. Catastrophes and steady state earth. *Nat. Hist.*, 84, No. 2, 14–18.

Gould, S.J. 1976. *Ontogeny and Phylogeny*. Cambridge, Massachusetts, Harvard University Press.

Gould, S.J., Sepkoski, J.J., Raup, D.M., Schopf, T.J.M., and Simberloff, D. 1977 (in preparation). The shape of evolution: a comparison of real and random clades, to be submitted to *Science*.

Grant, P.R. 1968. Bill size, body size and the ecological adaptations of bird species to the competitive situations on islands. *Syst. Zool.*, 17, 319–33.

Grant, P.R. 1972a. Convergent and divergent character displacement. *Biol. J. Linn. Soc.*, 4, 39–68.

Grant, P.R. 1972b. Interspecific competition between rodents. *A. Rev. Ecol. Syst.*, 3, 79–106.

Griffiths, K.J. 1969. Development and diapause in *Pleolophus basizonus* (Hymenoptera: Ichneumonidae). *Can. Ent.*, 101, 907–14.

Griffiths, K.J., and Holling, C.S. 1969. A competition sub-model for parasites and predators. *Can. Ent.*, 101, 785–818.

Grinnell, J. 1917. The niche relationships of the California thrasher. *Auk*, 21, 364–82.

Grinnell, J. 1924. Geography and evolution. *Ecology*, 5, 225–9.

Grinnell, J. 1928. The presence and absence of animals. *Univ. Calif. Chronicle*, 30, 429–50.

Guckenheimer, J., Oster, G.F., and Ipaktchi, A. 1976. The dynamics of density dependent population models, (in press).

Hairston, N.G. 1951. Interspecies competition and its probable influence upon the vertical distribution of Appalachian salamanders of the genus *Plethodon*. *Ecology*, 32, 266–74.

Hairston, N.G. 1962. Population ecology and epidemiological problems. In G.E.W. Wolstenholme and M. O'Connor (eds.), *CIBA Foundation Symposium on Bilharziasis*. London, J. & A. Churchill Ltd.

Hairston, N.G. 1965a. On the mathematical analysis of schistosome populations. *Bull. Wld. Hlth. Org.*, 33, 45–62.

Hairston, N.G. 1965b. An analysis of age-prevalence data by catalytic models. *Bull. Wld Hlth Org.*, 33, 163–75.

Hairston, N.G. 1973. The dynamics of transmission. In N. Ansari (ed.), *Epidemiology and Control of Schistosmiasis* (*Bilharaziasis*). Basel, University Park Press, Baltimore.

Hairston, N.G. *et al.*, 1968. The relationship between species diversity and stability: an experimental approach with protozoa and bacteria. *Ecology*, 49, 1091–101.

Hallam, A. 1969. Faunal realms and facies in the Jurassic. *Palaeontology*, 12, 1–18.

Hallam, A. 1975. *Jurassic Environments*. Cambridge, Cambridge University Press.

Halter, A.N., and Dean, G.W. 1971. *Decisions Under Uncertainty with Research Application*. South-Western Pub. Co., Cincinnati.

Hamilton, W.D. 1964. The genetical theory of social behaviour, I, II. *J. Theoret. Biol.*, 7, 1–52.

Hamilton, W.D. 1972. Altruism and related phenomena, mainly in social insects. *A. Rev. Ecol. Syst.*, 3, 193–232.

Hamilton, W.D. 1974. Evolution sozialer Verhaltensweisen bei sozialen Insekten. In G.H. Schmidt (ed.), *Sozialpoly-morphismus bei Insekten*. Wissenschaftliche Verlagsgesellschaft MBH, Stuttgart, pp. 60–93.

Hanes, T. L. 1971. Succession after fire in the chaparral of southern California. *Ecol. Monogr.*, 41, 27–52.

Hardin, G. 1960. The competitive exclusion principle. *Science*, 131, 1292–7.

Harper, J.L. 1969. The role of predation in vegetational diversity. *Brookhaven Symp. Biol.*, 22, 48–62.

Harper, J.L., and Ogden, J. 1970. The reproductive strategy of higher plants. [i] the concept of strategy with special reference to *Senecio vulgaris* L. *J. Ecol.*, 58, 681–98.

Harper, J.L., and White, J. 1974. The demography of plants. *A. Rev. Ecol. Syst.*, 5, 419–63.

Harrison, G. 1977. *The Curse of Caliban: mosquitoes, malaria and man.* New York, E. P. Dutton, (in press).

Harte, J., and Levy, D. 1975. On the vulnerability of ecosystems disturbed by man. In W.H. van Dobben and R.H. Lowe-McConnell (eds.), *Unifying Concepts in Ecology*, *op. cit.*, pp. 208–23.

Hassell, M.P. 1971. Mutual interference between searching insect parasites. *J. Anim. Ecol.*, 40, 473–86.

Hassell, M.P., and Huffaker, C.B. 1969. Regulatory processes and population cyclicity in laboratory populations of *Anagasta kuehniella* (Zeller) (Lepidoptera: Phycitidae). III. The development of population models. *Researches Popul. Ecol. Kyoto University*, 11, 186–210.

Hassell, M.P., and Varley, G.C. 1969. New inductive population model for insect parasites and its bearing on biological control. *Nature*, (*Lond.*), 223, 1133–6.

Hassell, M.P., and Rogers, D.J. 1972. Insect parasite responses in the development of population models. *J. Anim. Ecol.*, 41, 661–76.

Hassell, M.P., and May, R.M. 1973. Stability in insect host-parasite models. *J. Anim. Ecol.*, 42, 693–726.

Hassell, M.P., and May, R.M. 1974. Aggregation in predators and insect parasites and its effect on stability. *J. Anim. Ecol.*, 43, 567–94.

Hassell, M.P., Lawton, J.H., and Beddington, J.R. 1976. The components of arthropod predation. I. The prey death-rate. *J. Anim. Ecol.*, 45, 135–64. [Referred to as Hassell *et al.*, 1976b.]

Hassell, M.P., Lawton, J.H., and May, R.M. 1976. Patterns of dynamical behaviour in single-species populations. *J. Anim. Ecol.*, 45, (in press). [Referred to as Hassell *et al.*, 1976a.]

Hassell, M.P., Lawton, J.H., and Beddington, J.R. 1977. Sigmoid functional responses in invertebrate predators and parasitoids. (in preparation).

Haven, S.B. 1973. Competition for food between the intertidal gastropods *Acmaea scabra* and *Acmaea digitalis*. *Ecology*, 54, 143–51.

Heal, O.W., and MacLean, S.F. 1975. Comparative productivity in ecosystems: secondary productivity. In W.H. van Dobben and R.H. Lowe-McConnell (eds.), *Unifying Concepts in Ecology*, *op. cit.*, pp. 89–108.

Heatwole, H., and Levins, R. 1972. Trophic structure stability and faunal change during recolonization. *Ecology*, 53, 531–4.

Hedgpeth, J.W. (ed.) 1957. *Treatise on Marine Ecology and Paleoecology*, *Vol.* 1 *Ecology*. Mem. Geol. Soc. Am. No. 67. pp. 1296.

Heinrich, B., and Raven, P.H. 1972. Energetics and pollination ecology. *Science*, 176, 597–602.

Heinselman, M.L. 1963. Forest sites, bog processes and peatland types in the glacial Lake Agassiz region, Minnesota. *Ecol. Monogr.*, 33, 327–74.

Heinselman, M.L. 1973. Fire in the virgin forests of the Boundary Waters Canoe Area, Minnesota. *Quarternary Res.*, 3, 329–82.

Henry, J.D., and Swan, J.M.A. 1974. Reconstructing forest history from live and dead plant material—an approach to the study of forest succession in southwest New Hampshire. *Ecology*, 55, 772–83.

Heron, A.C. 1972. Population ecology of a colonizing species: the pelagic tunicate *Thalia democratica*. *Oecologia*, 10, 269–93 and 294–312.

Hickman, J.C. 1975. Environmental unpredictability and plastic energy allocation strategies in the annual *Polygonum cascadense* (Polygonaceae). *J. Ecol.*, 63, 689–701.

Hirsch, M.W., and Smale, S. 1974. *Differential Equations, Dynamical Systems, and Linear Algebra*. New York, Academic Press.

Hoffman, D.B. 1975. *Schistosomiasis Research: the strategic plan*. New York, The Edna McConnell Clark Foundation.

Holling, C.S. 1959a. The components of predation as revealed by a study of small mammal predation of the European pine sawfly. *Can. Ent.*, 91, 293–320.

Holling, C.S. 1959b. Some characteristics of simple types of predation and parasitism. *Can. Ent.*, 91, 385–98.

Holling, C.S. 1965. The functional response of predators to prey density and its role in mimicry and population regulation. *Mem. Ent. Soc. Can.*, 45, 3–60.

Holling, C.S. 1973. Resilience and stability of ecological systems. *A. Rev. Ecol. Syst.*, 4, 1–24.

Hooykaas, R. 1963. *The Principle of Uniformity*. Leiden, E.J. Brill.

Horn, H.S. 1971. *The Adaptive Geometry of Trees*. Princeton, Princeton University Press.

Horn, H.S. 1974. The ecology of secondary succession. *A. Rev. Ecol. Syst.*, 5, 25–37.

Horn, H.S. 1975a. Forest succession. *Scientific American*, 232 (5), 90–8.

Horn, H.S. 1975b. Markovian properties of forest succession. In M.L. Cody and J.M. Diamond (eds.), *Ecology and Evolution of Communities*, *op. cit.*, pp. 196–211.

Horn, H.S., and MacArthur, R.H. 1972. Competition among fugitive species in a harlequin environment. *Ecology*, 53, 749–52.

Horowitz, D.L. 1975. Ethnic identity. In N. Glazer and D.P. Moynihan (eds.), *Ethnicity: theory and experience*. Cambridge, Mass, Harvard University Press.

Houston, D.B. 1973. Wildfires in northern Yellowstone National Park. *Ecology*, 54, 1111–17.

Howell, D.J. 1976. Flock-foraging in pollinating bats: energetics and advantages to host plants, (in press).

Huey, R.B., and Pianka, E.R. 1974. Ecological character displacement in a lizard. *Am. Zool.*, 14, 1127–36.

Huey, R.B., Pianka, E.R., Egan, M.E., and Coons, L.W. 1974. Ecological shifts in sympatry: Kalahari fossorial lizards (*Typhlosaurus*). *Ecology*, 55, 304–16.

Huffaker, C.B. 1958. Experimental studies on predation: dispersion factors and predator-prey oscillations. *Hilgardia*, 27, 343–83.

Hunt, G.J., Jr. and Hunt, M.W. 1974. Trophic levels and turnover rates: the avifauna of Santa Barbara Island, California. *Condor*, 76, 363–9.

Hutchinson, G.E. 1948. Circular causal systems in ecology. *Ann. N.Y. Acad. Sci.*, 50, 221–46.

Hutchinson, G.E. 1951. Copepodology for the ornithologist. *Ecology*, 32, 571–7.

Hutchinson, G.E. 1953. The concept of pattern in ecology. *Proc. Acad. Nat. Sci., Philadelphia*, 105, 1–12.

Hutchinson, G.E. 1957. Concluding remarks. *Cold Spring Harbour Symp. Quant. Biol.*, 22, 415–27.

Hutchinson, G.E. 1959. Homage to Santa Rosalia, or why are there so many kinds of animals? *Am. Nat.*, 93, 145–59.

Hutchinson, G.E. 1961. The paradox of the plankton. *Am. Nat.*, 95, 137–45.

Hutchinson, G.E. 1975. Variations on a theme by Robert MacArthur. In M.L. Cody and J.M. Diamond (eds.), *Ecology and Evolution of Communities*, *op. cit.*, pp. 492–521.

Ivlev, V.S. 1961. *Experimental Ecology of the Feeding of Fishes*. New Haven, Yale University Press.

Jaeger, R.G. 1970. Potential extinction through competition between two species of terrestrial salamanders. *Evolution*, 24, 632–42.

Jaeger, R.G. 1971. Competitive exclusion as a factor influencing the distributions of two species of terrestrial salamanders. *Ecology*, 52, 632–7.

Janzen, D.H. 1971. Euglossine bees as long-distance pollinators of tropical plants. *Science*. 171, 203–5.

Jarvis, M.J.F. 1974. The ecological significance of clutch size in the South African Gannet (*Sula capensis* (Lichtenstein)). *J. Anim. Ecol.*, 43, 1–17.

Jobin, W.R., and Michelson, E.H. 1967. Mathematical simulation of an aquatic snail population. *Bull. Wld. Hlth. Org.*, 37, 657–64.

Johnson, C.G. 1969. *Migration and Dispersal of Insects by Flight*. London, Methuen, pp. 763.

Jones, E.W. 1945. Structure and reproduction of the virgin forest of the north temporate zone. *New Phytol.*, 44, 130–48.

Jones, H.L., and Diamond, J.M. 1976. Short-time-base studies of turnover in breeding bird population on the California Channel Islands. Submitted to *Condor*.

Jordan, P., and Webbe, G. 1969. *Human Schistosomiasis*. Springfield, C.C. Thomas.

Joule, J., and Jameson, D.L. 1972. Experimental manipulation of population density in three sympatric rodents. *Ecology*, 53, 653–60.

Joule, J., and Cameron, G.N. 1975. Species removal studies. I. Dispersal strategies of sympatric *Sigmodon hispidus* and *Reithrodontomys fulvescens* populations. *J. Mammal.*, 56, 378–96.

Kagan, I.G. 1968. Serologic diagnosis of schistosomiasis. *Bull. N.Y. Acad. Med.*, 44, 262–77.

Kauffman, E.G. 1972. Evolutionary rates and patterns of North American Cretaceous Mollusca. *24th Int. Geol. Cong. Section* 7, 174–89.

Kay, M. 1947. Analysis of stratigraphy. *Bull. Am. Ass. Petroleum Geologists*, 31, 162–8.

Kayll, A.J. 1974. Use of fire in land management. In T.T. Kozlowski and C.E. Ahlgren (eds.), *Fire and Ecosystems*, pp. 483–511. New York, Academic Press.

Kennedy, C.R. 1975. *Ecological Animal Parasitology*. Oxford, Blackwell Scientific Publications.

Kennedy, J.S. 1975. Insect dispersal. In D. Pimentel (ed.), *Insects, Science & Society*, pp. 103–119. New York, Academic Press.

Kiester, A.R., and Barakat, R. 1974. Exact solutions to certain stochastic differential equation models of population growth. *Theoret. Pop. Biol.*, 6, 199–216.

Kincade, R.T., Laster, M.L., and Brazzel J.R. 1970. Effect on cotton yield of various levels of simulated *Heliothis* damage to squares and bolls. *J. econ. Ent.*, 63, 613–15.

King, C.E., Gallaher, E.E., and Levin, D.A. 1975. Equilibrium diversity in plant-pollinator systems. *J. Theoret. Biol.*, 53, 263–75.

King, Mary-Claire, and Wilson, A.C. 1975. Evolution at two levels in humans and chimpanzees. *Science*, 188, 107–16.

Klein, D.R. 1968. The introduction, increase, and crash of reindeer on St Matthew Island. *J. Wildl. Manage.*, 32, 350–67.

Knight, D.H. 1975. A phytosociological analysis of species-rich tropical forest on Barro Colorado Island, Panama. *Ecol. Monogr.*, 45, 259–84.

Kolmogorov, A.N. 1936. Sulla teoria di Volterra della lotta per l'esisttenza. *Giorn. Instituto Ital. Attuari*, 7, 74–80.

Koplin, J.R., and Hoffman, R.S. 1968. Habitat overlap and competitive exclusion in voles (*Microtus*). *Am. Midl. Natur.*, 80, 494–507.

Kostitzen, V.A. 1939. *Mathematical Biology*. London, Harrap.

Kozlovsky, D.G. 1968. A critical evaluation of the trophic level concept: I, ecological efficiencies. *Ecology*, 49, 48–60.

Krebs, C.J. 1972. *Ecology: the experimental analysis of distribution and abundance.* New York, Harper and Row.

Kuhn, T.S. 1962. *The Structure of Scientific Revolutions.* Chicago, University of Chicago Press.

Lack, D. 1965. Evolutionary ecology. *J. Anim. Ecol.*, 34, 223–31.

Lack, D. 1966. *Population Studies of Birds.* Oxford, Clarendon Press, pp. 341.

Lack, D. 1968. *Ecological Adaptations for Breeding in Birds.* London, Methuen, pp. 409.

Lack, D. 1973. The numbers and species of hummingbirds in the West Indies. *Evolution*, 27, 326–37.

Lack, D. 1976. *Island Birds.* Oxford, Blackwell Scientific Publications.

Ladd, H.S. 1957. *Treatise on Marine Ecology and Paleoecology, Vol. 2 Paleoecology.* Mem. Geol. Soc. Am. No. 67. pp. 1077.

Lawlor, L.R., and Maynard Smith, J. 1976. Coevolution and stability of competing species. *Am. Nat.*, 110, (in press).

Lawton, J.H. 1974. The structure of the arthropod community on brackeı (*Pteridium aquilinium* (L.) Kuhn). In F.H. Perring (ed.), *The Biology of Bracken.* New York, Academic Press.

Leak, W.B. 1970. Successional change in northern hardwoods predicted by birth and death simulation. *Ecology*, 51, 794–801.

Lehman, J.S., Farid, Z., and Bassily, S. 1970. Mortality in urinary schistosomiasis. *The Lancet.* October 17, 822–3.

Leith, H. 1975. Primary productivity in ecosystems: comparative analysis of global patterns. In W.H. van Dobben and R.H. Lowe-McConnell (eds.), *Unifying Concepts in Ecology, op. cit.*, pp. 67–88.

Leon, J.A. 1974. Selection in contexts of interspecific competition. *Am. Nat.*, 108, 739–57.

Leopold, A., Sowls, L.K. and Spencer, D.L. 1947. A survey of over-populated deer ranges in the United States. *J. Wildl. Manage.*, 11, 162–77.

Leslie, P.H. 1948. Some further notes on the use of matrices in population mathematics. *Biometrika*, 35, 213–45.

LeVine, R.A., and Campbell, D.T. 1972. *Ethnocentrism: theories of conflict, ethnic attitudes, and group behavior.* New York, Wiley.

Levin, S.A. 1970. Community equilibria and stability, and an extension of the competitive exclusion principle. *Am. Nat.*, 104, 413–23.

Levin, S.A. 1974. Dispersion and population interactions. *Am. Nat.*, 108, 207–28.

Levins, R. 1968. *Evolution in Changing Environments.* Princeton, New Jersey, Princeton University Press.

Levins, R. 1969. The effects of random variations of different types on population growth. *Proc. Natl. Acad. Sci., U.S.*, 62, 1061–5.

Levins, R. 1975. Evolution in communities near equilibrium. In M.L. Cody and J.M. Diamond (eds.), *Ecology and Evolution of Communities, op. cit.* pp. 16–50.

Levins, R., and Culver, D. 1971. Regional coexistence of species and competition between rare species. *Proc. Natl. Acad. Sci., U.S.*, 68, 1246–8.

Levington, J.S. 1970. The paleoecological significance of opportunistic species. *Lethaia*, 3, 69–78.

Lewontin, R.C., and Cohen, D. 1969. On population growth in a randomly varying environment. *Proc. Natl. Acad. Sci., U.S.*, 62, 1056–60.

Leyton, M.D. 1968. Stochastic models in populations of helminthic parasites in the definitive host, II: Sexual mating functions. *Mathematical Biosciences*, 3, 413–19.

Li, T.-Y., and Yorke, J.A. 1975. Period three implies chaos. *Am. Math. Monthly*, 82, 985–92.

Liem, K.F. 1973. Evolutionary strategies and morphological innovations: cichlid pharyngeal jaws. In S.J. Gould (ed.), ICSEB. Symposium on Evolutionary, development of form and symmetry. *Syst. Zool.*, 22, 425–41.

Lin, N., and Michener, C.D. 1972. Evolution of sociality in insects. *Quart. Rev. Biol.*, 47, 131–59.

Linhart, H. 1968. On some bilharzia infection and immunisation models. *S. Afr. Statist. J.*, 2, 61–6.

Lotka, A.J. 1925. *Elements of Physical Biology*. Baltimore, Williams and Wilkins.

Loucks, O.L. 1970. Evolution of diversity, efficiency, and community stability. *Am. Zool.*, 10, 17–25.

Loya, Y. 1976. *Stylophora pistillata*: an r-strategist among the Red Sea Corals. *Nature*, (*Lond.*), (in press).

Luckinbill, L.S. 1973. Coexistence in laboratory populations of *Paramecium aurelia* and its predator *Didinium nasutum*. *Ecology*, 54, 1320–7.

Ludwig, D. 1975. Persistence of dynamical systems under random perturbations. *SIAM Review*, 17, 605–40.

MacArthur, R.H. 1961. Community. In P. Gray (ed.), *The Encyclopedia of the Biological Sciences*, pp. 262–4. New York, Reinhold.

MacArthur, R.H. 1968. The theory of the niche. In R.C. Lewontin (ed.), *Population Biology and Evolution*, pp. 159–76. Syracuse, Syracuse University Press.

MacArthur, R.H. 1969. Species packing and what competition minimizes. *Proc. Natl. Acad. Sci., U.S.*, 64, 1369–71.

MacArthur, R.H. 1970. Species packing and competitive equilibrium for many species. *Theoret. Pop. Biol.*, 1, 1–11.

MacArthur, R.H. 1972. *Geographical Ecology*. New York, Harper and Row.

MacArthur, R.H., and MacArthur, J.W. 1961. On bird species diversity. *Ecology*, 42, 594–8.

MacArthur, R.H., and Wilson, E.O. 1963. An equilibrium theory of insular zoogeography. *Evolution*, 17, 373–87.

MacArthur, R.H., and Pianka, E.R. 1966. On optimal use of a patchy environment. *Am. Nat.*, 100, 603–9.

MacArthur, R.H., and Levins, R. 1967. The limiting similarity, convergence and divergence of coexisting species. *Am. Nat.*, 101, 377–85.

MacArthur, R.H., and Wilson, E.O. 1967. *The Theory of Island Biogeography*. Princeton, Princeton University Press.

McCormick, J., and Buell, M.F. 1968. The plains: pigmy forests of the New Jersey pine barrens, a review and annotated bibliography. *Bull. N.J. Acad. Sci.*, 13, 20–34.

Macdonald, G. 1957. *The Epidemiology and Control of Malaria*. London, Oxford University Press.

Macdonald, G. 1961. Epidemiological models in studies of vector borne diseases. *Publ. Hlth. Rep.*, 76, 753–64.

Macdonald, G. 1965. The dynamics of helminth infections, with special reference to schistosomes. *Trans. R. Soc. Trop. Med. Hyg.*, 59(5), 489–506.

Macdonald, G., Cuellar, C.B., and Foll, C.V. 1968. The dynamics of malaria. *Bull. Wld Hlth Org.*, 38, 743–55.

McMurtrie, R.E. 1975. Determinents of stability of large, randomly connected systems. *J. Theoret. Biol.*, 50, 1–11.

McMurtrie, R.E. 1976a. Stability of communities supported by depletable resources. *J. Theoret. Biol.*, (in press).

McMurtrie, R.E. 1976b. On the limit to niche overlap for nonuniform niches. *J. Theoret. Biol.*, (in press).

McNaughton, S.J. 1975. r- & K-selection in *Typha*. *Am. Nat.*, 109, 251–61.

McNaughton, S.J., and Wolf, L.L. 1973. *General Ecology*. New York, Holt, Rinehart and Winston.

Maissurow, D.K. 1941. The role of fire in the perpetuation of virgin forests of Northern Wisconsin. *J. Forestry*, 39, 201–7.

Malek, E.A. 1961. The ecology of schistosomiasis. In J.M. May (ed.), *Studies in Disease Ecology*, pp. 261–327. New York, Hafner.

Malinowski, B. 1941. An anthropological analysis of war. *Am. J. Sociol.*, 46, 521–50.

Margalef, R. 1968. *Perspectives in Ecological Theory*. Chicago, University of Chicago Press.

Matthes, F.E., posthumously prep. by F. Fryxell. 1965. Glacial reconnaissance of Sequoia National Park California. *U.S. Geol. Surv. Prof. Paper* 504-A, 1–58.

May, J.M. 1958. *The Ecology of Human Disease*. New York, MD Publications.

May, J.M. (ed.) 1961. *Studies in Disease Ecology*. New York, Hafner.

May, R.M. 1972a. Limit cycles in predator-prey communities. *Science*, 177, 900–2.

May, R.M. 1972b. Will a large complex system be stable? *Nature*, (*Lond.*), 238, 413–14.

May, R.M. 1973a. Time-delay versus stability in population models with two and three trophic levels. *Ecology*, 54, 315–25.

May, R.M. 1973b. Stability in randomly fluctuating versus deterministic environments. *Am. Nat.*, 107, 621–50.

May, R.M. 1973c. Qualitative stability in model ecosystems. *Ecology*, 54, 638–41.

May, R.M. 1974a. Biological populations with nonoverlapping generations: stable points, stable cycles, and chaos. *Science*, 186, 645–7.

May, R.M. 1974b. Ecosystem patterns in randomly fluctuating environments. In R. Rosen and F. Snell (eds.), *Progress in Theoretical Biology*, pp. 1–50. New York, Academic Press.

May, R.M. 1974c. How many species: some mathematical aspects of the dynamics of populations. In J.D. Cowan (ed.), *Some mathematical problems in Biology*, Vol. 4, pp. 64–98. Providence, R.I., The American Mathematical Society.

May, R.M. 1974d. On the theory of niche overlap. *Theoret. Pop. Biol.*, 5, 297–332.

May, R.M. 1975a. *Stability and Complexity in Model Ecosystems*. (Second edition). Princeton, Princeton University Press.

May, R.M. 1975b. Biological populations obeying difference equations: stable points, stable cycles, and chaos. *J. Theoret. Biol.*, 49, 511–24.

May, R.M. 1975c. Stability in ecosystems: some comments. In W.H. van Dobben and R.H. Lowe-McConnell (eds.), *Unifying Concepts in Ecology*, *op. cit.*, pp. 161–8.

May, R.M. 1975d. Some notes on estimating the competition matrix, α. *Ecology*, 737–41.

May, R.M. 1975e. Group selection. *Nature*, (*Lond.*), 254, 485.

May, R.M. 1975f. Patterns of species abundance and diversity. In M.L. Cody and J.M. Diamond (eds.), *Ecology and Evolution of Communities*, *op. cit.* pp. 81–120.

May, R.M. 1975g. Island biogeography and the design of wildlife preserves. *Nature*, (*Lond.*), 254, 177–8.

May, R.M. 1976a. Mathematical aspects of the dynamics of animal populations. In S.A. Levin (ed.), *Studies in Mathematical Biology*. Providence, R.I., American Mathematical Society.

May, R.M. 1976b. Estimating r: a pedagogical note. *Am. Nat.*, 110, (in press).

May, R.M., and MacArthur, R.H. 1972. Niche overlap as a function of environmental variability. *Proc. Natl. Acad. Sci., U.S.*, 69, 1109–13.

May, R.M., Conway, G.R., Hassell, M.P., and Southwood, T.R.E. 1974. Time delays, density dependence, and single species oscillations. *J. Anim. Ecol.*, 43, 747–70.

May, R.M., and Leonard, W.J. 1975. Nonlinear aspects of competition between three species. *SIAM J. Appl. Math.*, 29, 243–53.

May, R.M., and Oster, G.F. 1976. Bifurcations and dynamic complexity in simple ecological models. *Am. Nat.*, 110, (in press).

Maynard Smith, J. 1971a. Group selection and kin selection: a rejoinder. In G.C. Williams (ed.), *Group Selection*. Chicago, Aldine Atherton.

Maynard Smith, J. 1971b. What use is sex? *J. Theoret. Biol.*, 30, 319–35.

Menge, B.A. 1972. Competition for food between two intertidal starfish species and its effect on body size and feeding. *Ecology*, 53, 635–44.

Menge, J.L., and Menge, B.A. 1974. Role of resource allocation, aggression and spatial heterogeneity in coexistence of two competing intertidal starfish. *Ecol. Monogr.*, 44, 189–209.

Michelakis, S. 1973. *A study of the laboratory interaction between* Coccinella septempunctata *larvae and its prey* Myzus persicae. Unpublished M.Sc. thesis. University of London.

Michener, C.D., and D.J. Brothers. 1974. Were workers of eusocial Hymenoptera initially altruistic or oppressed? *Proc. Natl. Acad. Sci. U.S.*, 71, 671–4.

Miller, C.A. 1959. The interaction of the spruce budworm, *Choristoneura fumiferana* (Clem.), and the parasite *Apanteles fumiferanae* Vier. *Can. Ent.*, 91, 457–77.

Miller, R.S. 1967. Pattern and process in competition. *Adv. Ecol. Res.*, 4, 1–74.

Milne, A. 1961. Definition of competition among animals. In F.L. Milthorpe (ed.), *Mechanisms in Biological Competition, Symp. Soc. Exp. Biol.*, 15, 40–61.

Milthorpe, F.L. (ed.) 1961. *Mechanisms in Biological Competition, Symp. Soc. Exp. Biol.* 15. Cambridge, Cambridge University Press.

Mitchell, W.C., and Mau, R.F.L. 1971. Response of the female Southern green stink bug and its parasite, *Trichopoda pennipes*, to male stink bug pheromones. *J. Econ. Ent.* 64, 856–9.

Mogi, M. 1969. Predation response of the larvae of *Harmonia axyridis* Pallas (Coccinellidae) to the different prey density. *Jap. J. appl. Ent. Zool.*, 13, 9–16.

Monro, J. 1967. The exploitation and conservation of resources by populations of insects. *J. Anim. Ecol.*, 36, 531–47.

Monro, J. 1975. Environmental variation and efficiency of biological control—*Cactoblastis* in the southern hemisphere. *Proc. Ecol. Soc. Australia*, 9, 204–12.

Morowitz, H.J. 1968. *Energy Flow in Biology*. New York, Academic Press.

Moser, J.W. 1972. Dynamics of an uneven-aged forest stand. *Forest Sci.*, 18, 184–91.

Mountford, M.D. 1973. The significance of clutch size. In M.S. Bartlett and R.W. Hiorns (eds.), *The Mathematical Theory of the Dynamics of Biological Populations*, pp. 315–23. London, Academic Press.

Muench, H. 1959. *Catalytic Models in Epidemiology*. Cambridge, Harvard University Press.

Muggleton, J., and Benham, B.R. 1975. Isolation and the decline of the Large Blue Butterfly (*Maculinea arion*) in Great Britain. *Biol. Conserv.*, 7, 119–28.

Murdie, G., and Hassell, M.P. 1973. Food distribution, searching success and predator-prey models. In M.S. Bartlett and R.W. Hiorns (eds.), *The Mathematical Theory of the Dynamics of Biological Populations*, pp. 87–101. London, Academic Press.

Murdoch, W.W., and Oaten, A. 1975. Predation and population stability. *Adv. Ecol. Res.*, 9, 2–131.

Murphy, G.I. 1967. Vital statistics of the Pacific sardine (*Sardinops caerulea*) and the population consequences. *Ecology*, 48, 731–6.

Mutch, R.W. 1970. Wildland fires and ecosystems—a hypothesis. *Ecology*, 51, 1046–51.

Nakasuji, F., Hokyo, N., and Kiritani, K. 1966. Assessment of the potential efficiency of parasitism in two competitive scelionid parasites of *Nezara viridula* L. (Hemiptera: Pentatomidae). *Appl. Ent. Zool.*, 1, 113–19.

Namkoong, G., and Roberts, J.H. 1974. Extinction probabilities and the changing age structure of redwood forests. *Am. Nat.*, 108, 355–68.

Nasell, I. Unpublished a. A mathematical model of schistosomiasis with snail latency.

Nasell, I. Unpublished b. A mathematical model of schistosomiasis with external infection.

Nasell, I., and Hirsch, W.M. 1973. The transmission dynamics of schistosomiasis. *Communications on Pure and Applied Mathematics*, 26, 395–453.

Neill, W.E. 1972. *Effects of size-selective predation on community structure in laboratory aquatic microcosms*. Ph.D. Thesis, University of Texas, Austin.

Neill, W.E. 1974. The community matrix and interdependence of the competition coefficients. *Am. Nat.*, 108, 399–408.

Neill, W.E. 1975. Experimental studies of microcrustacean competition, community composition and efficiency of resource utilization. *Ecology*, 56, 809–26.

Nevo, E., Gorman, G., Soule, M., Yang, S.Y., Clover, R., and Jovanociv, V. 1972. Competitive exclusion between insular *Lacerta* species (Sauria, Lacertidae). Notes on experimental introductions. *Oecologia*, 10, 183–90.

Newell, N.D. 1973. The very last moment of the Paleozoic Era. *Mem. Can. Soc. Petroleum Geologists*, 2, 1–10.

Newsome, A.E. 1969. A population study of house-mice temporarily inhabiting a South Australian wheatfield. *J. Anim. Ecol.*, 38, 341–59.

Neyman, J., Park, T., and Scott, E.L. 1956. Struggle for existence. The *Tribolium* model: biological and statistical aspects. In *Proc. 3rd Berkeley Symp. on Mathematical Statistics and Probability*, Vol. IV, pp. 41–79. Berkeley, University California Press.

Nicholson, A.J. 1933. The balance of animal populations. *J. Anim. Ecol.*, 2, 131–78.

Nicholson, A.J. 1947. Fluctuations of animal populations. *Rep. 26th Meeting Aust. N.Z. Assn. Advnt. Sci, Perth.*

Nicholson, A.J., & Bailey, V.A. 1935. The balance of animal populations. Part 1. *Proc. zool. Soc. Lond.*, 551–98.

Norton, G.A. 1975. Multiple cropping and pest control, an economic perspective. *Meded. Fac. Landbouww Rijks. Univ. Gent*, 40, 219–28.

Norton, G.A. 1976a. Pest control decision making—an overview. *Ann. Appl. Biol.*, (in press).

Norton, G.A. 1976b. The analysis of decision making in crop protection. *Agroecosystems*, (in press).

Noyes, J.S. 1974. *The Biology of the Leek Moth*, Acrolepia assectella (*Zeller*). Unpublished Ph.D. thesis, University of London.

Noy-Meir, I. 1975. Stability of grazing systems: an application of predator-prey graphs. *J. Ecol.*, 63, 459–81.

Odum, E.P. 1968. Energy flow in ecosystems: a historical review. *Am. Zool.*, 8, 11–18.

Odum, E.P. 1969. The strategy of ecosystem development. *Science*, 164, 262–70.

Oliver, C.D. 1975. *The development of northern red oak* (Quercus rubra L.) *in mixed species, even-aged stands in central New England.* Doctoral dissertation. New Haven, Yale University.

Olson, J.S. 1958. Rates of succession and soil changes on southern Lake Michigan sand dunes. *Bot. Gazette*, 119, 125–70.

Orians, G.H. 1975. Diversity, stability and maturity in natural ecosystems. In W.H. van Dobben and R.H. Lowe-McConnell (eds.), *Unifying Concepts in Ecology*, *op. cit.*, pp. 139–50.

Orians, G.H., and Horn, H.S. 1969. Overlap in foods and foraging of four species of blackbirds in the Potholes of central Washington. *Ecology*, 50, 930–8.

Orians, G.H., and Willson, M.F. 1964. Interspecific territories of birds. *Ecology*, 45, 736–45.

Osman, R.W. 1975. *The influence of seasonality and stability on the species equilibrium.* Ph.D. Dissertation, Dept. Geology, University of Chicago.

Oster, G.F., Auslander, D., and Allen, J. 1976. Deterministic and stochastic effects of population dynamics. (In preparation.)

Owen, D.F., and Chanter, D.O. 1972. Species diversity and seasonal abundance in *Charaxes* butterflies (Nymphalidae). *J. Ent.* (*A*), 46, 135–43.

Paine, R.T. 1966. Food web complexity and species diversity. *Am. Nat.*, 100, 65–75.

Paine, R.T. 1974. Intertidal community structure. *Oecologia*, 15, 93–120.

Pan, C.T. 1965. Studies on the host-parasite relationship between *Schistosoma mansoni* and the snail *Australorbis glabratus*. *Amer. J. Trop. Med. Hyg.*, 14, 931–76.

Park, T. 1948. Experimental studies of interspecific competition. I. Competition between populations of flour beetles *Tribolium confusum* Duval and *T. castaneum* Herbst. *Physiol. Zool.*, 18, 265–308.

Park, T. 1954. Experimental studies of interspecific competition. II. Temperature, humidity, and competition in two species of *Tribolium*. *Physiol. Zool.*, 27, 177–238.

Park, T. 1962. Beetles, competition, and populations. *Science*, 138, 1369–75.

Park, T., Leslie, P.H., and Mertz, D.B. 1964. Genetic strains and competition in populations of *Tribolium*. *Physiol. Zool.*, 37, 97–162.

Patrick, R. 1973. Use of algae, especially diatoms, in the assessment of water quality. *American Society for Testing and Materials*, Special Tech. Publ., 528, 76–95.

Patrick, R. 1975. Structure of stream communities. In M.L. Cody and J.M. Diamond (eds.), *Ecology and Evolution of Communities*, *op. cit.* pp. 445–59.

Patten, B.C. 1961. Competitive exclusion. *Science*, 134, 1599–601.

Peden, L.M., Williams, J.S., and Frayer, W.E. 1973. A Markov model for stand projection. *Forest Sci.*, 19, 303–14.

Peterkin, G.F., and Tubbs, C.R. 1965. Woodland regeneration in the New Forest, Hampshire, since 1650. *J. Appl. Ecol.*, 2, 159–70.

Petrusewicz, K. 1967. *Secondary Productivity in Terrestrial Ecosystems*. Warsaw, Polish Acad. Sciences.

Pianka, E.R. 1969. Sympatry of desert lizards (*Ctenotus*) in Western Australia. *Ecology*, 50, 1012–30.

Pianka, E.R. 1970. On *r*- and *K*-selection. *Am. Nat.*, 104, 592–7.

Pianka, E.R. 1972. *r* and *K*-selection or *b* and *d* selection? *Am. Nat.*, 106, 581–8.

Pianka, E.R. 1973. The structure of lizard communities. *A. Rev. Ecol. Syst.*, 4, 53–74.

Pianka, E.R. 1974. Niche overlap and diffuse competition. *Proc. Natl. Acad. Sci., U.S.*, 71, 2141–5.

Pianka, E.R. 1975. Niche relations of desert lizards. In M. Cody and J. Diamond (eds.), *Ecology and Evolution of Communities*, pp. 292–314. Cambridge, Harvard University Press.

Pianka, E.R. 1976. Natural selection of optimal reproductive tactics. *Am. Zool.*, 16, (in press).

Pianka, E.R., and Pianka, H.D. 1976. Comparative ecology of twelve species of nocturnal lizards (Gekkonidae) in the Western Australian desert. *Copeia*, (in press).

Pico, M.M., Maldonado, D., and Levins, R. 1965. Ecology and genetics of Puerto Rican *Drosophila*: I. Food preferences of sympatric species. *Carib. J. Sci.*, 5, 29–37.

Pielou, E.C. 1972. Niche width and niche overlap: a method of measuring them. *Ecology*, 53, 687–92.

Pielou, E.C. 1975. *Ecological Diversity*. New York, Wiley-Interscience.

Pimentel, D. 1973. Extent of pesticide use, food supply and pollution. *J.N.Y. Entomol Soc.*, 81, 13–37.

Pimentel, D., Levin, S.A., and Soans, A.B. 1975. On the evolution of energy balance in some exploiter-victim systems. *Ecology*, 56, 381–90.

Pitelka, F.A. 1967. Some characteristics of microtine cycles in the arctic. In H.P. Hanson (ed.), *Arctic Biology* (2nd, edn.), pp. 153–84. Corvallis, Oregon State University Press.

Poluektov, R.A. (ed.) 1974. *Dynamical Theory of Biological Populations*. Moscow, Science Pubs. (In Russian).

Pontin, A.J. 1969. Experimental transplantation of nest-mounds of the ant *Lasius flavus* (F.) in a habitat containing also *L. niger* (L.) and *Myrmica scabrinodis* Nyl. *J. Anim. Ecol.*, 38, 747–54.

Pratt, D.M. 1943. Analysis of population development in *Daphnia* at different temperatures. *Biol. Bull.*, 85, 116–40.

Preston, F.W. 1948. The commonness, and rarity, of species. *Ecology*, 29, 254–83.

Preston, F.W. 1962. The canonical distribution of commonness and rarity. *Ecology*, 43, 185–215 and 410–32.

Price, P.W. 1972. Parasitoids utilizing the same host: adaptive nature of differences in size and form. *Ecology*, 53, 190–5.

Price, P.W. 1973. Reproductive strategies in parasitoid wasps. *Am. Nat.*, 107, 684–93.

Pruszynski, S. 1973. The influence of prey density on prey consumption and oviposition of *Phytoseiulus persimilis* Athias-Henriot (Acarina: Phytoseiidae). SROP/WPRS Bulletin: Integrated Control in Glasshouses. 1973–74, pp. 41–6.

Pulliam, H.R. 1975. Coexistence of sparrows: a test of community theory. *Science*, 189, 474–6.

Rabinovich, J.E. 1974. Demographic strategies in animal populations: a regression analysis. In F.B. Golley and E. Medina (eds.), *Tropical Ecological Systems*, pp. 19–40. Ecological Studies 11. New York, Springer-Verlag.

Raup, D.M. 1972. Taxonomic diversity during the Phanerozoic. *Science*, 177, 1065–71.

Raup, D.M., Gould, S.J., Schopf, T.J.M., and Simberloff, D. 1973. Stochastic models of phylogeny and the evolution of diversity. *J. Geol.*, 81, 525–42.

Raup, D.M., and Gould, S.J. 1974. Stochastic simulation and evolution of morphology: towards a nomothetic paleontology. *Syst. Zool.*, 23, 305–22.

Raup, H.M. 1964. Some problems in ecological theory and their relation to conservation. *J. Ecol.*, 52 (Suppl.), 19–28.

Raup, H.M. 1975. Species versatility in shore habitats. *J. Arnold Arboretum Harvard University*, 56, 126–63.

Recher, H.F. 1969. Bird species diversity and habitat diversity in Australia and North America. *Am. Nat.*, 103, 75–80.

Reynolds, H.T., Adkisson, P.L., and Smith, R.F. 1975. Cotton insect pest management. In R.L. Metcalf and W.H. Luckmann (eds.), *Introduction to Insect Pest Management*, pp. 379–443. New York, Wiley-Interscience.

Richardson, B.J. 1975. *r*- and *K*-selection in kangaroos. *Nature*, (*Lond.*), 255, 323–4.

Richman, S. 1958. The transformation of energy by *Daphnia pulex*. *Ecol. Mongr.*, 28, 273–91.

Roberts, A.P. 1974. The stability of a feasible random ecosystem. *Nature*, (*Lond.*), 251, 607–8.

Rodin, L.E., *et al.* 1975. Primary productivity of the main world ecosystems. In *Proceedings of the First International Congress of Ecology*, pp. 176–81. Wageningen, Centre for Agricultural Pub.

Rogers, D.J. 1972a. Random search and insect population models. *J. Anim. Ecol.*, 41, 369–83.

Rogers, D.J. 1972b. The ichneumon wasp *Venturia canescens*: oviposition and avoidance of superparasitism. *Ent. exp. appl.*, 15, 190–4.

Rogers, D.J., and Hassell, M.P. 1974. General models for insect parasite and predator searching behaviour: interference. *J. Anim. Ecol.*, 43, 239–53.

Rogers, D.J., and Hubbard, S. 1974. How the behaviour of parasites and predators promotes population stability. In M.B. Usher and M.H. Williamson (eds.), *Ecological Stability*, pp. 99–119. London, Chapman & Hall.

Rollins, H.B., and Donahue, J. 1975. Towards a theoretical basis of paleoecology: concepts of community dynamics. *Lethaia*, 8, 255–70.

Rosenzweig, M.L. 1971. Paradox of enrichment: destabilization of exploitation ecosystems in ecological time. *Science*, 171, 385–7.

Rosenzweig, M.L., 1972. Stability of enriched aquatic ecosystems. *Science*, 175, 564–5.

Rosenzweig, M.L., and MacArthur, R.H. 1963. Graphical representation and stability condition of prey-predator interactions. *Am. Nat.*, 97, 209–23.

Ross, H.H. 1957. Principles of natural coexistence indicated by leafhopper populations. *Evolution*, 11, 113–29.

Ross, H.H. 1958. Further comments on niches and natural coexistence. *Evolution*, 12, 112–13.

Roughgarden, J. 1972. Evolution of niche width. *Am. Nat.*, 106, 683–718.

Roughgarden, J. 1974a. Species packing and the competition function with illustrations from coral reef fish. *Theoret. Pop. Biol.*, 5, 163–86.

Roughgarden, J. 1974b. Niche width: biogeographic patterns among *Anolis* lizard populations. *Am. Nat.*, 108, 429–42.

Roughgarden, J. 1974c. The fundamental and realized niche of a solitary population. *Am. Nat.*, 108, 232–5.

Roughgarden, J. 1975a. A simple model for population dynamics in stochastic environments. *Am. Nat.*, 109, 713–36.

Roughgarden, J. 1975b. Species packing and faunal build-up: an evolutionary approach based on density-dependent natural selection. *Theoret. Pop. Biol.*, (in press).

Roughgarden, J. 1976. Resource partitioning among competiting species: a coevolutionary approach, (in press).

Roughgarden, J., and Feldman, M. 1975. Species packing and predation pressure. *Ecology*, 56, 489–92.

Rowe, J.S., and Scotter, G.W. 1973. Fire in the boreal forest. *Quarternary Res.*, 3, 444–64.

Royama, T. 1971. A comparative study of models for predation and parasitism. *Res. Popul. Ecol.*, Suppl. 1, 1–91.

Rudwick, M.J.S. 1972. *The Meaning of Fossils*. New York, American Elsevier.

Rundel, P.W. 1971. Community structure and stability in the giant sequoia groves of the Sierra Nevada, California. *Am. Midl. Natur.*, 85, 478–92.

Rundel, P.W. 1972. Habitat restriction in giant sequoia: the environmental control of grove boundaries. *Am. Midl. Natur.*, 87, 81–99.

Russell, B. 1946. *The History of Western Philosophy*. London, Heinemann.

Salazar, J.A. 1976. *Analysis of the damage caused by the black bean aphid* (Aphis fabae) *to the field bean* (Vicia faba). Ph.D. Thesis, University of London.

Sanders, H.L. 1968. Marine benthic diversity: a comparative study. *Am. Nat.*, 102, 243–82.

Schaffer, W.M. 1974. Selection for optimal life histories: the effects of age structure. *Ecology*, 55, 291–303.

Schoener, T.W. 1965. The evolution of bill size differences among sympatric congeneric species of birds. *Evolution*, 19, 189–213.

Schoener, T.W. 1968. The *Anolis* lizards of Bimini: resource partitioning in a complex fauna. *Ecology*, 49, 704–26.

Schoener, T.W. 1969. Optimal size and specialization in constant and fluctuating environments. In *Diversity and Stability in Ecological Systems*. Springfield, Va., U.S. Department of Commerce, pp. 103–14.

Schoener, T.W. 1970. Nonsynchronous spatial overlap of lizards in patchy habitats. *Ecology*, 51, 408–18.

Schoener, T.W. 1971. Theory of feeding strategies. *A. Rev. Ecol. Syst.*, 2, 369–404.

Schoener, T.W. 1974. Resource partitioning in ecological communities. *Science*, 185, 27–39.

Schoener, T.W. 1975a. Competition and the form of habitat shift. *Theoret. Pop. Biol.*, 5, 265–307.

Schoener, T.W. 1975b. Presence and absence of habitat shift in some widespread lizard species. *Ecol. Monogr.*, 45, 232–58.

Schoener, T.W. 1976a. Competition and the niche. In D.W. Tinkle and W.W. Milstead (eds.), *Biology of the Reptilia*. New York, Academic Press.

Schoener, T.W. 1976b. The species-area relation within archipelagos: models and evidence from island land birds. *Proc. 16th International Ornith. Congress, Canberra, August* 1974. (In press.)

Schoener, T.W., and Gorman, G.C. 1968. Some niche differences among three species of Lesser Antillean anoles. *Ecology*, 49, 819–30.

Schopf, T.J.M. 1974. Permo-Triassic extinctions: relation to sea-floor spreading. *J. Geol.*, 82, 129–43.

Schopf, T.J.M., and Gooch, J.C. 1972. A natural experiment to test the hypothesis that loss of genetic variability was responsible for mass extinctions of the fossil record. *J. Geol.*, 80, 481–3.

Shelford, V.E. 1943. The relation of snowy owl migration to the abundance of the collared lemming. *Auk*, 62, 592–4.

Shoemaker, C. 1973. Optimisation of agricultural pest management: II, Formulation of a control model. *Math. Biosciences*, 17, 357–65.

Shugart, H.H.Jr., Crow, T.R., and Hett, J.M. 1973. Forest succession models: a rationale and methodology for modelling forest succession over large regions. *Forest Sci.*, 19, 203–12.

Simberloff, D. 1974. Permo-Triassic extinctions: effects of area on biotic equilibrium. *J. Geol.*, 82, 267–74.

Simberloff, D.S., and Wilson, E.O. 1969. Experimental zoogeography of islands: the colonisation of empty islands. *Ecology*, 50, 278–95.

Simberloff, D.S., and Abele, L.G. 1976. Island biogeography theory and conservation practice. *Science*, 191, 285–6.

Simon, C.A. 1975. The influence of food abundance on territory size in the Iguanid lizard *Sceloporus jarrovi*. *Ecology*, 56, 993–8.

Simpson, G.G. 1944. *Tempo and Mode in Evolution*. New York, Columbia University Press.

Simpson, G.G. 1949. *The Meaning of Evolution*. New Haven, Yale University Press.

Simpson, G.G. 1953. *The Major Features of Evolution*. New York, Columbia University Press.

Simpson, G.G. 1969. The first three billion years of community evolution. In *Diversity and Stability in Ecological Systems*. Springfield, Va., U.S. Department of Commerce, pp. 162–77.

Skellam, J.G. 1951. Random dispersal in theoretical populations. *Biometrika*, 38, 196–218.

Slatkin, M. 1974. Competition and regional ecoxistence. *Ecology*, 55, 128–34.

Slobodkin, L.B. 1961. *Growth and Regulation of Animal Populations*. New York, Holt, Rinehart and Winston.

Slobodkin, L.B. 1964. The strategy of evolution. *Am. Sci.*, 52, 342–57.

Smith, C.C., and Fretwell, S.D. 1974. The optimal balance between size and number of offspring. *Am. Nat.*, 108, 499–506.

Smith, F.E. 1976. Ecosystems and evolution. 1975 presidential address to Am. Ecol. Soc. (to be published).

Smith, R.F., and Reynolds, H.T. 1966. Principles, definitions and scope of integrated pest control. *Proc. of the FAO Symp. on Integrated Pest Control* 1, 11–17.

Snyder, J., and Bretsky, P.W. 1971. Life habits of diminutive bivalve mollusks in the Maquoketa Formation (Upper Ordovician). *Am. J. Sci.*, 271, 227–51.

Solomon, M.E. 1949. The natural control of animal populations. *J. Anim. Ecol.*, 18, 1–35.

Southern, H.N. 1970. The natural control of a population of Tawny Owls (*Strix aluco*). *J. Zool., Lond.*, 162, 197–285.

Southwood, T.R.E. 1962. Migration of terrestrial arthropods in relation to habitat. *Biol. Rev.*, 37, 171–214.

Southwood, T.R.E. 1970. The natural and manipulated control of animal populations. In L.R. Taylor (ed.), *The Optimum Population for Britain*. *Inst. Biol. Symp.* 19. London, Academic Press, pp. 87–102.

Southwood, T.R.E. 1975. The dynamics of insect populations. In D. Pimentel (ed.), *Insects, Science and Society*. New York, Academic Press, pp. 151–99.

Southwood, T.R.E. 1976. *Ecological Methods* (Second edition). London, Chapman and Hall.

Southwood, T.R.E., and Norton, G.A. 1973. Economic aspects of pest management strategies and decisions. In P.W. Geier, L.R. Clark, D.J. Anderson and H.A. Nix (eds.), *Insects: Studies in Pest Management*, pp. 168–84. Mem. Ecological Society of Australia, Canberra.

Southwood, T.R.E., May, R.M., Hassell, M.P., and Conway, G.R. 1974. Ecological strategies and population parameters. *Am. Nat.*, 108, 791–804.

Stanley, S.M. 1973a. An ecological theory for the sudden origin of multicellular life in the Late Precambrian. *Proc. Natl. Acad. Sci. U.S.*, 70, 1486–9.

Stanley, S.M. 1973b. Effects of competition on rates of evolution, with special reference to bivalve mollusks and mammals. In S.J. Gould (ed.), I.C.S.E.B. Symposium on Evolutionary development of form and symmetry. *Syst. Zool.*, 22, 486–506.

Stanley, S.M. 1975. A theory of evolution above the species level. *Proc. Natl. Acad. Sci. U.S.*, 72, 646–50.

Stephens, E.P. 1955. *The historical-developmental method of determining forest trends.* Doctoral dissertation. Harvard University, Cambridge, Massachusetts.

Stephens, G.R., and Waggoner, P.E. 1970. The forests anticipated from 40 years of natural transitions in mixed hardwoods. *Bull. Conn. Agric. Exp. Station, New Haven*, 707, 1–58.

Stern, K., and Roche, L. 1974. *Genetics of Forest Ecosystems.* New York, Springer-Verlag.

Stern, V.M. 1966. Significance of the economic threshold in integrated pest control. *Proc. FAO Symp. Integrated Control*, 2, 41–56.

Sternlicht, M. 1973. Parasitic wasps attracted by the sex pheromones of their coccid hosts. *Entomophaga*, 18, 339–43.

Stiles, F.G. 1975. Ecology, flowering phenology, and hummingbird pollination of some Costa Rican *Heliconia* species. *Ecology*, 56, 285–301.

Stimson, J.S. 1970. Territorial behaviour in the owl limpet, *Lottia gigantea. Ecology*, 51, 113–18.

Stimson, J.S. 1973. The role of the territory in the ecology of the intertidal limpet *Lottia gigantea* (Gray). *Ecology*, 54, 1020–30.

Sturrock, R.F., and Webbe, G. 1971. The application of catalytic models to schistosomiasis in snails. *Journal of Helminthology*, 45, 189–200.

Sturrock, R.F., Cohen, J.E., and Webbe, G. 1975. Catalytic curve analysis of schistosomiasis in snails. *Annals of Tropical Medicine and Parasitology*, 69, 133–4.

Sutherland, J.P. 1974. Multiple stable points in natural communities. *Am. Nat.*, 108, 859–73.

Tanner, J.T. 1975. The stability and the intrinsic growth rates of prey and predator populations. *Ecology*, 56, 855–67.

Tappan, H. 1971. Microplankton, ecological succession and evolution. *North Am. Paleont. Convention Chicago*, 1969, *Proc.*, *H*, 1058–1103.

Terborgh, J. 1974. Faunal equilibria and the design of wildlife preserves. In F. Golley and E. Medina (eds.), *Tropical Ecological Systems: Trends in Terrestrial and Aquatic Research.* New York, Springer.

Terborgh, J., and Faaborg, J. 1973. Turnover and ecological release in the avifauna of Mona Island, Puerto Rico. *Auk*, 90, 759–79.

Thayer, C.W. 1973. Taxonomic and environmental stability in the Paleozoic. *Science*, 182, 1242–3.

Thompson, D.J. 1975. Towards a predator-prey model incorporating age structure: the effects of predator and prey size on the predation of *Daphnia magna* by *Ichnura elegans*. *J. Anim. Ecol.*, 44, 907–16.

Thoreau, H.D. 1860. The succession of forest trees. In *Excursions* (1863). Boston, Houghton Mifflin & Co.

Trivers, R.L., and Hare, H. 1976. Haplodiploidy and the evolution of the social insects. *Science*, 191, 249–63.

Turnbull, A.L. 1962. Quantitative studies of the food of *Linyphia triangularis* Clerck (Araneae: Linyphiidae). *Can. Ent.*, 94, 1233–49.

Twight, P.A., and Minckler, L.S. 1972. Ecological forestry for the central hardwood forest. Washington, National Parks and Conservation Association, pp. 12.

Ullyett, G.C. 1949a. Distribution of progeny by *Chelonus texanus* Cress. (Hymenoptera: Braconidae). *Can. Ent.*, 81, 25–44.

Ullyett, G.C. 1949b. Distribution of progeny by *Cryptus inornatus* Pratt (Hymenoptera: Ichneumonidae). *Can. Ent.*, 81, 285–99.

Usher, M.B. 1966. A matrix approach to the management of renewable resources with special reference to selection forests. *J. Appl. Ecol.*, 3, 355–67.

Utida, S. 1967. Damped oscillation of population density at equilibrium. *Res. Pop., Ecol.*, 9, 1–9.

Valentine, J.W. 1970. How many marine invertebrate fossil species? A new approximation. *J. Paleontol.*, 44, 410–15.

Valentine, J.W. 1973. Phanerozoic taxonomic diversity: a test of alternate models. *Science*, 180, 1078–9.

Valentine, J.W., Hedgecock, D., Zumwalt, G., and Ayala, F.J. 1973. Mass extinctions and genetic polymorphism in the 'killer clam', *Tridacna*. *Bull. Geol. Soc. Am.*, 84, 3411–14.

Valentine, J.W., and Ayala, F.J. 1974. On scientific hypotheses, killer clams, and extinctions. *Geology*, 69–71.

Vance, R.R. 1972. Competition and mechanism of coexistence in three sympatric species of intertidal hermit crabs. *Ecology*, 53, 1062–74.

Vandermeer, J.H. 1972. Niche theory. *A. Rev. Ecol. Syst.*, 3, 107–32.

Van Oye, E. 1964. *The World Problem of Salmonellosis*. The Hague, W. Junk.

Van Valen, L. 1965. Morphological variation and the width of the ecological niche. *Am. Nat.*, 100, 377–89.

Viereck, L.A. 1966. Plant succession and soil development on gravel outwash of the Muldrow Glacier, Alaska. *Ecol. Monogr.*, 36, 181–99.

Volterra, V. 1926. Variations and fluctuations of the number of individuals in animal species living together. *J. Cons. perm. int. Ent. Mer*. 3, 3–51. (Also reprinted in Chapman, R.N. 1931. *Animal Ecology*. New York and London.)

Vuilleumier, F. 1970. Insular biogeography in continental regions: the northern Andes of South America. *Am. Nat.*, 104, 373–88.

Walker, K.R., and Alberstadt, L.P. 1975. Ecological succession as an aspect of structure in fossil communities. *Paleobiology*, 1, 238–57.

Wallace, B. 1973. Misinformation, fitness, and selection. *Am. Nat.*, 107, 1–7.

Wangersky, P.J., and Cunningham, W.J. 1957. Time lag in prey-predator population models. *Ecology*, 38, 136– 9.

Warren, K.S. 1973. Regulation of the prevalence and intensity of schistosomiasis in man: immunology or ecology. *J. Inf. Diseases*, 127, 595–609.

Warren, K.S. and Newill, V.A. 1967. *Schistosomiasis—A Bibliography of the World's Literature from 1852 to 1962*. Cleveland, Ohio: Press of Western Reserve University.

Watt, K.E.F. 1963. Dynamic programming, 'look head programming', and the strategy of insect pest control. *Can. Ent.*, 95, 525–636.

Watt, K.E.F. 1968. *Ecology and Resource Management.* (And references therein.) New York, McGraw-Hill.

Way, M.J. 1973. Objectives, methods and scope of integrated control. In P.W. Geier, L.R. Clark, D.J. Anderson and H.A. Nix (eds.), *Insects: Studies in Pest Management.* Mem. 1, pp. 137–52, Ecological Society of Australia, Canberra.

Weaver, H. 1974. Effects of fire on temperate forests: Western United States. In T.T. Kozlowski and C.E. Ahlgren (eds.), *Fire and Ecosystems*, pp. 279–319. New York, Academic Press.

Webb, L.J., Tracey, J.G., and Williams, W.T. 1972. Regeneration and pattern in the subtropical rain forest. *J. Ecol.*, 60, 675–95.

Weisbrod, B.A., Andreano, R.L., Baldwin, R.E., Epstein, E.H., and Kelley, A.C. 1973. *Disease and Economic Development; the Impact of Parasitic Diseases in St. Lucia.* Madison, Wisconsin, The University of Wisconsin Press.

Whittaker, R.H. 1957. Recent evolution of ecological concepts in relation to the eastern forests of North America. *Am. J. Bot.*, 44, 197–206.

Whittaker, R.H. 1972. Evolution and measurement of species diversity. *Taxon*, 21, 213–51.

Whittaker, R.H. 1975. *Communities and Ecosystems.* (Second edition). New York, Macmillan.

Wilbur, H.M. 1972. Competition, predation, and the structure of the *Ambystoma-Rana sylvatica* community. *Ecology*, 53, 3–21.

Wildbotz, T., and Meier, W. 1973. Integrated control: critical assessment of case histories in affluent economies. In P.W. Geier, L.R. Clark, D.J. Anderson and H.A. Nix (eds.), *Insects: Studies in Pest Management.* Mem. 1, pp. 137–52, Ecological Society of Australia, Canberra.

Williams, G.C. 1975. *Sex and Evolution.* Princeton, New Jersey, Princeton University Press.

Williams, W.T., Lance, G.N., Webb, L.J., Tracey, J.G., and Dale, M.B. 1969. Studies in the numerical analysis of complex rain-forest communities III. The analysis of successional data. *J. Ecol.*, 57, 515–35.

Williamson, M. 1972. *The Analysis of Biological Populations.* London, Edward Arnold.

Williamson, M. 1973. Species diversity in ecological communities. In M.S. Bartlett and R.W. Hiorns (eds.), *The Mathematical Theory of the Dynamics of Biological Populations*, pp. 325–35. New York, Academic Press.

Willis, E.O., 1974. Populations and local extinction of birds on Barro Colorado Island, Panama. *Ecol. Monographs*, 44, 153–69.

Wilson, A.G.L., Hughes, R.D., and Gilbert, N.E. 1972. The response of cotton to pest attack. *Bull. ent. Res*, 61, 405–14.

Wilson, D.S. 1975. A theory of group selection. *Proc. Natl. Acad. Sci. U.S.*, 72, 143–6.

Wilson, E.O. 1969. The new population biology. *Science*, 163, 1184–5.

Wilson, E.O. *et al.* 1973. *Life on Earth.* Stamford, Conn., Sinauer.

Wilson, E.O. 1975. *Sociobiology: the new synthesis.* Cambridge, Mass., Belknap Press of Harvard University Press.

Wilson, E.O., and Willis, E.O. 1975. Applied biogeography. In M.L. Cody and J.M. Diamond (eds.), *Ecology and Evolution of Communities*, *op. cit.*, pp. 522–34.

Wood, B.J. 1971. Development of integrated control programs for pests of tropical perennial crops in Malaysia. In C.B. Huffaker (ed.), *Biological Control*, pp. 422–57. New York, Plenum.

Wood, D.L., Browne, L.E., Bedard, W.D., Tilden, P.E., Silverstein, R.M., and Rodin, J.O. 1968. Response of *Ips confusus* to synthetic sex pheromones in Nature. *Science*, 159, 1373–4.

Wratten, S.D. 1973. The effectiveness of the coccinellid beetle, *Adalia bipunctata* (L.), as a predator of the lime aphid, *Eucallipterus tiliae* L. *J. Anim. Ecol.*, 42, 785–802.

Wright, H.E., Jr. 1974. Landscape development, forest fires, and wilderness management. *Science*, 186, 487–95.

Wright, H.E., and Heinselman, M.L. 1973. The ecological role of fire in natural conifer forests in western and northern North America—Introduction. *Quarternary Res.*, 3, 319–28.

Wynne-Edwards, V.C. 1962. *Animal Dispersion in Relation to Social Behaviour.* Edinburgh, Oliver & Boyd. pp. 653.

Young, A.M. 1974. On the biology of *Hamadryas februa* [Lepidoptera: Nymphalidae] in Guanacuste, Costa Rica. *Z. ang. Ent.*, 76, 380–93.

Young, A.M., and Muyshondt, A. 1972. Biology of *Morpho polyphemus* (Lepidoptera: Morphidae) in El Salvador. *J. New York ent. Soc.*, 80, 18–42.

Organism Index

Subject Index